ELECTROANALYTICAL CHEMISTRY

VOLUME 1

ELECTROANALYTICAL CHEMISTRY

A SERIES OF ADVANCES

Edited by

ALLEN J. BARD

DEPARTMENT OF CHEMISTRY
UNIVERSITY OF TEXAS
AUSTIN, TEXAS

VOLUME 1

1966

MARCEL DEKKER, Inc., NEW YORK

INTRODUCTION TO THE SERIES

This series is designed to provide authoritative reviews in the field of modern electroanalytical chemistry defined in its broadest sense. Coverage will be comprehensive and critical. Enough space will be devoted to each chapter of each volume so that derivations of fundamental equations, detailed descriptions of apparatus and techniques, and complete discussions of important articles can be provided, so that the chapters may be useful without repeated reference to the periodical literature. Chapters will vary in length and subject area. Some will be reviews of recent developments and applications of well-established techniques, whereas others will contain discussions of the background and problems in areas still being investigated extensively and in which many statements may still be tentative. Finally, chapters on techniques generally outside the scope of electroanalytical chemistry, but which can be applied fruitfully to electrochemical problems, will be included.

Electroanalytical chemists and others are concerned not only with the application of new and classical techniques to analytical problems, but also with the fundamental theoretical principles upon which these techniques are based. Electroanalytical techniques are proving useful in such diverse fields as electro-organic synthesis, fuel cell studies, and radical ion formation, as well as with such problems as the kinetics and mechanism of electrode reactions, and the effects of electrode surface phenomena, adsorption, and the electrical double layer on electrode reactions.

It is hoped that the series will prove useful to the specialist and non-specialist alike—that it will provide a background and a starting point for graduate students undertaking research in the areas mentioned, and that it will also prove valuable to practicing analytical chemists interested in learning about and applying electroanalytical techniques. Furthermore, electrochemists and industrial chemists concerned with problems of electrosynthesis, electroplating, corrosion, and fuel cells, as well as other chemists wishing to apply electrochemical techniques to chemical problems, may find useful materials in these volumes.

A. J. B.

June 1966

CONTRIBUTORS TO VOLUME 1

DONALD E. SMITH, *Department of Chemistry, Northwestern University, Evanston, Ill.*

DONALD G. DAVIS, *Department of Chemistry, Louisiana State University in New Orleans, New Orleans, Louisiana*

THEODORE KUWANA, *Department of Chemistry, Case Institute of Technology, Cleveland, Ohio*

DAVID M. MOHILNER, *Department of Chemistry, University of Texas, Austin, Texas*

CONTENTS OF VOLUME 2

Electrochemistry of Aromatic Hydrocarbons and Related Substances, MICHAEL E. PEOVER, *Ministry of Technology, Basic Physics Division, National Physical Laboratory, Teddington, Middlesex, England*

Stripping Voltammetry, E. BARENDRECHT, *Central Laboratory, Staatsmijnen in Limburg, Geleen, the Netherlands*

The Anodic Film on Platinum Electrodes, S. GILMAN, *General Electric Research and Development Center, Schenectady, New York*

Oscillographic Polarography at Controlled Alternating Current, M. HEYROVSKÝ AND K. MICKA, *J. Heyrovský Institute of Polarography, Czechoslovakia, Academy of Sciences, Prague, Czechoslovakia*

CONTENTS OF VOLUME 1

Introduction to the Series   v
Contributors to Volume 1   vi
Contents of Volume 2   vii

AC Polarography and Related Techniques: Theory and Practice   1

    DONALD E. SMITH

  I.   Introduction   2
  II. The AC Polarographic Experiment   3
 III. Theory   13
 IV. Instrumentation   102
  V. Data Analysis   132
      References   148

Applications of Chronopotentiometry to Problems
in Analytical Chemistry   157

    DONALD G. DAVIS

  I.   Introduction   157
  II. Theory   161
 III. Experimental Methods   175
 IV. Applications   184
      References   193

Photoelectrochemistry and Electroluminescence   197

    THEODORE KUWANA

  I.   Introduction and Scope   197
  II. Nomenclature   198
 III. Some Photochemical Fundamentals   201
 IV. Photopotentials from Organic Systems   206
  V. Irradiation of Metal Electrode Surfaces in
        "Nonabsorbing" Solutions   215

VI.	Metal–Metal Oxide Surfaces and Semiconductors	219
VII.	Thermal Effects	222
VIII.	Electroluminescence	223
	References	235

The Electrical Double Layer, Part I:
Elements of Double-Layer Theory   241

DAVID M. MOHILNER

I.	Introduction	242
II.	Double-Layer Thermodynamics	249
III.	Adsorption and Double-Layer Models	298
	Appendixes	391
	References	405

| *Author Index* | 411 |
| *Subject Index* | 422 |

AC POLAROGRAPHY

AND RELATED TECHNIQUES:

Theory and Practice

Donald E. Smith

DEPARTMENT OF CHEMISTRY
NORTHWESTERN UNIVERSITY
EVANSTON, ILLINOIS

I. Introduction 2
II. The AC Polarographic Experiment 3
III. Theory 13
   A. Background 13
   B. The Reversible AC Polarographic Wave 15
   C. The Quasi-Reversible AC Polarographic Wave 26
   D. Systems with Coupled Chemical Reactions 42
   E. Systems with Multistep Charge Transfer 71
   F. Adsorption 77
   G. Double-Layer Effects 87
   H. Time Dependence of the AC Polarographic Wave 92
   I. Contributions of Electrode Growth and Geometry 93
   J. Stationary Electrodes 98
   K. Theory for Some Related AC Techniques 100
   L. Faradaic Rectification and Intermodulation Polarography .. 101
IV. Instrumentation 102
   A. General 102
   B. The Operational Amplifier 104
   C. The Follower Amplifier 107
   D. The Controlled-Potential Configuration 108
   E. Signal Sources 110
   F. Tuned Amplifiers 111
   G. Phase-Sensitive Detectors 114
   H. Conventional Rectifiers 119
   I. Phase Shifters 119
   J. Trigger Circuits 119
   K. Holding Circuits 120

L. Fundamental-Harmonic AC Polarography   121
M. Phase-Selective AC Polarography   121
N. Second-Harmonic and Intermodulation AC Polarography   ..   122
O. Phase-Selective Second-Harmonic and Intermodulation   ..
        AC Polarography   125
P. Phase-Angle Measurement   125
Q. Possible Future Trends   128
V. Data Analysis   132
    A. Vectorial Correction for Ohmic Resistance and Double-Layer
            Charging Current   132
    B. Analysis of Faradaic-Component Data   135
    C. Plotting AC Polarographic Data in the Complex-Impedance or
            Admittance Plane   140
Notation Definitions   146
References   148

I. INTRODUCTION

Alternating current polarography became an active area of research in the middle and late 1940s as a result of the pioneering efforts of Breyer and Gutmann (*1–3*), Grahame (*4*), Ershler (*5*), Randles (*6*), and others. The technique found increasing interest through the 1950s and presently ranks high in importance among electrochemical techniques. The reader may find support for the latter statement in the amount of discussion devoted to ac polarography and related techniques in two recent review articles devoted to developments in electrochemistry (*7,8*). The technique has shown advance, not just in volume of application, but in the increasing fraction of quantitative kinetic studies.

The purpose of this chapter is to provide a detailed discussion of modern concepts in ac polarographic theory, experimental approach, data analysis, and instrumentation. No attempt will be made to provide in detail a historical development or discussion of the literature. Such material is found in the recently published book by Breyer and Bauer (*9*), to which the reader is referred for much material supplementary to the present discussion.

Relative to their importance in this area, topics such as faradaic rectification, tensammetry, and the influence of the electrical double layer have been slighted to a certain extent: The former two because of recent excellent reviews of the topics (*9,10*); the latter because it is the subject of another chapter in this volume (*11*).

It is assumed that the reader commands a working knowledge of basic concepts regarding theory of electrode reactions and ac circuitry. Those unfamiliar with such topics are urged to consult suitable texts as a source of background material [for example, (*12,13*)].

II. THE AC POLAROGRAPHIC EXPERIMENT

Basically, the ac polarographic experiment amounts to determining the impedance of an electrolytic cell under polarographic conditions while employing small-amplitude (< 10 mV) alternating potentials. The normal objectives are to relate this impedance to electrode-reaction mechanism, kinetics, and/or concentration of an electroactive material. Physical processes contributing to the cell impedance may be divided into three categories: (a) ionic migration which, among other things, determines the ohmic resistance of the cell; (b) charging of the electrical double layer at the electrode-solution interface, which provides a path for ac flow independent of electrolysis; (c) the faradaic process which encompasses all phenomena related to electrolysis (charge transfer, mass transfer, adsorption, and chemical reaction). Ac polarographic measurements have been employed to study the double-layer capacity, particularly the tensammetric process which is related to the influence of adsorption on the double-layer capacity (*9,14–18*). However, the faradaic process is most frequently the primary subject of the ac experiment. As a result, considerable effort in instrumental and experimental design has been devoted to minimizing contributions of the double-layer charging current and ohmic resistance. Although some of these efforts have met notable success, these "extraneous" factors always represent the ultimate limitation of the ac method when investigating the faradaic process.

For determining the impedance of the polarographic cell to fundamental harmonic alternating current, two experimental approaches are evident. One may employ an impedance bridge, thus determining the resistive and capacitive components of the cell impedance through the precise null method. Alternatively, the direct measurement of alternating current flowing as a result of the applied alternating potential is possible. To effect correction for ohmic-potential drop and double-layer charging current (which require vectorial subtraction), as well as to obtain more complete knowledge of the electrical nature of the faradaic impedance, two parameters related to cell impedance must be measured (*6,12,19,20*). These may be the resistive and capacitive components of the impedance or alternating current, the total alternating current plus phase angle, the total current plus resistive component, etc. The impedance-bridge method was the earliest instrumental technique employed for precision kinetic studies (*6,21,22,23*), and it remains ultimately the most accurate approach, because of its reliance on the null signal. It yields as readout the equivalent series (or parallel) resistive and capacitive components of the cell impedance.

The earliest measurements with the impedance bridge employed solutions containing both oxidized and reduced forms of the redox couple of interest, with measurements confined to the equilibrium potential (zero direct current). If the investigator was interested in studying the faradaic process as a function of dc potential, a series of solutions was prepared containing different ratios of oxidized and reduced forms so that the equilibrium direct potential varied. The desire to obtain bridge measurements as a function of direct potential when one of the redox forms was unstable and/or to avoid the inconvenience of preparing a series of solutions led to simple modification of the impedance bridge so that dc polarization could be effected (*24,25*). Impedance measurements in the presence of direct current can be considered a means of accomplishing *in situ* variation of the "mean" or dc concentrations of reactants at the electrode-solution interface and its immediate vicinity (*10*). Some workers abandoned the impedance bridge with its associated point-by-point measurement capability in favour of direct measurement of alternating current as a function of dc potential. This approach, which is compatible with automatic recording methods, permits one to obtain information on the faradaic process more rapidly, at some loss in accuracy. A small amplitude alternating potential is superimposed on the dc potential which is varied linearly with time. The resulting fundamental harmonic alternating current is recorded as a function of direct potential. This latter approach was entitled *AC polarography* and was pioneered by the Australian school (*1–3,9,14–16,18,19*). Conventional usage led to the term *faradaic impedance measurement* being associated with studies employing the impedance bridge. The reader should recognize the artificiality of the difference in terminology, since the titles refer to two different instrumental approaches to obtaining the same data. The form of the fundamental-harmonic ac polarographic wave (the *ac polarogram*) obtained upon recording the alternating current as a function of direct potential is illustrated in Fig. 1. Faradaic alternating current flows only when both oxidized and reduced forms are present simultaneously at the interface, that is, on the rising portion of the dc polarographic wave. Thus, one obtains a polarogram with a maximum corresponding roughly to the shape of the derivative of the dc wave. The details regarding shape and magnitude of the ac polarographic wave are influenced markedly by kinetics of the various rate processes associated with the electrode reaction. One may obtain results ranging from the well-formed symmetric wave shown in Fig. 1 to no detectable ac wave for irreversible processes (*9*) Herein lies the basis for application of ac polarography in kinetic studies. The wave shown in Fig. 1 was obtained at a dropping mercury electrode

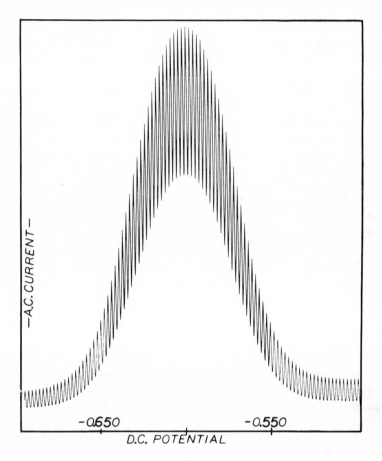

Fig. 1. Fundamental harmonic ac polarograms [from (*34*)]. System: 3×10^{-3} *M* Cd(II), 1.0 *M* Na$_2$SO$_4$. Applied potential: 10 mV peak-to-peak, 320 cps ac signal; 25 mV min^{-1} dc scan rate. (Courtesy of *Analytical Chemistry*.)

(DME); thus drop oscillations are observed. An ac wave obtained at a stationary electrode has essentially the same shape, without the drop oscillations (*26,27*). No drastic alteration in the *qualitative* shape of the wave is observed resulting from a change from the DME to a stationary electrode (neglecting variation in kinetics due to change in electrode material). This contrasts to the situation in dc polarization studies, where the dc polarogram and the linear scan chronoamperogram differ significantly in qualitative characteristics (*12,28*). Direct measurement of the funda-mental-harmonic alternating current, applied extensively by Breyer, Bauer, and co-workers, led to applications in analysis and qualitative studies of electrode-reaction kinetics (*1–3,9,29–31*). Tensammetric waves arising from adsorption phenomena were elucidated (*14–18*). The realiz-ation that a second parameter, in addition to total fundamental-harmonic current, must be measured to correct properly for ohmic losses and charging current in quantitative kinetic studies made point-by-point measurement of phase angle the lot of some experimentalists employing automatic recording of currents (*32,33*). This detracted from the advantage of speed and convenience associated with direct measurement of currents. However, instrumentation providing for automatic recording of phase angle is readily obtained (*34*), or, alternatively, one can combine automatic recording of total alternating current with phase-selective measurement of the resistive (or capacitive) current component to provide the necessary experimental information. Thus, the advantage of the direct method relative to the bridge method regarding the speed and convenience associated with automatic recording applies, even for quantitative kinetic measurements. In addition, modern instrumentation is sufficiently well developed that automatic recording equipment with precision and accuracy approaching ~ 0.1 per cent (with care and frequent calibration) is available, and automatic simultaneous recording of two or more variables is no problem. The direct measurement of current is readily adapted to modern data-processing equipment (analog-to-digital converters, digital readout to punched tape, magnetic tape, IBM card punch, and others). Because it is debatable whether the average electrochemical system can be reproduced to 0.1 per cent or better, the greater inherent accuracy of the bridge method [0.01 per cent bridges are available (*35*)] might be considered a purely academic point. These considerations, coupled with the need for extensive data to effect understanding of many electrochemical systems, appear to make the direct approach the choice of the future, despite the significantly larger initial investment involved.

Regardless of the instrumental method, or the quality and/or speed of the instrument, problems associated with double-layer charging current and ohmic losses remain, limiting the range of application of fundamental harmonic ac polarography in both analysis and kinetic investigation. At concentrations much below 10^{-4} M, analysis is difficult and inaccurate because charging current predominates over the small faradaic current signal. The upper range of usable frequencies is limited by the double-layer charging current, which increases directly with frequency, whereas the faradaic current normally increases with the square root of frequency or less. In fundamental-harmonic measurements, frequencies much above 5 to 10 kc result in prohibitively low signal-to-noise (faradaic-to-charging current) ratios. The advantage of ac polarography in studying very rapid rate processes lies in one's ability to observe the response of the faradaic process to the rapidly changing potential. This restriction imposed on the frequency range limits the upper magnitude of rate constants accessible to fundamental-harmonic ac study. An additional problem arises because of the low impedance of the cell at high frequencies. This leads to significant decrease in accuracy because of ohmic losses. Fortunately, a major instrumental advance has significantly reduced the latter problem. Application of operational-amplifier circuitry with a three-electrode configuration (*26,27,32,34,36–40*) drastically reduces the sources of ohmic resistance. Not only has this advance effected reduction of the problem for work in aqueous solution, but it has made relatively low conducting solvents such as acetonitrile and dimethylformamide accessible to quantitative ac polarographic study. Operational amplifier instrumentation does not eliminate all sources of ohmic resistance (*34,40*), for example, in the DME capillary, so the effect still exists. However, the frontiers have been moved back considerably.

Limitations imposed by double-layer charging current have led to the appearance of a variety of modifications of the ac polarographic method. Phase-selective detection of the fundamental-harmonic current takes advantage of the phase difference between faradaic and charging current. By measuring the component in phase with the applied alternating potential, one discriminates against the charging current which is 90° out of phase with the alternating potential across the interface, whereas the faradaic component is usually 45° or less out of phase. Perfect selectivity would be realized if ohmic resistance did not exist, since charging current would then be exactly 90° out of phase with the applied potential (*34*). Insofar as ohmic resistance is not zero, resolution of double-layer charging current and faradaic current is imperfect. Nevertheless, significant improvement is realized and *phase-selective ac polarography* has been applied frequently as a means of enhancing the sensitivity of fundamental-harmonic measurements for analytical applications (*9,27,32,34,41–51*). As already indicated, phase-sensitive detection in combination with conventional ac polarography represents a means to obtaining sufficient data for kinetic studies. Another approach to enhancing the sensitivity of fundamental-harmonic measurements is *square-wave polarography*, first developed by Barker and Jenkins (*52,53*). A square-wave voltage is applied in place of the normal sinusoidal wave form. Because the double-layer charging process takes place in a very short period of time at the beginning of each square-wave half-cycle, whereas faradaic current decays much more slowly [cf., for example p. 109 of (9)], one can discriminate against the charging current by confining measurement to near the end of each half-cycle with the aid of appropriate gating circuits. The technique has been eminently successful in trace-analysis applications, permitting analysis at concentrations as low as 10^{-8} M for some metal ions in aqueous solutions (54). It has been applied widely for analysis (9,43,52–56), but kinetic applications are rare. Square-wave polarography operates on essentially the same basic principle as phase-selective ac polarography. Both techniques discriminate against charging current on the basis of the smaller RC time constant associated with the double-layer charging process relative to the faradaic process. A phase angle of nearly 90° in a sinusoidal technique and a decay of current early in the half-cycle of a square wave manifest the same phenomena: a small RC time constant. On this basis the two techniques should be equally effective. However, perusal of the literature seems to indicate greater sensitivity for square-wave polarography [$\sim 10^{-8}$ M limit as opposed to $\sim 10^{-6}$ M limit (45) for phase-selective detection at the DME]. It would appear that the disparity arises from the difference in readout technique conventionally employed with the two methods. The readout employed in square-wave polarography yields essentially the current at a particular instant in the life of the mercury drop. Phase-selective methods have employed continuous monitoring of current during drop life. The former approach eliminates the drop oscillation and thus provides a more easily suppressed base line. Application of the same readout technique in phase-selective ac polarography should substantially improve the sensitivity. This subject is considered in more detail in Sec. IV.

The impedances presented by the faradaic and double-layer charging processes are not perfectly linear. As a result, current flowing under ac polarographic conditions contains higher harmonic components as well as a dc "rectification" component generated by application of the alternating potential. The latter dc component arises together with the normal dc polarographic current. Because nonlinearity associated with the faradaic process is considerably greater than for the double-layer charging process, the measurement of higher-order current components has been studied extensively as a means of reducing the contribution of charging current (8–10,34,53,57–81). Most frequently utilized has been measurement of the dc rectification component (10,57–61,63–72). Because evaluation of this relatively small dc component is difficult in the presence of the much larger normal dc polarographic current, special experimental conditions are required (*10,57*) which entail use of solutions containing both forms of the electroactive redox couple. Under these conditions, measurements of the dc rectification component can be made at the equilibrium potential, where the normal direct current is zero. Alternatively, one can employ a high impedance to direct current ($I_{dc} = 0$), in which case a dc rectification voltage is generated (*10,57*). Alternating current or voltage can be controlled with either approach. This type of experiment is referred to as *faradaic rectification measurement* and has been developed extensively by Delahay and co-workers (*57,65–72*) and by Barker (*61*), primarily for kinetic studies. Adaptation to analytical work is inconvenient, owing to the requirement that both redox forms be present in solution. Faradaic rectification measurements have been applied up to the megacycle-frequency range (*10*). At these very high frequencies, the over-all cell impedance is dependent almost solely on the *RC* impedance presented by the double-layer capacity and ohmic resistance. At the same time, the rectification component is influenced only slightly by the double-layer charging process. This ideal situation permits convenient and accurate calculation of the effective voltage applied across the electrode-solution interface (effective voltage equals charging current times impedance of double-layer capacity), provided stray capacitance and inductance are negligible. Even more convenient is a method developed by Imai (*70*) which does not require knowledge of the effective applied potential. Despite these innovations, which reduce or eliminate the problem of correcting adequately for ohmic-potential drop, the upper range of frequencies remains limited by this factor. At frequencies in excess of a megacycle, high currents are passed and energy dissipation in the ohmic resistance effects a significant temperature rise (*10*). Thus, as the frequency is increased, observation times must be decreased. To reduce this problem, Senda, Imai, and Delahay devised a double-pulse method which has been applied with frequencies up to 50 Mc (*69,71*). To adapt the measurement of the faradaic-rectification effect to analytical work, Barker devised a technique he called *rf polarography* (*53,61,62*). Superimposed on the dc polarographic potential scan is a small amplitude rf (100 kc to 6·4 Mc.) signal modulated at 225 cps. A 225-cps signal is measured. Although Barker presented the technique as a method of measuring the dc rectification component in the presence of the normal dc polarographic current, it is more properly considered a form of *intermodulation polarography*, as indicated by Reinmuth (*8*). In one form of intermodulation polarography, two sinusoidal potential signals of differing frequency are superimposed on the direct potential, and the ac signal corresponding to the difference frequency is measured (*8,73–75*). The difference frequency component, like the higher harmonics and dc rectification components, is a manifestation of the nonlinearity of the faradaic and double-layer charging processes. Paynter and Reinmuth (*8,73*) have investigated in detail the possibilities of intermodulation polarography for kinetic application. ·As in measurement of the dc rectification component, this method appears ideally suited to studies at high frequencies, because of the relatively small charging-current contribution. In addition, readout is obtained conveniently as a function of direct potential and yields an *intermodulation wave* which is readily adapted to analytical work. Neeb (*75*) has used intermodulation methods at low frequencies for analytical purposes. This technique also suffers the ultimate limitation on frequency range associated with energy dissipation in the ohmic resistance at multi-megacycle frequencies. The problem is more serious than with faradaic rectification measurements because the intermodulation method is based on a steady-state readout. Second harmonic measurements have been considered for purposes of kinetic and analytical applications (*73,74,76–82*). The technique appears promising for either endeavour. Paynter (*73*) also presented a detailed analysis of *second-harmonic ac polarography*, comparing it with the intermodulation approach. It appears that the latter has a significant advantage in kinetic studies because double-layer charging currents are significantly smaller at frequencies above a kilocycle (*73*). For either second-harmonic or intermodulation polarography, the contribution of charging current can be reduced by phase-selective detection (*34*). Both second-harmonic and intermodulation waves normally show two peaks, except under extreme conditions. Phase-selective detection yields a wave often having the qualitative shape of the second derivative of the dc polarogram. Some examples are shown in Fig. 2(*a*) and (*b*). As with the fundamental-harmonic wave, the form of second-harmonic and intermodulation polarograms is strongly dependent on the kinetics of charge transfer, chemical reaction, etc. In some cases the polarograms may show only a single peak (*61,73*).

A characteristic of many results reported from these "second-order" ac methods is that kinetic data is often at variance with results obtained by other techniques (*8*). This seemingly unfortunate situation possibly can be explained on the basis of two considerations: (a) As indicated by theoretical work, the second-order currents are expected to be more sensitive to kinetic effects. Thus, subtle kinetic contributions may significantly influence these currents, whereas their effects go unnoticed in dc or fundamental-harmonic ac polarography; (b) the higher frequencies normally employed with the second-order techniques introduce kinetic control by processes too rapid to be kinetically important at lower frequencies or longer observation times (for non-ac methods).

Ac polarography and its numerous variations represent a relatively untapped methodological storehouse, not to mention the variety of transient relaxation methods which have also been devised. It is true that much useful quantitative and qualitative information regarding electrode processes already has resulted from the study of the fundamental harmonic component [cf., for example, Breyer and Bauer (9)]. However, the past

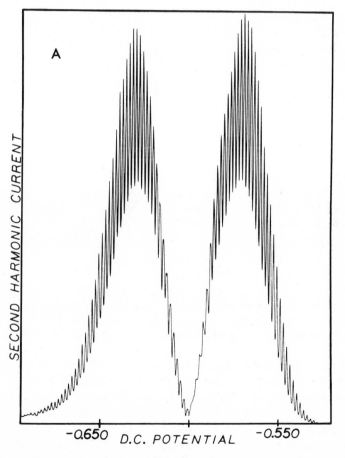

Fig. 2a

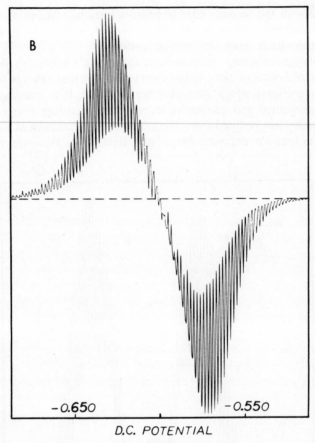

D.C. POTENTIAL

Fig. 2b

Fig. 2. Second harmonic ac polarograms [from (*34*)]. System: $3 \times 10^{-3} M$ Cd(II), 1.0 M Na$_2$SO$_4$. Applied potential: 80 cps, 10-mV peak-to-peak ac signal; 25 mV min^{-2} dc scan rate. (a) "Conventional" second-harmonic ac polarogram. (b) Phase-selective second-harmonic ac polarogram. (Courtesy of *Analytical Chemistry*.)

decade has seen an inordinate amount of effort directed toward overcoming limitations of existing methods by applying electronic "trickery" to evolve more sophisticated techniques, such as the higher-order ac methods just described and a variety of transient techniques such as the double-pulse galvanostatic (*10,83*) and coulostatic methods (*84–90*). Many of these recent innovations have been investigated only briefly. It is likely that this emphasis on technique improvement will soon be replaced by emphasis on utilization of existing techniques within their useful realm. In this regard, it is well to reflect on the rewards in the form of knowledge of electrode processes and useful analytical methods which have resulted and are resulting from widespread and judicious application of dc polarography. One envisions a bright future for electrochemistry if one extrapolates this consideration to possible rewards in the offing if, for example, the ac polarographic experiment and its many variations receive similar treatment.

The small-amplitude ac polarographic approach represents a single category in a vast array of electrochemical methodology. Normally, this situation would make desirable some discussion directed toward a comparison of the ac approach with other competing techniques. However, this need has been fulfilled in a recent review by Reinmuth (8), who gives a detailed and realistic comparison of ac polarography and its various modifications with other electrochemical methods. The reader is referred to this review. It should be mentioned that the present author looks upon applications of ac polarography in analysis and the study of kinetics of coupled homogeneous chemical reactions in a somewhat brighter vein than Reinmuth. Some reasons for this feeling will become apparent in the forthcoming discussion. Perhaps the most important advantage of the ac approach which may make it survive the rigors of the future was indicated by Reinmuth (8) in the statement; "The application of the faster techniques is bringing to light a great number of anomalous results due presumably to reaction mechanisms more complicated than those assumed in the simplest theoretical treatments. In cases of complicated mechanisms for which several kinetic parameters must be derived, the accuracy of the impedance technique makes it the method of choice."

III. THEORY

A. Background

Because methods to measure directly concentrations of electroactive species in the diffusion layer have not been devised, they must be inferred from a theoretical model relating them to experimentally measurable parameters such as current, impedance, or phase angle. Thus, whatever success ac polarography and most other electrochemical techniques provide in quantitative investigations of electrode-reaction kinetics depends, to a large extent, on the ability of the theoretician to account adequately for various factors controlling the electrode reaction. The problem is not simple. Ac polarographic measurements are normally performed at the DME and difficulties associated with solving the complicated MacGillavry-Rideal partial-differential equation present themselves. In ac polarography, the potential term is sinusoidal and enters into equations to be solved as the argument of the exponential function, yielding rather complicated non-linear forms. Effects due to shielding (*91,92*) and depletion (*93,94*) might be important and are difficult to handle theoretically. The electrical double layer, in addition to providing a source of undesirable background current, presents a unique ionic atmosphere in the region of the electrode-solution interface, whose properties differ considerably from the bulk of the solution. In this region, mass-transfer and chemical-reaction processes may be altered considerably. At first glance these considerations appear sufficient to discourage even the most experienced applied mathematician. Notwith-standing these obstacles, ac polarographers, armed with the body of knowledge derived from the theoretical treatment of the dc polarographic problem, combined judicious choice of experimental conditions with mathe-matical agility to show considerable progress in obtaining a quantitative theory. It was realized early that use of sufficiently small amplitudes of applied alternating potential would permit linearization of the complex exponential forms. Gerischer (*95*) and Koutecky (*96*) showed that, at least for many situations, the influence of curvature and motion relative to the solution of the electrode surface had a negligible influence on the ac wave. Advances have been made regarding our understanding of specific influences of the electrical double layer. It appears that equations derived neglecting double-layer effects can be readily corrected when such effects are important. The net results of over a decade of effort have shown that the ac polarographic theoretical problem is often relatively simple. Even when one is interested in complex kinetic schemes and/or higher-order terms in the ac equations (second harmonics, higher-amplitude fundamental-harmonic terms, etc.), derivation of the appropriate expressions can be accomplished readily if one possesses sufficient fortitude to wade through the cumbersome algebraic expressions. Indeed, it will become apparent in the forthcoming discussion that solution of the *dc* polarographic problem is often the most significant obstacle to obtaining ac wave equations for a given kinetic scheme which are applicable for any combination of rate parameters. One reason why the ac polarographic problem yields readily to solution is that the periodicity of the ac signal makes it subject to the powerful mathematical tool of Fourier analysis.

Derivation of equations for the diffusion-controlled ac wave (reversible case) will be given in some detail. It is emphasized that these equations represent a limiting case in which diffusion is the sole rate-determining step. Such conditions are rather infrequently encountered in practice, particularly at higher frequencies (> 1 kc). Perhaps the reader will find the consideration given to the reversible case exceeds that warranted on the basis of its practical importance. However, the reversible case represents an "ideal" or reference system to which all "real" systems encountered might be compared. Such practice is common in all areas of science (for example, the ideal gas or the ideal solution). In addition, the derivation serves to illustrate a generally applicable theoretical approach without encumbering these pages with the greater algebraic complexity required in derivation for more complex kinetic schemes. Derivation for the latter will be outlined only briefly. The planar-diffusion mass-transfer model will be employed through-out most of the theoretical discussion for simplicity and because such equations are often sufficiently accurate. The influence of electrode geometry and growth are discussed separately. Also discussed separately are double-layer and activity corrections, which will be neglected in the derivation.

The theoretical method employed is essentially modeled after Matsuda's treatment of the quasi-reversible case (*97*).

A list of notation definitions is given at the end of this chapter.

B. The Reversible AC Polarographic Wave

For the system represented by the reaction scheme

$$O + ne \rightleftharpoons R$$

the partial-differential equations, initial and boundary conditions for diffusion to a stationary planar electrode are (*12*)

$$\frac{\partial C_O}{\partial t} = D_O \frac{\partial^2 C_O}{\partial x^2} \tag{1}$$

$$\frac{\partial C_R}{\partial t} = D_R \frac{\partial^2 C_R}{\partial x^2} \tag{2}$$

For $t = 0$, any x,

$$C_O = C_O^* \tag{3}$$

$$C_R = C_R^* \tag{4}$$

For $t > 0$, $x \to \infty$,

$$C_O \to C_O^* \tag{5}$$

$$C_R \to C_R^* \tag{6}$$

For $t > 0$, $x = 0$,

$$D_O \frac{\partial C_O}{\partial x} = -D_R \frac{\partial C_R}{\partial x} = \frac{i(t)}{nFA} \tag{7}$$

$$C_O = C_R \left(\frac{D_R}{D_O}\right)^{\frac{1}{2}} \exp\left\{\frac{nF}{RT} [E(t) - E_{\frac{1}{2}}^r]\right\} \tag{8}$$

The assumptions that Fick's law is applicable to each species independently, that coupled chemical reactions exert no influence, and that electrode curvature and motion relative to the solution have a negligible effect are incorporated in Eqs. (1) and (2). Equation (7) assumes that each reacting species is soluble either in the solution or electrode phase, that is, there is no accumulation of electroactive material at the interface. Equation (8) states the Nernst equation in terms of the reversible dc polarographic half-wave potential, $E_{\frac{1}{2}}^r$, where

$$E_{\frac{1}{2}}^r = E^0 - \frac{RT}{nF} \ln \left(\frac{f_R}{f_O}\right)\left(\frac{D_O}{D_R}\right)^{\frac{1}{2}} \tag{9}$$

The assumption of nernstian behavior is obeyed if heterogeneous charge-transfer kinetics are very rapid. Conditions for nernstian behavior are stated more precisely later.

Application of the method of Laplace transformation to Eqs. (1) through (7) yields, for surface concentrations,

$$C_{O_{x=0}} = C_O^* - \int_0^t \frac{i(t-u)\,du}{nFA(D_O\pi u)^{\frac{1}{2}}} \tag{10}$$

$$C_{R_{x=0}} = C_R^* + \int_0^t \frac{i(t-u)\,du}{nFA(D_R\pi u)^{\frac{1}{2}}} \tag{11}$$

Derivation of Eqs. (10) and (11) is well known (66,91,98–100) and contains no features specific to ac polarography. For simplicity, we shall assume at this stage that the reduced form is initially absent from the solution; that is, $C_R^* = 0$. Substituting Eq. (10) and (11) in Eq. (8) and rearranging yields the integral expression (given in a form most convenient for subsequent steps)

$$e^{-j(t)} - e^{-j(t)} \int_0^t \frac{\psi(t-u)\,du}{(\pi u)^{\frac{1}{2}}} = \int_0^t \frac{\psi(t-u)\,du}{(\pi u)^{\frac{1}{2}}} \tag{12}$$

where

$$\psi(t) = \frac{i(t)}{nFAC_O^* D_O^{\frac{1}{2}}} \tag{13}$$

$$j(t) = \frac{nF}{RT} [E(t) - E_{\frac{1}{2}}^r] \tag{14}$$

In ac polarography the applied potential is given by

$$E(t) = E_{\text{dc}} - \Delta E \sin \omega t \tag{15}$$

The dc potential term is considered constant. This assumes the rate of scan is slow relative to the rate of change of alternating potential and that the dc potential does not change significantly over the life of a single mercury drop. The latter restriction applies also to dc polarography. Substitution of Eq. (15) in the exponential term gives

$$e^{-j(t)} = e^{-j} \exp\left[\frac{nF \Delta E}{RT} \sin \omega t\right] \tag{16}$$

where

$$j = \frac{nF}{RT} (E_{\text{dc}} - E_{\frac{1}{2}}^r) \tag{17}$$

One then develops the power series

$$\exp\left[\frac{nF \Delta E}{RT} \sin \omega t\right] = \sum_{p=0}^{\infty} \left(\frac{nF \Delta E}{RT}\right)^p \frac{(\sin \omega t)^p}{p!} \tag{18}$$

$$\psi(t) = \sum_{p=0}^{\infty} \psi_p(t) \left(\frac{nF \Delta E}{RT}\right)^p \tag{19}$$

$$p = 0, 1, 2, 3 \ldots \tag{20}$$

Substituting Eq. (18) and (19) in Eq. (12) and equating coefficients of equal powers of $(nF\Delta E/RT)^p$, one obtains a system of integral equations:

$$\frac{e^{-j}(\sin \omega t)^p}{p!} - \sum_{r=0}^{p} \frac{e^{-j}(\sin \omega t)^r}{r!} \int_0^t \frac{\psi_{p-r}(t-u) \, du}{(\pi u)^{\frac{1}{2}}} = \int_0^t \frac{\psi_p(t-u) \, du}{(\pi u)^{\frac{1}{2}}} \tag{21}$$

Solutions of Eq. (21) for different values of p represent the various faradaic current components in the following manner. If k is the order of the current harmonic ($k = 0$ for dc, $k = 1$ for fundamental-harmonic ac, etc.), contributions to a particular harmonic k are obtained from values of p given by

$$p = 2q + k \tag{22}$$

where

$$q = 0, 1, 2, 3, \ldots \tag{23}$$

Thus, dc components are obtained for $p = 2q = 0, 2, 4, \ldots$ and fundamental-harmonic ac terms for $p = 2q + 1 = 1, 3, 5, 7, \ldots$. The contribution to a given harmonic decreases with increasing p for $\Delta E < \sim 0.1$ volt, a condition always obeyed in ac polarography. At sufficiently small amplitudes (see below) only the term corresponding to the lowest value of p need be considered.

For $p = 0$, one obtains

$$\int_0^t \frac{\psi_0(t - u) \, du}{(\pi u)^{\frac{1}{2}}} = \frac{1}{1 + e^j} \tag{24}$$

The solution for $\psi_0(t)$ is obtained most conveniently by application of the method of Laplace transformation to Eq. (24). Applying Eqs. (13) and (19) to the solution for $\psi_0(t)$, we get

$$i_{dc}(t) = \frac{nFAC_O^* D_O^{\frac{1}{2}}}{(1 + e^j)(\pi t)^{\frac{1}{2}}} \tag{25}$$

This is the well-known expression for the potentiostatic direct current at a planar electrode with a reversible process (*12*).

For $p = 1$, one has

$$e^{-j}\left[1 - \int_0^t \frac{\psi_0(t - u) \, du}{(\pi u)^{\frac{1}{2}}}\right] \sin \omega t = (1 + e^{-j}) \int_0^t \frac{\psi_1(t - u) \, du}{(\pi u)^{\frac{1}{2}}} \tag{26}$$

Note that the integral equation for the small-amplitude fundamental harmonic contains a term in the dc component $\psi_0(t)$. This represents a general feature of the theoretical method. As will be seen, the integral equation for the second harmonic contains terms in $\psi_1(t)$ and $\psi_0(t)$. In general, the integral equation for any higher-order term contains all the lower-order terms. For the reversible case, explicit solution of the lower-order terms is not required, since the integral equations provide the required relationships. Thus, substitution of Eq. (24) in Eq. (26) eliminates $\psi_0(t)$, yielding

$$\int_0^t \frac{\psi_1(t - u) \, du}{(\pi u)^{\frac{1}{2}}} = \frac{1}{4 \cosh^2 (j/2)} \sin \omega t \tag{27}$$

However, for most electrode-reaction mechanisms, explicit solution for $\psi_0(t)$ is required to obtain a general solution for $\psi_1(t)$, solution of $\psi_0(t)$ and

$\psi_1(t)$ is required for $\psi_2(t)$, etc. The need to obtain an explicit solution for the dc component often presents the most serious obstacle to obtaining completely general ac polarographic wave equations for a particular kinetic scheme. It is apparent from this discussion that the theoretical approach is iterative, as pointed out by Reinmuth (*101*).

The form of Eq. (27) is such that $\psi_1(t)$ can contain only fundamental–harmonic terms. Thus we may write

$$\psi_1(t) = A \sin \omega t + B \cos \omega t \qquad (28)$$

Substituting this relation in Eq. (27), employing the trigonometric identities

$$\sin \omega(t-u) = \sin \omega t \cos \omega u - \cos \omega t \sin \omega u \qquad (29)$$

$$\cos \omega(t - u) = \cos \omega t \cos \omega u + \sin \omega t \sin \omega u \qquad (30)$$

applying the steady-state approximation (*10*)

$$\left(\int_0^t = \int_0^\infty \right)$$

and the relations (*102*)

$$\int_0^\infty \frac{\cos \omega u \, du}{(\pi u)^{\frac{1}{2}}} = \int_0^\infty \frac{\sin \omega u \, du}{(\pi u)^{\frac{1}{2}}} = \frac{1}{(2\omega)^{\frac{1}{2}}} \qquad (31)$$

we obtain the result

$$(A + B) \sin \omega t - (A - B) \cos \omega t = \frac{(2\omega)^{\frac{1}{2}}}{4 \cosh^2 (j/2)} \sin \omega t \qquad (32)$$

The steady-state approximation neglects transient behaviour in the ac concentration profile. As shown by Berzins and Delahay (*103*), transients are negligible when $(\omega t)^{\frac{1}{2}} \gg 1$, where t is time elapsed after application of the alternating potential.

Equating coefficients of $\sin \omega t$ and $\cos \omega t$ on both sides of Eq. (32), and solving for A and B, we obtain

$$A = B = \frac{(2\omega)^{\frac{1}{2}}}{8 \cosh^2 (j/2)} \qquad (33)$$

Application of the trigonometric identity

$$a \sin \omega t + b \cos \omega t = (a^2 + b^2)^{\frac{1}{2}} \sin\left(\omega t + \cot^{-1} \frac{a}{b}\right) \qquad (34)$$

yields

$$\psi_1(t) = \frac{\omega^{\frac{1}{2}}}{4 \cosh^2 (j/2)} \sin\left(\omega t + \frac{\pi}{4}\right) \tag{35}$$

Applying Eqs. (13) and (19), one obtains as the solution for the small-amplitude fundamental-harmonic current:

$$I(\omega t) = \frac{n^2 F^2 A C_0^*(\omega D_0)^{\frac{1}{2}} \Delta E}{4RT \cosh^2 (j/2)} \sin\left(\omega t + \frac{\pi}{4}\right) \tag{36}$$

This equation is discussed below.

One may proceed in a similar manner to obtain higher-order current components. Equation (21) has the form, for $p = 2$,

$$\frac{e^{-j}}{2}\left[1 - \int_0^t \frac{\psi_0(t-u)\,du}{(\pi u)^{\frac{1}{2}}}\right]\sin^2 \omega t - \left[e^{-j}\int_0^t \frac{\psi_1(t-u)\,du}{(\pi u)^{\frac{1}{2}}}\right]\sin \omega t$$

$$= (1 + e^{-j})\int_0^t \frac{\psi_2(t-u)\,du}{(\pi u)^{\frac{1}{2}}} \tag{37}$$

Substituting Eqs. (24) and (27) in Eq. (37), rearranging, and applying the trigonometric identity

$$\sin^2 \omega t = \tfrac{1}{2}(1 - \cos 2\omega t) \tag{38}$$

we get

$$\int_0^t \frac{\psi_2(t-u)\,du}{(\pi u)^{\frac{1}{2}}} = \frac{\sinh(j/2)}{16 \cosh^3(j/2)} (1 - \cos 2\omega t) \tag{39}$$

The form of Eq. (39) indicates that $\psi_2(t)$ is made up of dc and second-harmonic components, so that one can write

$$\psi_2(t) = A_0(t) + A_2 \sin 2\omega t + B_2 \cos 2\omega t \tag{40}$$

Substituting Eq. (40) in Eq. (39) generates expressions for the dc and ac components [applying Eqs. (29), (30), and (31) to the ac portion] given by

$$\int_0^t \frac{A_0(t-u)\,du}{(\pi u)^{\frac{1}{2}}} = \frac{\sinh(j/2)}{16 \cosh^3(j/2)} \tag{41}$$

and

$$(A_2 - B_2)\cos 2\omega t - (A_2 + B_2)\sin 2\omega t = \frac{\omega^{\frac{1}{2}} \sinh(j/2)}{8 \cosh^3(j/2)} \cos 2\omega t \tag{42}$$

Solution of Eq. (41) is accomplished readily by the method of Laplace transformation and, after application of Eqs. (13) and (19), the dc component corresponding to $A_0(t)$ is shown to be

$$I_2(\text{dc}) = \frac{n^3 F^3 A C_O^* D_O^{\frac{1}{2}} \Delta E^2 \sinh(j/2)}{16 R^2 T^2 (\pi t)^{\frac{1}{2}} \cosh^3(j/2)} \tag{43}$$

This represents the small-amplitude faradaic-rectification dc component flowing under the experimental conditions described. Addition of this term to the normal direct current [Eq. (25)] flowing in absence of alternating current yields an expression for the Fournier polarogram (*66,104,105*) within the framework of the planar-diffusion model. As will be pointed out below, the planar-diffusion model sometimes appears accurate for ac components observed with a reversible process at a DME, but yields rather inaccurate expressions for the dc components. Influence of electrode growth and geometry must be considered in deriving the dc terms.

Proceeding as in the derivation of the fundamental-harmonic component, one obtains solutions for A_2 and B_2 in Eq. (42), corresponding to a second-harmonic component given by

$$I(2\omega t) = \frac{2^{\frac{1}{2}} n^3 F^3 A C_O^* (\omega D_O)^{\frac{1}{2}} \Delta E^2 \sinh(j/2)}{16 R^2 T^2 \cosh^3(j/2)} \sin\left(2\omega t - \frac{\pi}{4}\right) \tag{44}$$

One may solve the integral equation for larger values of p to obtain expressions for third, fourth, etc., harmonics and larger amplitude contributions to all current components. Table I gives solutions of these integral equations up to $p = 5$. All current expressions shown in Table I have been given in the literature. The fundamental-harmonic component expression was derived independently by a number of workers (*12,30,66, 91,106*). The second-harmonic and higher-order terms were given by Senda and Tachi (*66*). Matsuda derived an expression for the entire complex wave form applicable to very large amplitudes (*91*).

The small amplitude expression given by Eq. (36) corresponds to a faradaic impedance (magnitude)

$$Z_f = \frac{4RT \cosh^2(j/2)}{n^2 F^2 A C_O^* (\omega D_O)^{\frac{1}{2}}} \tag{45}$$

which is equivalent to a series RC circuit with

$$R_f = \frac{1}{\omega C_f} = \frac{4RT \cosh^2(j/2)}{n^2 F^2 A C_O^* (2\omega D_O)^{\frac{1}{2}}} \tag{46}$$

Equations (45) and (46) could have been obtained from the classical faradaic-impedance equations for a reversible process by replacing bulk

TABLE I

Contributions to Faradaic Current Amplitudes from Solutions of Eq. (21) for Various Values of p^a

Current components	$p=0$	$p=1$	$p=2$	$p=3$	$p=4$	$p=5$
I_{dc}	$\dfrac{K}{(1+e')(\pi t)^{1/2}}$	0	$\dfrac{K\delta^2 \sinh(j/2)}{16(\pi t)^{1/2}\cosh^3(j/2)}$	0	$\dfrac{K\delta^4 \sinh(j/2)[\cosh^2(j/2)-3]}{256(\pi t)^{1/2}\cosh^5(j/2)}$	0
$I(\omega t)$	0	$\dfrac{K\delta\omega^{1/2}}{4\cosh^2(j/2)}$	0	$\dfrac{K\delta^3\omega^{1/2}[2\cosh^2(j/2)-3]}{64\cosh^4(j/2)}$	0	$\dfrac{K\delta^5\omega^{1/2}[2\cosh^4(j/2)-15\sinh^2(j/2)]}{1536\cosh^6(j/2)}$
$I(2\omega t)$	0	0	$\dfrac{2^{1/2}K\delta^2\omega^{1/2}\sinh(j/2)}{16\cosh^3(j/2)}$	0	$\dfrac{2^{1/2}K\delta^4\omega^{1/2}\sinh(j/2)[\cosh^2(j/2)-3]}{192\cosh^5(j/2)}$	0
$I(3\omega t)$	0	0	0	$\dfrac{3^{1/2}K\delta^3\omega^{1/2}[3-2\cosh^2(j/2)]}{192\cosh^4(j/2)}$	0	$\dfrac{3^{1/2}K\delta^5\omega^{1/2}[15\sinh^2(j/2)-2\cosh^4(j/2)]}{3072\cosh^6(j/2)}$
$I(4\omega t)$	0	0	0	0	$\dfrac{K\delta^4\omega^{1/2}\sinh(j/2)[3-\cosh^2(j/2)]}{384\cosh^5(j/2)}$	0
$I(5\omega t)$	0	0	0	0	0	$\dfrac{5^{1/2}K\delta^5\omega^{1/2}[2\cosh^4(j/2)-15\sinh^2(j/2)]}{15360\cosh^6(j/2)}$

$^a\delta = nF\Delta E/RT$; $K = nFAC_O^* D_O^{1/2}$; phase angle $= (-1)^{p+1}(\pi/4)$

concentration terms for species O and R with dc surface concentrations and substituting the Nernst equation for the latter (*10*). The success of this procedure for this case and for more complex mechanisms implies that the *small-amplitude* ac impedance is responsive to the dc concentrations primarily in the vicinity of the interface. This concept is important in gaining an intuitive understanding of theoretical results for more complex systems.

Expressions for the fundamental-harmonic alternating current (or impedance) suggest a number of criteria characteristic of a reversible ac polarographic wave. The most common of these are outlined in Table II.

TABLE II

CRITERIA FOR REVERSIBLE FUNDAMENTAL-HARMONIC AC POLAROGRAPHIC WAVE

Experimental parameter	Criteria (for all frequencies, dc potentials, and amplitudes, except where noted)
1. Phase angle	$45°$
2. Impedance	Equal resistive and capacitive components
3. Frequency dependence of $I(\omega t)$	Linear $I(\omega t) - \omega^{\frac{1}{2}}$ plot with intercept at origin
4. Peak potential of $I(\omega t)$	$[E_{dc}]_{peak} = E_{\frac{1}{2}}^r$
5. Width of wave	Half-width $= 90/n$ mV at 25°C (only for $\Delta E \leq 8/n$ mV)
6. Shape of wave	Linear plot of E_{dc} versus $\log [(I_p/I)^{\frac{1}{2}} - [(I_p - I)/I]^{\frac{1}{2}}]$ of slope $120/n$ mV at 25°C (only for $\Delta E \leq 8/n$ mV); has the form of the first derivative of the dc wave

The fact that the ac wave has the form of the first derivative of the dc polarographic wave applies only at small amplitudes and only for reversible processes. One can obtain from Eq. (36) the expression

$$E_{dc} = E_{\frac{1}{2}}^r + \frac{2RT}{nF} \ln\left[\left(\frac{I_p}{I}\right)^{\frac{1}{2}} - \left(\frac{I_p - I}{I}\right)^{\frac{1}{2}}\right] \qquad (47)$$

where I_p is the peak current corresponding to cosh $(j/2) = 1$, $[I \equiv I(\omega t)]$. Thus, one obtains criterion 6 in Table II. Examination of terms in ΔE^3 and ΔE^5 for the fundamental harmonic indicates that the ΔE^3 term is insignificant (at the 1 per cent level) for $\Delta E \leq 8/n$ mV and the term in ΔE^5 is insignificant for $\Delta E \leq 35/n$ mV.

Measurement of the second harmonic has held some interest and probably will find more application in the future. Although small, second-harmonic currents are accessible to reasonably precise experimental measurement. Criteria characteristic of a reversible second-harmonic wave are given in Table III. The second-derivative form of the second-harmonic wave is

TABLE III

CRITERIA FOR REVERSIBLE SECOND-HARMONIC AC POLAROGRAPHIC WAVE

Experimental parameter	Criteria (for all frequencies, dc potentials, and amplitudes, except where noted)
1. Phase angle	45° and 225° relative to fundamental harmonic on positive and negative peaks, respectively
2. Frequency dependence	Linear $I(2\omega t) - \omega^{\frac{1}{2}}$ plot with intercept at origin
3. Minimum potential	Minimum is a true null located at $[E_{dc}]_{min} = E_{\frac{1}{2}}^r$
4. Peak potentials	$[E_{dc}]_{peak} = E_{\frac{1}{2}}^r \pm (0.034/n)$ volt
5. Shape of wave	Second derivative of dc polarographic wave when $\Delta E \leq 8/n$ mV; linear $I(2\omega t) - E_{dc}$ dependence in vicinity of null of slope $2^{\frac{1}{2}}nFK\delta^2/32RT$ (see Table I for definitions of K and δ)

observed experimentally only if one employs a phase-sensitive detector. Conventional measurement of alternating current gives current amplitude without regard to sign, and thus one simply obtains two peaks. Second-harmonic ac polarograms obtained by phase-selective and conventional means were illustrated in Fig. 2. Criterion 5 of Table III is obtained by introducing the relations

$$\sinh\left(\frac{j}{2}\right) \sim \frac{j}{2} \tag{48}$$

$$\cosh\left(\frac{j}{2}\right) \sim 1.0 \tag{49}$$

for $j/2 < 0.30$, ($\pm 5\%$). This linear variation of the second-harmonic current with potential for a reversible process at potentials in the vicinity of the null point is seen in Fig. 2(b).

Little advantage is apparent in measurement of harmonics beyond the second, so these components will not be discussed here. It should be noted that the shape of third, fourth, etc., harmonic ac polarograms with a reversible process should have the form of the third, fourth, etc., derivatives of the dc polarographic wave.

The dc faradaic rectification component [Eq. (43)] shows exactly the same form of variation with potential as the second harmonic. In addition to this characteristic, it is frequency independent with a reversible process. As mentioned earlier, direct measurement of this component for quantitative purposes is difficult unless experimental conditions differ from those under consideration.

In general, theoretical literature agrees with the above equations for a reversible ac wave with planar diffusion. A few dissenting ideas have been presented, but appear invalid. Bauer (*107*) presented calculations indicating that the peak of the reversible fundamental harmonic ac polarographic wave would coincide with $E_{\frac{1}{2}}^{r}$ only if $D_O = D_R$. However, it was shown that an inconsistency in derivation rendered his equations valid only for equality of diffusion coefficients (*108–110*). Bauer also has contested the equation of Tachi and Senda (*66,78*) for the reversible second-harmonic wave [Eq. (44)], particularly regarding the shape of the wave. Bauer is led to this conclusion on the basis of equations he derived for the quasi reversible case which are seriously in error, except at high frequencies. This matter is discussed further in the next section. Bauer has since corrected this work (*78a*).

Experimental work also supports the validity of these equations. Although ac polarographic waves controlled solely by diffusion are relatively rare, sufficient examples of nearly ideal systems exist to test the theory. Results are encouraging regarding current magnitudes and phase angles for the small-amplitude fundamental harmonic (*21,27,30,81,111*) as well as the magnitude of the second harmonic (*81*). The ferric-ferrous redox system in oxalate media is one example of an electrode process showing essentially reversible ac polarographic behavior at audio frequencies. Several authors have found very close agreement between the above equations and a variety of experimental observations with this system, including (a) phase relations (*21,111*), (b) amplitude of fundamental harmonic at both the DME (*81,111*) and the hanging mercury drop (*27*), (c) magnitude of the second harmonic (*81*), (d) shape of the phase-selective second-harmonic polarogram (*34*), and (e) the separation of peaks for third-harmonic polarograms (*73,112*).

Bauer and Foo (*113*) have recently reported a comparison of experimental results with Matsuda's equation (*91*) for the reversible ac polarographic current at large amplitudes. Only fair agreement was found. These results may not be conclusive. Matsuda's equation gives the amplitude of the complex periodic signal including higher harmonics. However, Bauer and Foo employed a frequency-selective detector, and it appears that these authors measured primarily the fundamental harmonic rather than the amplitude of the combined harmonics. Bauer and Foo do not indicate that Matsuda's equations were revised to give only the fundamental-harmonic contribution or that the wave analyzer was modified to provide a broad-band pass. If neither step was taken, deviations between theory and experiment are to be expected. If the deviations between theory and experiment are not due to such experimental artefact, they likely arise from nonideality of the systems studied and not the effect described by these authors who propose that the discrepancies arise because the effective alternating potential is not the potential across the interface region, but some fraction of this potential. A concept of this nature is applicable to waves influenced by charge-transfer kinetics as is well known in the theory (*10,114*) for the effect of the structure of the double layer on rates of electrode reactions. To apply the same idea to the theory for reversible systems in which the electrode reaction exerts only a thermodynamic influence represents a considerable departure from existing thermodynamic concepts and is probably untenable (*115*). Potential terms in equations for reversible systems represent nernstian (thermodynamic) potentials, that is, the potential across the interface (relative to some reference).

C. The Quasi-Reversible AC Polarographic Wave

Theoretical treatment of the ac polarographic wave kinetically controlled by both charge transfer and diffusion (the "quasi-reversible" case) requires minor modification in the mathematical formulation of the reversible case. One deletes the assumption of nernstian behavior [Eq. (8)] and replaces it by the absolute rate expression (*116*)

$$\frac{i(t)}{nFAk_s} = C_{O_{x=0}} \exp\left[\frac{-\alpha nF}{RT}\left(E(t) - E^0\right)\right]$$

$$- C_{R_{x=0}} \exp\left[\frac{(1-\alpha)nF}{RT}\left(E(t) - E^0\right)\right] \quad (50)$$

One substitutes the expression for surface concentrations [Eqs. (10) and (11)] in Eq. (50) and, proceeding as for the reversible case, one obtains the system of integral equations

$$\psi_p(t) = \frac{k_s}{D^{\frac{1}{2}}}\left\{\frac{\alpha^p e^{-\alpha j}(\sin \omega t)^p}{p!}\right.$$

$$\left. - \sum_{r=0}^{p}\left[(\alpha^r e^{-\alpha j} + (-1)^r \beta^r e^{\beta j})\frac{(\sin \omega t)^r}{r!}\int_0^t \frac{\psi_{p-r}(t-u)\,du}{(\pi u)^{\frac{1}{2}}}\right]\right\} \tag{51}$$

where

$$\beta = 1 - \alpha \tag{52}$$

$$D = D_O^\beta D_R^\alpha \tag{53}$$

$\psi(t)$ and j have the same significance as above. The significance of solutions of Eq. (51) for various values of p is the same as in Eq. (21). The fundamental-harmonic ac wave is the solution of the integral equation

$$\psi_1(t) = \frac{k_s}{D^{\frac{1}{2}}}\left\{G_0(t)\sin \omega t - (e^{-\alpha j} + e^{\beta j})\int_0^t \frac{\psi_1(t-u)\,du}{(\pi u)^{\frac{1}{2}}}\right\} \tag{54}$$

where

$$G_0(t) = \alpha e^{-\alpha j} - (\alpha e^{-\alpha j} - \beta e^{\beta j})\int_0^t \frac{\psi_0(t-u)\,du}{(\pi u)^{\frac{1}{2}}} \tag{55}$$

Solving Eq. (54), as with the reversible case, one obtains

$$\psi_1(t) = \frac{(2\omega)^{\frac{1}{2}}G_0(t)}{(e^{-\alpha j} + e^{\beta j})[1 + (1 + ((2\omega)^{\frac{1}{2}}/\lambda))^2]^{\frac{1}{2}}}\sin[\omega t + \cot^{-1}(1 + ((2\omega)^{\frac{1}{2}}/\lambda))] \tag{56}$$

where

$$\lambda = \frac{k_s}{D^{\frac{1}{2}}}(e^{-\alpha j} + e^{\beta j}) \tag{57}$$

This result may be rearranged to a more convenient form through the relationship provided by the integral equation for the dc process:

$$\psi_0(t) = \frac{k_s}{D^{\frac{1}{2}}}\left\{e^{-\alpha j} + (e^{-\alpha j} + e^{\beta j})\int_0^t \frac{\psi_0(t-u)\,du}{(\pi u)^{\frac{1}{2}}}\right\} \tag{58}$$

followed by algebraic rearrangement and incorporation of Eq. (13) and (19) to yield

$$I(\omega t) = I_{rev}F(t)G(\omega^{\frac{1}{2}}\lambda^{-1})\sin\left[\omega t + \cot^{-1}\left(1 + \frac{(2\omega)^{\frac{1}{2}}}{\lambda}\right)\right] \tag{59}$$

where

$$F(t) = 1 + \frac{(\alpha e^{-j} - \beta)D^{\frac{1}{2}}\psi_0(t)}{k_s e^{-\alpha j}} \tag{60}$$

$$G(\omega^{\frac{1}{2}}\lambda^{-1}) = \left[\frac{2}{1 + (1 + [(2\omega)^{\frac{1}{2}}/\lambda])^2}\right]^{\frac{1}{2}} \tag{61}$$

and I_{rev} is the amplitude term for the reversible wave [Eq. (36)]. One completes the solution by incorporating the expression for $\psi_0(t)$:

$$\psi_0(t) = \frac{k_s e^{-\alpha j}}{D^{\frac{1}{2}}} e^{\lambda^2 t} \operatorname{erfc}(\lambda t^{\frac{1}{2}}) \tag{62}$$

obtained by solving the potentiostatic problem for planar diffusion [Eq. (58)] (12,117–119). From a theoretical standpoint, a critical test of Eq. (59) is that it reduces to the reversible wave equation when charge-transfer kinetics become very rapid, that is, when the conditions

$$(2\omega)^{\frac{1}{2}}/\lambda \ll 1 \tag{63}$$

and

$$\lambda t^{\frac{1}{2}} > 50 \tag{64}$$

apply. It is easily shown the Eq. (59) meets this requirement. In terms of an impedance, Eq. (59) is expressed as

$$Z_f = \frac{Z_{rev}}{F(t)}\left[\frac{1 + (1 + [(2\omega)^{\frac{1}{2}}/\lambda])^2}{2}\right]^{\frac{1}{2}} \tag{65}$$

which is equivalent to a series RC circuit represented by

$$R_f = R_{rev}\left(1 + \frac{(2\omega)^{\frac{1}{2}}}{\lambda}\right)\bigg/ F(t) \tag{66}$$

$$1/\omega C_f = R_{rev}/F(t) \tag{67}$$

Z_{rev} and R_{rev} represent the reversible impedances given by Eqs. (45) and (46). Equations (59) through (62) indicate that the influence of charge-transfer kinetics on the ac polarographic wave can effect a number of significant differences relative to the behavior predicted for the reversible wave. These differences are outlined in Table IV. The amplitude of the ac wave is written as the product of three terms: the amplitude of the reversible wave and two "correction" terms which differ from unity when significant deviations from nernstian behavior exist. $F(t)$ is responsive to conditions in the dc process [note the $\psi_0(t)$ term] and deviates from unity when the mean (or dc) surface concentrations are nonnernstian ($k_s < 10^{-2} \text{cm sec}^{-1}$, or $\lambda t^{\frac{1}{2}} < 50$). $G(\omega^{\frac{1}{2}}\lambda^{-1})$ accounts for conditions in the ac concentration profile and differs from unity when the fundamental harmonic ac concentration components are nonnernstian [$(2\omega)^{\frac{1}{2}}/\lambda > 0.01$]. One of the significant aspects of Eq. (59) lies in the relative simplicity of the phase-angle relation,

TABLE IV

Criteria for Quasi-Reversible Fundamental-Harmonic ac Polarographic Wave

Experimental parameter	Criteria (apply only for $\Delta E \leq 8/n$ mV)
1. Phase angle	Cot ϕ varies linearly with $\omega^{\frac{1}{2}}$; intercept at 1.00; slope of $2^{\frac{1}{2}}/\lambda$; cot ϕ versus E_{dc} plot exhibits a maximum with peak potential and magnitude given by Eqs. (68) and (69)
2. Impedance	Resistive and capacitive components differ by $[R_{rev}/F(t)](2\omega)^{\frac{1}{2}}/\lambda$
3. Frequency dependence of $I(\omega t)$	Nonlinear, approaching linearity at low frequencies with intercept at origin and approaching limiting amplitude of $$\frac{n^2 F^2 A C_O^* D_O^{\frac{1}{2}} \Delta E \lambda F(t)}{4RT \cosh^2(j/2)}$$ at high frequencies
4. Peak potential of $I(\omega t)$	Varies with frequency, approaching $E_{\frac{1}{2}}^r$ at low frequency and, for $k_s > 10^{-2}$ cm sec^{-1}, the high-frequency limit is given by Eq. (70)
5. Shape of wave	Does not have form of derivative of dc wave except when $\alpha = 0.5$ and $k_s > 10^{-2}$ cm sec^{-1}; for other conditions, shape is strongly dependent on k_s, α, and ω; when $k_s < 10^{-2}$ cm sec^{-1} a secondary peak or shoulder may be exhibited with small α

particularly its independence of $F(t)$. At any point along the ac wave, cot ϕ varies linearly with $\omega^{\frac{1}{2}}$ (see Table IV). A plot of cot ϕ versus E_{dc} at any frequency has a maximum at the potential

$$[E_{dc}]_{max} = E_{\frac{1}{2}}^r + \frac{RT}{nF} \ln\left(\frac{\alpha}{\beta}\right) \tag{68}$$

with a magnitude

$$[\cot \phi]_{max} = 1 + \frac{(2\omega D)^{\frac{1}{2}}}{k_s[(\alpha/\beta)^{-\alpha} + (\alpha/\beta)^{\beta}]} \tag{69}$$

Thus, one can determine α and, if $D^{\frac{1}{2}}$ is known, k_s from measurement of position and magnitude of maximum cot ϕ. Use of these simple relationships to obtain charge-transfer kinetic parameters was first proposed by Tamamushi and Tanaka (*24*) and later applied by others (*32,33*). Taking the difference between the resistive and capacitive components of the faradaic impedance [Eqs. (66) and (67)] represents another relatively simple approach to data analysis derived from classical faradaic-impedance measurements (*6,10,12,21,22*). However, when the dc process is nonnernstian [$F(t) \neq 1$], this method loses its simplicity unless one employs the equilibrium method [$i_{dc} = 0$, $F(t) = 1$].

Differentiation of the expression for current amplitude with respect to E_{dc} to obtain similar expressions for position and magnitude of peak current yields intractable forms. This is true, even if one can assume that the dc process is nernstian. However, for the latter case the high-frequency limit ($(2\omega)^{\frac{1}{2}}/\lambda \gg 1$) can be treated, giving the results (*9,120*)

$$[E_{dc}]_{peak} = E^r_{\frac{1}{2}} + \frac{RT}{nF} \ln \frac{\beta}{\alpha} \tag{70}$$

$$[I(\omega t)]_{peak} = \frac{n^2 F^2 A C^*_O \, \Delta E k_s \beta^\beta \alpha^\alpha}{RT} \left(\frac{D_O}{D_R}\right)^{\alpha/2} \sin \omega t \tag{71}$$

Influence on the predicted ac wave of various values of k_s and α are illustrated in Fig. 3, 4, and 5. An interesting aspect of the ac polarographic wave observed under conditions of nonnernstian dc behavior is the time-dependence introduced by the $F(t)$ term. The net result is that the ac wave obtained at a DME will depend on mercury-column height (*121–126*). The influence of drop life under these conditions is illustrated in Fig. 5. Note the dependence of the position and magnitude of the peak current on column height (drop life). The phase angle with a quasi-reversible process is not subject to $F(t)$ and, thus, no time-dependence is predicted. The cross-over point seen in Fig. 5 is predicted to occur at the dc potential

$$E^*_{dc} = E^r_{\frac{1}{2}} + \frac{RT}{nF} \ln\left(\frac{\alpha}{\beta}\right) \tag{72}$$

Measurement of this potential has been proposed for the determination of the charge-transfer coefficient (*121*).

The complexities introduced into the ac wave equation by nonnernstian dc behavior go well beyond that indicated in Eqs. (60) and (62). Accurate calculation of $F(t)$ for such systems requires incorporation of contributions of drop growth and geometry, as will be shown later. The I_{rev} and $G(\omega^{\frac{1}{2}}\lambda^{-1})$ terms do not require such modification. The significance of the fact that the phase angle is independent of $F(t)$ is enhanced by the latter considerations.

For kinetic investigations, the study of the quasi-reversible fundamental harmonic ac process has received considerably greater interest than any

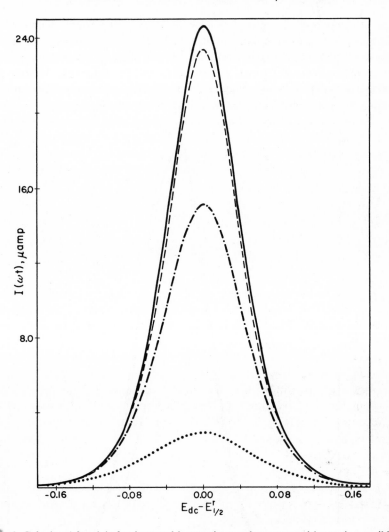

Fig. 3. Calculated faradaic fundamental-harmonic ac polarograms with quasi-reversible systems. Parameters: $\omega = 2500$ sec^{-1}, $\alpha = 0.500$; $D = 9 \times 10^{-6}$ cm^2 sec^{-1}, $A = 0.035$ cm^2; $C_0^* = 1.00 \times 10^{-3}M$, $T = 298°$K; $\Delta E = 5.00$ mV, $n = 1$. —, $k_S = \infty$; ---, $k_S = 1$; — · —, $k_S = 0.1$; ···, $k_S = 0.01$. Shows faradaic ac at end of drop life.

other type of kinetic scheme. Much of this work was carried out by employing the classical faradaic-impedance measurement at the equilibrium potential ($i_{dc} = 0$). Considerable kinetic data for the charge-transfer process has been the reward of such efforts [cf., Breyer and Bauer (*9*) for review]. Verification of many important predictions of the above equations have resulted, particularly with regard to the frequency dependence of the impedance components and phase angles and the difference between the resistive- and capacitive-impedance components. More recently, studies have been carried out under actual ac polarographic conditions (net dc

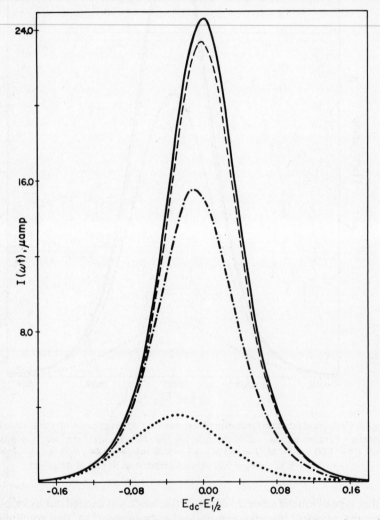

Fig. 4. Calculated faradaic fundamental-harmonic ac polarograms with quasi-reversible systems. Parameters: Same as Fig. 3, except $\alpha = 0.800$.

current flow). Predictions regarding phase angle have been subject to careful test, and the above equations adequately account for the behaviour of several systems (24,32,33,111). Figure 6 shows some experimental data obtained with Ti^{4+}/Ti^{3+}. Randles' measurements of impedance components under ac polarographic conditions appear in fair accord with theory (127). Recent studies of the time dependence of ac polarographic waves indicate that the phenomena appears when expected (121–125). In this regard,

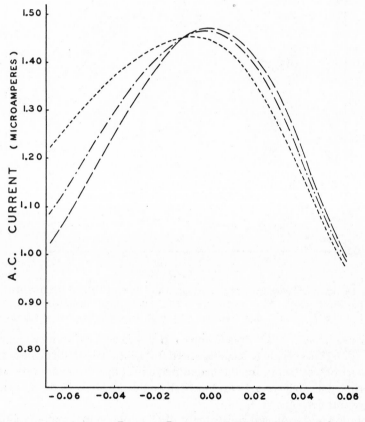

Fig. 5. Calculated faradaic fundamental-harmonic ac polarograms with quasi-reversible systems [from (121)]. Parameters: $\omega = 40\pi$ sec^{-1}; $\alpha = 0.400$; $D = 5.0 \times 10^{-6}$ cm^2 sec^{-1}; $A = 0.035$ cm^2; $C_O^* = 5.00 \times 10^{-3}$ M; $T = 298°$K; $\Delta E = 5.00$ mV; $n = 1$; $k_s = 1.0 \times 10^{-3}$ cm sec^{-1}. ——, $\tau = 12.0$ sec; — — —, $\tau = 9.0$ sec; – – –, $\tau = 5.0$ sec.; $\tau =$ drop life, calculation based on expanding-plane model. (Courtesy of *Analytical Chemistry*.)

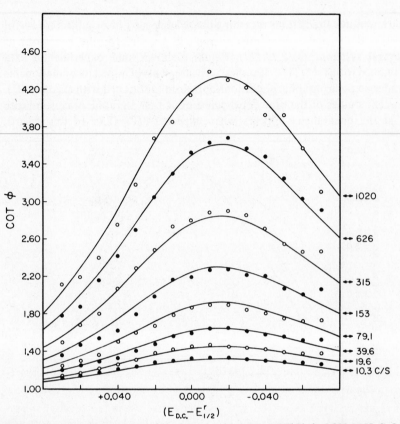

Fig. 6. Dc potential dependence of phase angle with 3.36 mM TiCl$_4$ in 0.200 M H$_2$C$_2$O$_4$ solution [from *(33)*]. $\Delta E = 5.00$ mV, $T = 25°$C. Solid lines represent theory for $k_s = 4.6 \times 10^{-2}$, $\alpha = 0.35$, and $D = 0.66 \times 10^{-5}$. (Courtesy of *Analytical Chemistry.*)

results with the V^{3+}/V^{2+} system appear in accord with the above equations (*111,121*). However, there remains a need for additional careful quantitative comparisons of theory for the magnitude and shape of the ac wave with experimental results on systems kinetically influenced only by diffusion and charge transfer.

Solution of the integral equation for $p = 2$ to obtain the small-amplitude second-harmonic and dc rectification components is readily accomplished and contains no features different from the treatment of the reversible case. For the dc rectification component, one obtains

$$I_2(\text{dc}) = \frac{n^3 F^3 A C_O^* D_O^{\frac{1}{2}} \, \Delta E^2 \lambda J e^{\lambda^2 t} \operatorname{erfc}(\lambda t^{\frac{1}{2}})}{2R^2 T^2} \qquad (73)$$

and, for the second-harmonic component,

$$I(2\omega t) = \frac{n^3 F^3 A C_O^*(\omega D_O)^{\frac{1}{2}} \Delta E^2}{R^2 T^2} \left[\frac{J^2 + H^2}{1 + (1 + (2\omega^{\frac{1}{2}}/\lambda))^2} \right]^{\frac{1}{2}}$$

$$\times \sin\left[2\omega t + \cot^{-1} \frac{J + (1 + (2\omega^{\frac{1}{2}}/\lambda))H}{H - (1 + (2\omega^{\frac{1}{2}}/\lambda))J} \right] \quad (74)$$

where

$$J = \frac{1}{8\cosh^3(j/2)} \left\{ \frac{\alpha^2}{2} (e^{j/2} + 2e^{-j/2} + e^{-3j/2}) \right.$$

$$- \frac{(\alpha e^{-j/2} - \beta e^{j/2})(2 + ((2\omega)^{\frac{1}{2}}/\lambda))}{[1 + (1 + ((2\omega)^{\frac{1}{2}}/\lambda))^2]} \left[1 + \frac{(\alpha e^{-j} - \beta)D^{\frac{1}{2}}\psi_0(t)}{k_s e^{-\alpha j}} \right]$$

$$\left. - \frac{(\alpha^2 e^{-j} + \beta^2)(e^{-j/2} + e^{j/2})}{2} \left[1 - \frac{(1 + e^j)}{\lambda} \psi_0(t) \right] \right\} \quad (75)$$

$$H = \frac{1}{8\cosh^3(j/2)} \left\{ \frac{(\alpha e^{-j/2} - \beta e^{j/2})((2\omega)^{\frac{1}{2}}/\lambda)}{[1 + (1 + ((2\omega)^{\frac{1}{2}}/\lambda))^2]} \left[1 + \frac{(\alpha e^{-j} - \beta)D^{\frac{1}{2}}\psi_0(t)}{k_s e^{-\alpha j}} \right] \right\}$$

$$76)$$

and $\psi_0(t)$ is given by Eq. (62). The expression for the second-harmonic current was derived by Paynter et al. (*73,128*). It has been discussed by Reinmuth (*8*), but has not been published at the time of this writing. As is expected, both Eqs. (73) and (74) reduce to the corresponding expressions for the reversible case [Eqs. (43) and (44)] when charge-transfer kinetics are rapid relative to diffusion [Eqs. (63) and (64)]. Also, when the dc process is nernstian, terms containing $\psi_0(t)$ are negligible. The algebraic complexity of the second-harmonic expression precludes the derivation of simple relationships for the separation of peaks, position of minima, etc. However, it appears that difficulties associated with the cumbersome nature of Eq. (74) can be overcome through the use of high-speed computers. Detailed calculations of the predictions of Eq. (74) can be effected rapidly, the results of which may then be expressed in graphic form and compared with experimental data. Efforts along these lines are in progress (*78,129,130*).

Some conventional second-harmonic polarograms predicted by Eq. (74) are given in Figs. 7, 8, and 9. One can see that the symmetry of the second-harmonic polarogram (or the second-derivative form in phase-selective measurements) is lost when charge-transfer kinetics play an important role. The wave loses its symmetry under such conditions when $\alpha \neq 0.50$

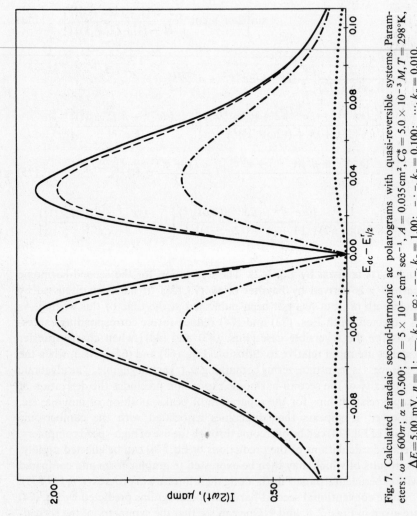

Fig. 7. Calculated faradaic second-harmonic ac polarograms with quasi-reversible systems. Parameters: $\omega = 600\pi$; $\alpha = 0.500$; $D = 5 \times 10^{-5}\ \text{cm}^2\ \text{sec}^{-1}$, $A = 0.035\ \text{cm}^2$, $C_0^* = 5.0 \times 10^{-3}\ M$, $T = 298°\text{K}$, $\Delta E = 5.00\ \text{mV}$, $n = 1$; —, $k_s = \infty$; — —, $k_s = 1.00$; — · —, $k_s = 0.100$; · · ·, $k_s = 0.010$.

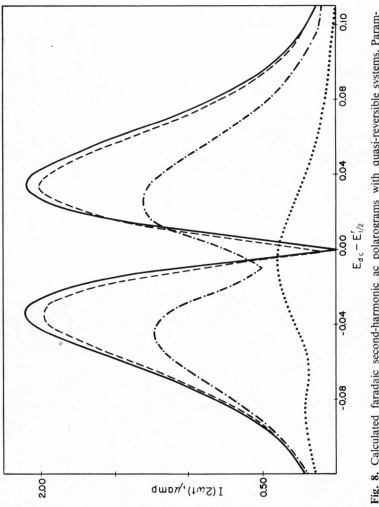

Fig. 8. Calculated faradaic second-harmonic ac polarograms with quasi-reversible systems. Parameters: Same as Fig. 7 except $\alpha = 0.800$.

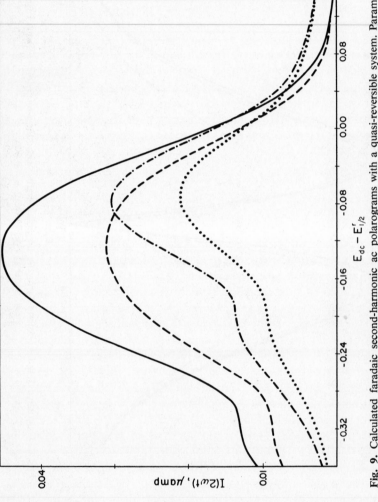

Fig. 9. Calculated faradaic second-harmonic ac polarograms with a quasi-reversible system. Parameters: $\alpha = 0.500$, $k_s = 1.00 \times 10^{-3}$ cm sec^{-1}, $D = 1.0 \times 10^{-5}$ cm^2 sec^{-1}, $A = 0.035$ cm^2, $C_0^* = 5.0 \times 10^{-3}$ M, $T = 298°$K, $n = 1.00$, $\Delta E = 5.00$ mV; —, $\omega^{1/2} = 90$ sec$^{-1/2}$, $\tau = 4$ sec; - - -, $\omega^{1/2} = 90$ sec$^{-1/2}$, $\tau = 9$ sec; - · -, $\omega^{1/2} = 10$ sec$^{-1/2}$, $\tau = 4$ sec; · · ·, $\omega^{1/2} = 10$ sec$^{-1/2}$, $\tau = 9$ sec. τ = drop life; calculation based on expanding plane model.

and even when $\alpha = 0.50$, if $\lambda t^{\frac{1}{2}} < 50$ ($k_s < 10^{-2}$cm sec^{-1}). The second-harmonic ac polarogram with a nonnernstian dc process is predicted to be extremely sensitive to mercury column height, as seen in Fig. 9.

Recently, Bauer (78) has published equations for the second-harmonic current with a quasi-reversible process which are at variance with Eq. (74). Bauer assumed that the dc polarographic process is nernstian and that second-harmonic concentration changes may be neglected. The former assumption is not serious, because it is expected that most interest in second-harmonic measurements will be concerned with rapid charge-transfer processes. However, neglect of second-harmonic concentration changes yields equations which are accurate only at relatively high frequencies, as pointed out by Reinmuth and Paynter (8,73). In this respect Bauer's result is similar to van Cakenberghe's equation (79) which neglects both fundamental- and second-harmonic concentration terms, but is somewhat more accurate. It can be shown that if one assumes nernstian dc behavior, Equation (74) differs from Bauer's less-exact expression by the factor (130)

$$\left[\left(\frac{\lambda}{2\omega^{\frac{1}{2}}}\right)^2 + \left(\frac{\lambda}{2\omega^{\frac{1}{2}}} + 1\right)^2\right]^{\frac{1}{2}}$$

and, thus, the results are the same only when $2\omega^{\frac{1}{2}}/\lambda \gg 1$.

Despite the potentialities of second-harmonic measurements and the sensitivity of the second harmonic to kinetic influence, no detailed quantitative kinetic studies have been reported.

The equations for the fundamental-harmonic component with a quasi-reversible system have been extended to include larger amplitudes ($\Delta E \leq 35/n$ mV, approx.) by calculating the fundamental-harmonic term arising from the solution to Eq. (51) for $p = 3$ (131). In view of difficulties in derivation and application associated with the extremely cumbersome nature of the equations, one may question whether there is any practical basis for undertaking this task. It is conceded that measurements at sufficiently small amplitudes may be accomplished readily with good quality instrumentation and that dc, fundamental harmonic and/or second harmonic data should suffice in most any kinetic investigation. However, motivations to carry out these calculations were supplied by the following considerations. (a) The magnitude of the large-amplitude terms (terms in ΔE^3, etc.) relative to the small-amplitude term should be a function of the charge-transfer-rate parameters. Thus, the guide provided by the theory for the reversible case, which indicates that the small-amplitude expression is sufficiently accurate for $\Delta E \leq 8/n$ mv may not apply when charge-transfer kinetics are important. (b) The phase angle between the fundamental-harmonic and applied potential is predicted to be independent of amplitude for the reversible case. Although this is not expected to occur with the quasi-reversible case, it is possible that the phase angle will be relatively insensitive to ΔE so that measurements made at amplitudes larger than $\Delta E = 8/n$ mV, where instrument-noise levels are a less serious problem, may still obey Eqs. (59), (68), and (69) with acceptable accuracy. (c) Possibly some advantages of larger-amplitude measurements, not apparent from intuitive concepts, may become evident upon examination of the equations. (d) High-speed computers enable extensive calculation, even from the most cumbersome algebraic relationships.

The expression for the quasi-reversible fundamental-harmonic ac wave for amplitudes up to $\Delta E = 35$ mV and $k_s > 10^{-2}$ is given by (*131*)

$$I(\omega t) = \frac{n^2 F^2 A C_O^* D_O^{\frac{1}{2}} \, \Delta E}{RT} \left\{ \left[A_1 + \left(\frac{nF \, \Delta E}{RT} \right)^2 A_2 \right]^2 + \left[B_1 + \left(\frac{nF \, \Delta E}{RT} \right)^2 B_2 \right]^2 \right\}^{\frac{1}{2}}$$

$$\times \sin \left[\omega t + \cot^{-1} \left(\frac{A_1 + (nF \, \Delta E/RT)^2 A_2}{B_1 + (nF \, \Delta E/RT)^2 B_2} \right) \right] \tag{77}$$

where

$$A_1 = \frac{(2\omega)^{\frac{1}{2}}(1 + 2^{\frac{1}{2}}Z)}{4 \cosh^2(j/2)[1 + (1 + 2^{\frac{1}{2}}Z)^2]} \tag{78}$$

$$B_1 = \frac{(2\omega)^{\frac{1}{2}}}{4 \cos \mathrm{h}^2(j/2)[1 + (1 + 2^{\frac{1}{2}}Z)^2]} \tag{79}$$

$$A_2 = \left\{ (2\omega)^{\frac{1}{2}} \left[256 \cosh^3 \left(\frac{j}{2} \right) [1 + 3.414Z + 3.828Z^2 + 4.828Z^3 + 2.00Z^4] \right. \right.$$

$$\left. \times [1 + (1 + 2^{\frac{1}{2}}Z)^2] \right]^{-1} \right\}$$

$$\times \left\{ 8(\alpha^3 + \beta^3)(1 + 4.828Z + 8.655Z^2 + 10.241Z^3 + 8.827Z^4 + 2.828Z^5) \right.$$

$$\times \cosh \left(\frac{j}{2} \right) - 12(\alpha e^{-j/2} - \beta e^{j/2})(2\alpha - 1)$$

$$\times (1 + 4.828Z + 9.517Z^2 + 10.517Z^3 + 6.356Z^4 + 1.885Z^5)$$

$$+ \frac{12(\alpha e^{-j/2} - \beta e^{j/2})^2}{\cosh(j/2)} (1 + 4.357Z + 6.104Z^2 + 4.161Z^3 + 1.333Z^4)$$

$$\left. - 12(\alpha^2 e^{-j/2} + \beta^2 e^{j/2})(1 + 4.357Z + 7.713Z^2 + 6.713Z^3 + 2.000Z^4) \right\}$$

$$\tag{80}$$

$$B_2 = \left\{ (2\omega)^{\frac{1}{2}} \left[256 \cosh^3\left(\frac{j}{2}\right) [1 + 3.414Z + 3.828Z^2 + 4.828Z^3 + 2.00Z^4] \right.\right.$$

$$\left.\left. \times [1 + (1 + 2^{\frac{1}{2}}Z)^2] \right]^{-1} \right\}$$

$$\times \left\{ 8(\alpha^3 + \beta^3)(1 + 3.414Z + 3.828Z^2 + 4.828Z^3 + 2.00Z^4)\cosh\left(\frac{j}{2}\right) \right.$$

$$- 12(\alpha e^{-j/2} - \beta e^{j/2})(2\alpha - 1)$$

$$\times (1 + 2.747Z + 3.747Z^2 + 2.552Z^3 + 0.862Z^4)$$

$$+ \frac{12(\alpha e^{-j/2} - \beta e^{j/2})^2}{\cosh(j/2)}(1 + 1.805Z + Z^2 + 0.276Z^3)$$

$$\left. - 12(\alpha^2 e^{-j/2} + \beta^2 e^{j/2})(1 + 2.471Z + 2.609Z^2 + 0.276Z^3 - 0.667Z^4) \right\}$$

$$\tag{81}$$

$$Z = \omega^{\frac{1}{2}}/\lambda \tag{82}$$

It can be shown that Eq. (77) reduces to the corresponding expression for the reversible case when $\omega^{\frac{1}{2}}/\lambda \to 0$. The extremely cumbersome nature of these expressions obviously dictates against their direct application in kinetic studies. Although we are uncertain that the rewards justified the efforts in deriving these expressions, some insight into the importance and behavior of the large-amplitude correction was gained. Figures 10 and 11 illustrate some predictions of the equations regarding the influence of ΔE on current amplitude and phase angle for a quasi-reversible system. Results of detailed calculations show that the relative importance of the term in ΔE^3 is dependent on k_s and α, as expected. However, the effect is not sufficiently pronounced to invalidate the statement, based on theory for the reversible system, that the small-amplitude expression is accurate for $\Delta E \le 8/n$ mV. Indeed, it appears that the small-amplitude equations can be applied to data obtained at significantly larger amplitudes without introducing significant error for many types of measurements with quasi-reversible systems. For example, application of Eqs. (68) and (69) appears sufficiently accurate up to amplitudes of $\Delta E = 20/n$ mV (cf., Figs. 10 and 11). Extension of these equations to $k_s < 10^{-2}$ (*131*) indicates that time-dependent terms appear in both amplitude and phase-angle relationships at large amplitudes, whereas only in the former at small amplitudes. One must keep this effect in mind when attempting to draw mechanistic conclusions based on time-dependent studies.

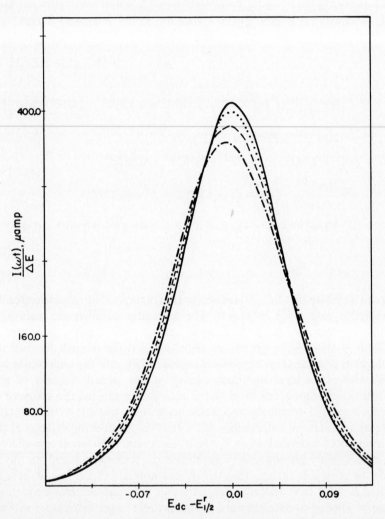

Fig. 10. Calculated faradaic fundamental-harmonic ac polarograms with a quasi-reversible system and varying ΔE. Parameters: $\alpha = 0.200$, $k_s = 1.0 \times 10^{-2}$ cm sec^{-1}, $\omega = 40\pi$, $D = 5.00 \times 10^{-6}$ cm^2 sec^{-1}, $A = 0.035$ cm^2, $C_O^* = 1.0 \times 10^{-3}$ M, $T = 298°$K, $n = 1.00$; —, $\Delta E = 5.00$ mV, $\cdots$, $\Delta E = 20.0$ mV, – –, $\Delta E = 30.0$ mV, — $\cdot$ —, $\Delta E = 40.0$ mV.

D. Systems with Coupled Chemical Reactions

Electrode processes in which currents are influenced by kinetics of coupled homogeneous chemical reactions are well known from studies of

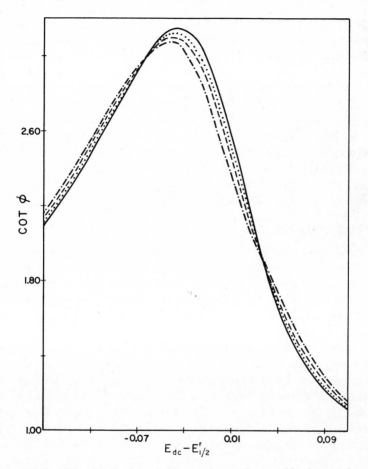

Fig. 11. Calculated faradaic fundamental-harmonic ac polarographic phase angles with a quasi-reversible system and varying ΔE. Parameters: Same as Fig. 10.

the dc polarographic wave. Considerable work, both theoretical and experimental, has been devoted to this class of electrode reactions in dc polarography, particularly by the Czechoslovakian school [cf., for example, (*132–136*)]. Although few studies have been reported, it is likely that this type of process will prove even more prevalent in ac polarography. One is led to this conclusion because ac polarography, like the other electro-chemical-relaxation techniques, involves measurements over effective periods of time considerably shorter than in dc polarography. The enhanced rates of mass transfer associated with short time measurements

(*10*) increase the likelihood of kinetic influence by other rate processes such as homogeneous chemical reactions. Chemical rate processes exhibiting only thermodynamic influence on the dc process may show a decided kinetic effect in ac polarography. The same reasoning suggests that kinetic effects due to multistep chemical kinetic schemes (*132–145*) may prove important in ac polarography, whereas they appear only infrequently in dc polarography. Therefore, consideration of the theory of ac polarographic waves influenced by coupled homogeneous chemical reactions appears important, not only because ac polarography shows promise as a technique for obtaining rate parameters for rapid chemical reactions, but also because proper interpretation of much ac polarographic data may require consideration of influence of coupled chemical reactions.

1. First-Order Reactions

Derivation of ac wave equations for systems with coupled chemical reactions is accomplished for systems with first-order or pseudo-first-order reactions in essentially the same manner as the cases treated above. The problem of handling the nonlinear partial-differential equations associated with second- and higher-order chemical kinetics involves special difficulties and will be discussed separately. Development of a theory for the ac polarographic wave with coupled chemical reactions has been rather recent. Gerischer (*95*) gave expressions for the faradaic impedance at the equilibrium potential (zero direct current) for systems with coupled preceding reactions of any order. Matsuda et al. (*146*) later considered the first-order preceding reaction in more detail. Delahay (*10*) suggested a method for extending Gerischer's equations to ac polarographic conditions. Delahay's approach contains the assumption that the dc polarographic process is diffusion controlled. A communication by Satyanarayana and co-workers (*147*) presented equations without derivation for the same system. However, these equations appear inconsistent with the work of others. More recently Aylward and co-workers (*122,123,150*) and Smith and co-workers (*33, 121,129,137,148,149*) have extended the theory.

The method of derivation will be outlined for the system with a preceding first-order chemical reaction designated by the scheme

$$Y \underset{k_2}{\overset{k_1}{\rightleftharpoons}} O + ne \rightleftharpoons R$$

Reactants present in large excess are neglected for simplicity. Their concentrations are incorporated in the rate and equilibrium constants. For this process, the system of partial-differential equations, initial, and boundary conditions is (*12,148,151–153*)

$$\frac{\partial C_Y}{\partial t} = D_Y \frac{\partial^2 C_Y}{\partial x^2} - k_1 C_Y + k_2 C_O \tag{83}$$

$$\frac{\partial C_O}{\partial t} = D_O \frac{\partial^2 C_O}{\partial x^2} + k_1 C_Y - k_2 C_O \tag{84}$$

$$\frac{\partial C_R}{\partial t} = D_R \frac{\partial^2 C_R}{\partial x^2} \tag{85}$$

For $t = 0$, any x,

$$C_Y + C_O = C_Y^* + C_O^* = C^0 \tag{86a}$$

$$\frac{C_O}{C_Y} = \frac{C_O^*}{C_Y^*} = \frac{k_1}{k_2} = K \tag{86b}$$

$$C_R = 0 \tag{86c}$$

For $t > 0$, $x \to \infty$,

$$C_Y + C_O \to C^0 \tag{87a}$$

$$\frac{C_O}{C_Y} \to K \tag{87b}$$

$$C_R \to 0 \tag{87c}$$

For $t > 0$, $x = 0$,

$$D_O \frac{\partial C_O}{\partial x} = -D_R \frac{\partial C_R}{\partial x} = \frac{i(t)}{nFA} \tag{88a}$$

$$D_Y \frac{\partial C_Y}{\partial x} = 0 \tag{88b}$$

The effect of the chemical reaction is incorporated into the boundary-value problem by introducing chemical rate terms into the Fick's law equations, chemical equilibrium conditions for $t = 0$ and $x \to \infty$ and a statement that species Y is not electroactive [Eq. (88b)]. Introducing the linear transformations

$$\Psi = C_Y + C_O \tag{89}$$

$$\theta = \left[C_Y - \frac{1}{K} C_O \right] e^{kt} \tag{90}$$

where

$$k = k_1 + k_2 \tag{91}$$

and, assuming $D_Y = D_O$, a new set of equations is obtained in terms of the variables Ψ and θ (*132,148,151,152*)

$$\frac{\partial \Psi}{\partial t} = D_O \frac{\partial^2 \Psi}{\partial x^2} \tag{92}$$

$$\frac{\partial \theta}{\partial t} = D_O \frac{\partial^2 \theta}{\partial x^2} \tag{93}$$

$$\frac{\partial C_R}{\partial t} = D_R \frac{\partial^2 C_R}{\partial x^2} \tag{94}$$

For $t = 0$, any x,

$$\Psi = C^0 \tag{95a}$$

$$\theta = 0 \tag{95b}$$

$$C_R = 0 \tag{95c}$$

For $t > 0$, $x \to \infty$,

$$\Psi \to C^0 \tag{96a}$$

$$\theta \to 0 \tag{96b}$$

$$C_R \to 0 \tag{96c}$$

For $t > 0$, $x = 0$,

$$\frac{\partial \Psi}{\partial x} = \frac{i(t)}{nFAD_O} \tag{97a}$$

$$\frac{\partial \theta}{\partial x} = -\frac{i(t)e^{kt}}{KnFAD_O} \tag{97b}$$

$$\frac{\partial C_R}{\partial x} = -\frac{i(t)}{nFAD_R} \tag{97c}$$

The problem has been reduced to one of solving simple Fick's second-law differential equations with initial and boundary conditions different from those for the cases considered earlier. Routine application of the method of Laplace transformation yields, for surface concentrations of the electroactive forms (*148,151*),

$$C_{O_{x=0}} = C_O^* - \frac{1}{(1 + K)} \int_0^t \frac{e^{-ku} i(t - u)\, du}{nFA(D_O \pi u)^{\frac{1}{2}}} - \frac{K}{1 + K} \int_0^t \frac{i(t - u)\, du}{nFA(D_O \pi u)^{\frac{1}{2}}} \tag{98}$$

$$C_{R_{x=0}} = \int_0^t \frac{i(t - u)\, du}{nFA(D_R \pi u)^{\frac{1}{2}}} \tag{99}$$

These relationships are then substituted in the absolute rate expression [Eq. (50)], to include also effects of charge-transfer kinetics. Proceeding as with the reversible and quasi-reversible cases, one obtains the system of integral equations

$$\psi_p(t) = \frac{k_s}{D^{\frac{1}{2}}} \left\{ \alpha^p e^{-\alpha j} \frac{(\sin \omega t)^p}{p!} \right.$$

$$- \sum_{r=0}^{p} \left[\frac{\alpha^r e^{-\alpha j}}{1+K} \frac{(\sin \omega t)^r}{r!} \int_0^t \frac{e^{-ku} \psi_{p-r}(t-u)\, du}{(\pi u)^{\frac{1}{2}}} \right.$$

$$\left. \left. + \left(\frac{K\alpha^r e^{-\alpha j}}{1+K} + (-1)^r \beta^r e^{\beta j} \right) \frac{(\sin \omega t)^r}{r!} \int_0^t \frac{\psi_{p-r}(t-u)\, du}{(\pi u)^{\frac{1}{2}}} \right] \right\} \quad (100)$$

where $\psi_p(t)$ and p are defined by Eqs. (19) and (20). Solution to the integral equations for $p = 1, 2$, etc., is accomplished as before with the aid of the additional relationships (*148*)

$$\int_0^\infty \frac{e^{-ku} \cos \omega u\, du}{(\pi u)^{\frac{1}{2}}} = \left[\frac{(\omega^2 + k^2)^{\frac{1}{2}} + k}{2(\omega^2 + k^2)} \right]^{\frac{1}{2}} \quad (101)$$

$$\int_0^\infty \frac{e^{-ku} \sin \omega u\, du}{(\pi u)^{\frac{1}{2}}} = \left[\frac{(\omega^2 + k^2)^{\frac{1}{2}} - k}{2(\omega^2 + k^2)} \right]^{\frac{1}{2}} \quad (102)$$

For the fundamental-harmonic component, one obtains (*148,149*)

$$I(\omega t) = I_{rev} F_{dc}(t) G_{ac}(\omega) \sin\left(\omega t + \cot^{-1} \frac{V}{S} \right) \quad (103)$$

where

$$F_{dc}(t) = (1 + e^{-j}) \left[\frac{\alpha D^{\frac{1}{2}} \psi_0(t)}{k_s e^{-\alpha j}} + e^j \int_0^t \frac{\psi_0(t-u)\, du}{(\pi u)^{\frac{1}{2}}} \right] \quad (104)$$

$$\psi_0(t) = \frac{i_{dc}(t)}{nFAC_O^* D_O^{\frac{1}{2}}} \quad (105)$$

$$G_{ac}(\omega) = \left[\frac{2}{V^2 + S^2} \right]^{\frac{1}{2}} \quad (106)$$

$$V = \frac{(2\omega)^{\frac{1}{2}}}{\lambda} + \frac{1}{(1+e^j)} \left\{ e^j + \frac{K}{1+K} + \frac{1}{1+K} \left[\frac{(1+g^2)^{\frac{1}{2}} + g}{1+g^2} \right]^{\frac{1}{2}} \right\} \quad (107)$$

$$S = \frac{1}{(1+e^j)} \left\{ e^j + \frac{K}{1+K} + \frac{1}{1+K} \left[\frac{(1+g^2)^{\frac{1}{2}} - g}{1+g^2} \right]^{\frac{1}{2}} \right\} \quad (108)$$

$$g = \frac{k}{\omega} = \frac{k_1 + k_2}{\omega} \quad (109)$$

and the other terms have the same significance as above. Note that the $\psi_0(t)$ term is given in the general form of Eq. (105) and not by Eq. (62).

The latter is the specific form of $\psi_0(t)$ for the quasi-reversible dc system only. The significance of the $F_{dc}(t)$ and $G_{ac}(\omega)$ terms is the same as for the analogous terms in the quasi-reversible ac wave equation, $F(t)$ and $G(\omega^{\frac{1}{2}}\lambda^{-1})$. $F_{dc}(t)$ is responsive to kinetic influence of either chemical reaction or charge transfer on the *dc* process. $G_{ac}(\omega)$ accounts for similar effects on the ac process. One impasse arises in attempting to give completely general expressions for this system. Closed-form solutions for the dc polarographic current with consideration of simultaneous kinetic contribution of chemical reaction and charge transfer have not been effected. Thus, a general expression for $\psi_0(t)$ and $F_{dc}(t)$ cannot be given. Solution is possible for the following special cases: (a) the dc process is diffusion controlled; (b) the dc process is influenced kinetically only by charge transfer; (c) the dc process is influenced kinetically only by the chemical reaction. The forms of $F_{dc}(t)$ for these special cases are (*130,148,149*):

Case a

$$F_{dc}(t) = \frac{(1 + e^j)(1 + K)}{M} \tag{110}$$

where

$$M = K + (1 + K)e^j \tag{111}$$

Case b

$$F_{dc}(t) = \frac{(1 + e^j)(1 + K)}{M}\left[1 + \left(\frac{\alpha K e^{-j}}{1 + K} - \beta\right)e^{\lambda_K^2 t}\operatorname{erfc}(\lambda_K t^{\frac{1}{2}})\right] \tag{112}$$

where

$$\lambda_K = \frac{k_s}{D^{\frac{1}{2}}}\left(\frac{K e^{-\alpha j}}{1 + K} + e^{\beta j}\right) \tag{113}$$

Case c

$$F_{dc}(t) = \frac{(1 + e^j)(1 + K)}{1 - M^2}\left\{\exp\left(\frac{-kt}{2}\right)I_0\left(\frac{kt}{2}\right)\right.$$

$$+ \frac{k\exp[M^2 kt/(1 - M^2)]}{1 - M^2}\int_0^t \exp\left[\frac{k(M^2 + 1)u}{2(M^2 - 1)}\right]I_0\left(\frac{ku}{2}\right)du$$

$$\left. - \frac{1}{M}\left[\exp\left(\frac{M^2 kt}{1 - M^2}\right) + M^2 - 1\right]\right\} \tag{114}$$

where I_0 represents a Bessel function of imaginary argument.

The failure to obtain a general solution is probably not serious, since these special cases should fit most systems encountered. For fast electrode processes, the simple expression based on the assumption that the dc process is diffusion controlled will suffice. Even more encouraging is the fact that the phase angle is independent of $F_{dc}(t)$. Thus, the phase-angle expression is applicable for any combination of rate parameters, in addition to being considerably less cumbersome than the current-amplitude expression. It should be noted that $F_{dc}(t)$ has a time-dependent form for the cases in which the dc process is influenced by charge-transfer or chemical-reaction kinetics.

Derivation of equations for the fundamental-harmonic ac polarographic current for other mechanisms involving first-order chemical reactions coupled to a single charge-transfer step is effected as above. Linear transformations analogous to Eqs. (89) and (90) can be obtained to aid in solution, even for the most complex multistep chemical reaction schemes (*137*). A general solution to the dc problem [and thus to $F_{dc}(t)$] has proved inaccessible for all chemical-reaction mechanisms except the simple catalytic case (*131*). However, solution for cases *a* and *b* are always obtainable and also case *c* for many systems. The phase angle proves independent of the functions responsive to the dc process for all such mechanisms and, thus, general expressions for phase angle are readily accessible.

In general, the ac wave equations for systems with coupled first-order chemical reactions can be written in the form of Eq. (103), where $G_{ac}(\omega)$ can be expressed as in Eq. (106) with (*137*)

$$V = \frac{(2\omega)^{\frac{1}{2}}}{\lambda} + \frac{1}{(1 + e^j)}\left\{W + Ye^j + \sum_{d=1}^{N_O} X_d\left[\frac{(1 + g_d^2)^{\frac{1}{2}} + g_d}{1 + g_d^2}\right]^{\frac{1}{2}}\right.$$
$$\left. + e^j \sum_{m=1}^{N_R} Z_m\left[\frac{(1 + h_m^2)^{\frac{1}{2}} + h_m}{1 + h_m^2}\right]^{\frac{1}{2}}\right\} \tag{115}$$

$$S = \frac{1}{(1 + e^j)}\left\{W + Ye^j + \sum_{d=1}^{N_O} X_d\left[\frac{(1 + g_d^2)^{\frac{1}{2}} - g_d}{1 + g_d^2}\right]^{\frac{1}{2}}\right.$$
$$\left. + e^j \sum_{m=1}^{N_R} Z_m\left[\frac{(1 + h_m^2)^{\frac{1}{2}} - h_m}{1 + h_m^2}\right]^{\frac{1}{2}}\right\} \tag{116}$$

and $F_{dc}(t)$ may be written

$$F_{dc}(t) = (1 + e^{-j})\left\{\frac{\alpha\psi_0(t)D^{\frac{1}{2}}}{k_s e^{-\alpha j}} + e^j\left[Y\int_0^t \frac{\psi_0(t - u)\,du}{(\pi u)^{\frac{1}{2}}}\right.\right.$$
$$\left.\left. + \sum_{m=1}^{N_R} Z_m\int_0^t \frac{e^{-k_m u}\psi_0(t - u)\,du}{(\pi u)^{\frac{1}{2}}}\right]\right\} \tag{117}$$

where

$$g_d = \frac{k_d}{\omega} \tag{118}$$

$$h_m = \frac{k_m}{\omega} \tag{119}$$

W, X_d, Y, Z_m, k_d, and k_m are constants determined by the nature of the chemical reaction scheme and $\psi_0(t)$ is given by Eq.(105). N_O and N_R represent the number of chemical reactions coupled with species O and R, respectively. Table V lists values of W, X_d, Y, Z_m, k_d, and k_m for various coupled chemical reaction mechanisms. Table VI gives $F_{dc}(t)$ with the same mechanisms for the cases where solutions have been obtained. For all systems given in Tables V and VI, it can be shown that the equations reduce to the expressions for the quasi-reversible case [Eqs. (59) through (62)] when the chemical-reaction rate is so slow as to be ineffective or when K is so large that the chemical reaction is nonexistent. When reversible chemical reactions are so rapid that they exert only thermodynamic influence, the equations reduce to essentially the same form as the quasi-reversible case, but contain additional terms accounting for shift in position of the wave along the dc potential axis. This manifests the thermodynamic effect of the chemical reaction. Thus, these equations for ac waves influenced by coupled first-order chemical reactions are entirely consistent with theory for the quasi-reversible and reversible systems.

Figures 12 through 16 show predicted ac polarographic current and phase-angle behavior for some systems with coupled chemical reactions. One can see that significant influence of the chemical reaction is usually predicted to be readily apparent from *qualitative* inspection of phase-angle data, whereas similar examination of ac amplitudes may show nothing definitely suggestive of the chemical reaction. One interesting exception is the case of a relatively slow chemical reaction, where the chemical rate constants are much smaller than the frequency, but sufficiently rapid to influence the dc polarographic process. Here the phase angle is unaffected by the chemical reaction, but the current amplitude is influenced through the $F_{dc}(t)$ term. Aylward and others (122,150) have experimentally examined this type of system in some detail. Plots of cot ϕ versus $\omega^{\frac{1}{2}}$ at a given dc potential will show a "hump" when a preceding or following chemical reaction is operative, and a double "hump" may appear with two-step mechanisms. Systems with catalytic reductions show cot $\phi \to \infty$ as $\omega \to 0$. When preceding or following reactions are involved, plots of cot ϕ versus E_{dc} show a marked asymmetry about $E_{\frac{1}{2}}^r$, whereas a catalytic reaction introduces no asymmetry. Preceding reactions lead to larger values of cot ϕ at negative potentials; the opposite occurs with following reactions. It is interesting to note that with a preceding or following reaction, the plot of cot ϕ versus $\omega^{\frac{1}{2}}$ is predicted to show a linear segment at low frequencies ($\omega \to 0$) and high frequencies ($\omega \to \infty$). The high-frequency linear segment is determined by charge-transfer kinetics and is given by Eq. (59) (phase-angle term). The low-frequency linear segment is determined by the combined effects of charge transfer and chemical reaction. The equation for this linear segment is obtained by introducing the condition, $h^2, g^2 \gg 1$. For example, with the single-step preceding reaction mechanism, this condition yields the relationship

$$\cot \phi = 1 + \frac{1}{M}\left[\frac{1}{k^{\frac{1}{2}}} + \frac{M+1}{\lambda}\right](2\omega)^{\frac{1}{2}} \qquad (120)$$

with M given by Eq. (111). The corresponding expression for the single step following reaction mechanism is obtained by replacing M by M_f in Eq. (120) M_f is defined in Part II of Table VI.

General relationships for position and magnitude of peak current and cot ϕ appear inaccessible for systems with coupled chemical reactions, except for cot ϕ with the catalytic case (*33*).

Experimental data constituting a test of these equations are relatively scarce. Good quantitative agreement between theory for phase angle and experiment was found for the catalytic process occurring when Ti^{4+} is reduced in the presence of chlorate ion (*33*). Figures 17 and 18 show some comparisons of theory and experiment for this system. The chemical-rate constant obtained by dc and ac polarography were substantially the same. Phase-angle data on systems with preceding and following reactions have been found to be in qualitative agreement with theory for the reduction of Ti^{4+} in mixed oxalic acid-ammonium sulfate electrolyte (*111*). Some data are shown in Fig. 19. For this system, the existence of preceding-reaction control also is evident from the dc polarographic data. The larger values of cot ϕ at negative potentials with the lower frequencies is characteristic of the predictions of theory. Aylward and Hayes obtained values of the rate constant for the reaction of cadmium ion with EDTA through ac current-amplitude measurements (*150*). The rate constant agreed favorably with dc polarographic measurements. A similar study of the reaction between calcium ion and the EDTA complex of divalent europium ion gave less satisfactory agreement between the ac and dc methods (*150*).

TABLE V

VALUES OF PARAMETERS IN EQS. (115) TO (119) FOR VARIOUS

Reaction mechanism	W	Y	X_d
$Y \underset{k_2}{\overset{k_1}{\rightleftharpoons}} O$ $O + ne \rightleftharpoons R$	$\dfrac{K}{1+K}$ $K = k_1/k_2$	1	$X_1 = (1+K)^{-1}$
$O + ne \rightleftharpoons R$ $R \underset{k_1}{\overset{k_2}{\rightleftharpoons}} Y$	1	$\dfrac{K}{1+K}$ $K = k_1/k_2$	
$O + ne \rightleftharpoons R$ $\uparrow \!\!\boxed{\quad k \quad}$	0	0	$X_1 = 1$
$Y \underset{k_2}{\overset{k_1}{\rightleftharpoons}} O$ $O + ne \rightleftharpoons R$ $R \underset{k_3}{\overset{k_4}{\rightleftharpoons}} Z$	$\dfrac{K_p}{1+K_p}$ $K_p = k_1/k_2$	$\dfrac{K_f}{1+K_f}$ $K_f = k_3/k_4$	$X_1 = (1+K_p)^{-1}$
$Z \underset{k_2}{\overset{k_1}{\rightleftharpoons}} Y \underset{k_4}{\overset{k_3}{\rightleftharpoons}} O$ $O + ne \rightleftharpoons R$	$\dfrac{K_1 K_2}{1+K_1+K_1 K_2}$ $K_1 = k_1/k_2$ $K_2 = k_3/k_4$	1	$X_1 =$ $[k_4(k_1+k_2)(J+k_1+k_2-k_4)$ $-k_3 k_4(k_1-k_2)]/$ $2J(k_1 k_3 + k_2 k_4 + k_1 k_4)$ $X_2 =$ $[k_4(k_1+k_2)(J-k_1-k_2+k_4)$ $+k_3 k_4(k_1-k_2)]/$ $2J(k_1 k_3 + k_2 k_4 + k_1 k_4)$ $J = [(k_1+k_2+k_3+k_4)^2 + 4k_2 k_3]^{\frac{1}{2}}$
$O + ne \rightleftharpoons R$ $R \underset{k_3}{\overset{k_4}{\rightleftharpoons}} Y \underset{k_1}{\overset{k_2}{\rightleftharpoons}} Z$	1	$\dfrac{K_1 K_2}{1+K_1+K_1 K_2}$ $K_1 = k_1/k_2$ $K_2 = k_3/k_4$	

COUPLED FIRST-ORDER CHEMICAL-REACTION SCHEMES

Z_m	k_d	k_m
	$k_{d-1} = k_1 + k_2$	
$Z_1 = (1 + K)^{-1}$		$k_{m-1} = k_1 + k_2$
$Z_1 = 1$	$k_{d-1} = k$	$k_{m-1} = k$
$Z_1 = (1 + K_f)^{-1}$	$k_{d-1} = k_1 + k_2$	$k_{m-1} = k_3 + k_4$
	$k_{d-1} = (k_1 + k_2 + k_3 + k_4 - J)/2$	
	$k_{d-2} = (k_1 + k_2 + k_3 + k_4 + J)/2$	

$Z_1 =$
$[k_4(k_1 + k_2)(J + k_1 + k_2 - k_4)$
$- k_3 k_4(k_1 - k_2)]/$
$2J(k_1 k_3 + k_2 k_4 + k_1 k_4)$
$Z_2 =$
$[k_4(k_1 + k_2)(J - k_1 - k_2 + k_4)$
$+ k_3 k_4(k_1 - k_2)]/$
$2J(k_1 k_3 + k_2 k_4 + k_1 k_4)$
$J = [(k_1 + k_2 - k_3 - k_4)^2 + 4k_2 k_3]^{\frac{1}{2}}$

$k_{m-1} = (k_1 + k_2 + k_3 + k_4 - J)/2$
$k_{m-2} = (k_1 + k_2 + k_3 + k_4 + J)/2$

TABLE V—*continued*

Reaction mechanism	W	Y	X_d
$\begin{array}{cc} k_1 & k_3 \\ Z \rightleftharpoons O \rightleftharpoons Y \\ k_2 & k_4 \end{array}$ $O + ne \rightleftharpoons R$	$\dfrac{K_1K_2}{K_1K_2 + K_1 + K_2}$ $K_1 = k_1/k_2$ $K_2 = k_3/k_4$	1	$X_1 =$ $[k_2k_3(k_3 - k_4 - k_1 - k_2 + J)$ $+ k_1k_4(k_1 - k_2 - k_3 - k_4 + J)]/$ $2J(k_1k_3 + k_2k_3 + k_1k_4)$ $X_2 =$ $[k_2k_3(k_1 + k_2 - k_3 + k_4 + J)$ $+ k_1k_4(k_3 + k_4 + k_2 - k_1 + J)]/$ $2J(k_1k_3 + k_2k_3 + k_1k_4)$ $J = [(k_1 + k_2 - k_3 - k_4)^2 + 4k_2k_4]^{\frac{1}{2}}$
$O + ne \rightleftharpoons R$ $\begin{array}{cc} k_3 & k_2 \\ Z \rightleftharpoons R \rightleftharpoons Y \\ k_4 & k_1 \end{array}$	1	$\dfrac{K_1K_2}{K_1K_2 + K_1 + K_2}$ $K_1 = k_1/k_2$ $K_2 = k_3/k_4$	
$\begin{array}{c} k_1 \\ Y \rightleftharpoons O + ne \rightleftharpoons R \\ \uparrow k_2 \uparrow \;\; k_{CO} \quad \rceil \\ \lfloor \underline{\quad\quad\quad} \rceil \\ k_{CY} \end{array}$	0	0	$X_1 = \dfrac{k_1 - k_{CO}}{k_1 + k_2 - k_{CO} - k_{CY}}$ $X_2 = \dfrac{k_2 - k_{CY}}{k_1 + k_2 - k_{CO} - k_{CY}}$
$\begin{array}{c} k_2 \\ O + ne \rightleftharpoons R \rightleftharpoons Y \\ \uparrow \;\; k_{CR} \;\rceil\; k_1 \\ \lfloor \underline{\quad\quad\quad} \rceil \\ k_{CY} \end{array}$	0	0	$X_1 = \dfrac{k_{CR} - k_{CY} - k_1 - k_2 + J}{2J}$ $X_2 = \dfrac{k_1 + k_2 + k_{CY} - k_{CR} + J}{2J}$ $J = [(k_2 + k_{CR} - k_{CY} - k_1)^2 + 4k_1k_2]^{\frac{1}{2}}$

Z_m	k_d	k_m
	$k_{d=1} = (k_1 + k_2 + k_3 + k_4 - J)/2$ $k_{d=2} = (k_1 + k_2 + k_3 + k_4 + J)/2$	
$Z_1 =$ $[k_2 k_3 (k_3 - k_4 - k_1 - k_2 + J)$ $+ k_1 k_4 (k_1 - k_2 - k_3 - k_4 + J)]/$ $2J(k_1 k_3 + k_2 k_4 + k_1 k_4)$ $Z_2 =$ $[k_2 k_3 (k_1 + k_2 - k_3 + k_4 + J)$ $+ k_1 k_4 (k_3 + k_4 + k_2 - k_1 + J)]/$ $2J(k_1 k_3 + k_2 k_4 + k_1 k_4)$ $J = [(k_1 + k_2 - k_3 - k_4)^2$ $+ 4k_2 k_4]^{\frac{1}{2}}$		$k_{m=1} =$ $(k_1 + k_2 + k_3 + k_4 - J)/2$ $k_{m=2} =$ $(k_1 + k_2 + k_3 + k_4 + J)/2$
$Z_1 = 1$	$k_{d=1} = k_{CO} + k_{CY}$ $k_{d=2} = k_1 + k_2$	$k_{m=1} = k_{CO} + k_{CY}$
$Z_1 =$ $(k_2 - k_1 + k_{CR} - k_{CY} + J)/2J$ $Z_2 =$ $(k_1 - k_2 + k_{CY} - k_{CR} + J)/2J$	$k_{d=1} =$ $(k_1 + k_2 + k_{CY} + k_{CR} + J)/2$ $k_{d=2} =$ $(k_1 + k_2 + k_{CY} + k_{CR} - J)/2$	$k_{m=1} = k_{d=1}$ $k_{m=2} = k_{d=2}$

TABLE VI

$F_{dc}(t)$ FUNCTION FOR VARIOUS COUPLED CHEMICAL-REACTION SCHEMES[a]

$$I. \quad Y \underset{k_2}{\overset{k_1}{\rightleftharpoons}} O + ne \rightleftharpoons R; \ k = k_1 + k_2; \ K = k_1/k_2$$

Case a: Dc process is controlled solely by diffusion (reversible dc process):

$$F_{dc}(t) = \frac{(1 + e^J)(1 + K)}{M}$$

where

$$M = K + (1 + K)e^J$$

Case b: Dc process is controlled kinetically by diffusion and chemical reaction:

$$F_{dc}(t) = \frac{(1 + e^J)(1 + K)}{1 - M^2} \left\{ \exp\left(\frac{-kt}{2}\right) I_0\left(\frac{kt}{2}\right) - \frac{1}{M}\left[\exp\left(\frac{M^2 kt}{1 - M^2}\right) + M^2 - 1\right] \right.$$

$$\left. + \frac{k \exp[M^2 kt/(1 - M^2)]}{1 - M^2} \int_0^t \exp\left[\frac{k(M^2 + 1)u}{2(M^2 - 1)}\right] I_0\left(\frac{ku}{2}\right) du \right\}$$

Case c: Dc process is controlled kinetically by diffusion and charge transfer:

$$F_{dc}(t) = \frac{(1 + e^J)(1 + K)}{M} \left\{ 1 + \left(\frac{\alpha K e^{-J}}{1 + K} - \beta\right) e^{\lambda_K^2 t} \operatorname{erfc}(\lambda_K t^{\frac{1}{2}}) \right\}$$

where

$$\lambda_k = \frac{k_s}{D^{\frac{1}{2}}} \left(\frac{K e^{-\alpha J}}{1 + K} + e^{\beta J}\right)$$

$$II. \quad O + ne \rightleftharpoons R \underset{k_1}{\overset{k_2}{\rightleftharpoons}} Y; \ k = k_1 + k_2; \ K = k_1/k_2$$

Case a: Dc process is controlled solely by diffusion:

$$F_{dc}(t) = [(1 + e^{-J})K]/M_f$$

where

$$M_f = (1 + K)e^{-J} + K$$

Case b: Dc process is controlled kinetically by diffusion and chemical reaction:

$$F_{dc}(t) = (1 + e^{-J}) \left\{ 1 - \frac{e^{-J}(1 + K)}{(1 - M_f^2)} \left[\exp\left(\frac{-kt}{2}\right) I_0\left(\frac{kt}{2}\right) \right. \right.$$

$$+ \frac{k \exp[M_f^2 kt/(1 - M_f^2)]}{1 - M_f^2} \int_0^t \exp\left[\frac{k(M_f^2 + 1)u}{2(M_f^2 - 1)}\right] I_0\left(\frac{ku}{2}\right) du$$

$$\left. \left. - \frac{1}{M_f}\left[\exp\left(\frac{M_f^2 kt}{1 - M_f^2}\right) + M_f^2 - 1\right] \right] \right\}$$

TABLE VI—*continued*

Case c: Dc process is controlled kinetically by diffusion and charge transfer

$$F_{dc}(t) = \frac{(1 + e^{-j})(1 + K)}{M_f}\left\{\frac{K}{1 + K} + \left(\alpha e^{-j} - \frac{K\beta}{1 + K}\right)e^{\lambda_f^2 t}\,\text{erfc}(\lambda_f t^{\frac{1}{2}})\right\}$$

where

$$\lambda_f = \frac{k_s}{D^{\frac{1}{2}}}\left(e^{-\alpha j} + \frac{K e^{\beta j}}{1 + K}\right)$$

$$\text{III.} \qquad \begin{array}{c} O + ne \rightleftharpoons R \\ \uparrow \quad k \quad \mid \end{array}$$

Case a: Dc process is controlled kinetically by diffusion, charge-transfer, and chemical reaction (the general case)

$$F_{dc}(t) = 1 + \frac{(\alpha e^{-j} - \beta)}{(k - \lambda^2)}\,[k - \lambda k^{\frac{1}{2}}\,\text{erf}(kt)^{\frac{1}{2}}$$

$$-\lambda^2\exp[(\lambda^2 - k)t]\,\text{erfc}(\lambda t^{\frac{1}{2}})]$$

Case b: Dc process is controlled kinetically by diffusion and/or chemical reaction:

$$\dot{F}_{dc}(t) = 1$$

Case c: Dc process is controlled kinetically by diffusion and charge transfer (chemical reaction nonexistent):

$$F_{dc}(t) = 1 + (\alpha e^{-j} - \beta)e^{\lambda^2 t}\,\text{erfc}(\lambda t^{\frac{1}{2}})$$

$$\text{IV.} \quad Y\underset{k_2}{\overset{k_1}{\rightleftharpoons}}O + ne \rightleftharpoons R\underset{k_3}{\overset{k_4}{\rightleftharpoons}}Z; \qquad k_f = k_3 + k_4;\quad K_f = k_3/k_4$$

$$k_p = k_1 + k_2;\quad K_p = k_1/k_2$$

Case a: Dc process is controlled solely by diffusion:

$$F_{dc}(t) = \frac{(1 + e^j)K_f(1 + K_p)}{K_p(1 + K_f) + e^j K_f(1 + K_p)}$$

Case b: Dc process is controlled kinetically by diffusion and charge transfer:

$$F_{dc}(t) = \frac{(1 + e^j)}{K_p/(1 + K_p) + K_f e^j/(1 + K_f)}\left\{\frac{K_f}{1 + K_f} + \left(\frac{\alpha K_p e^{-j}}{1 + K_p} - \frac{K_f\beta}{1 + K_f}\right)e^{\lambda_{pf}^2 t}\,\text{erfc}(\lambda_{pf}t^{\frac{1}{2}})\right\}$$

where

$$\lambda_{pf} = \frac{k_s}{D^{\frac{1}{2}}}\left(\frac{K_p e^{-\alpha j}}{1 + K_p} + \frac{K_f e^{\beta j}}{1 + K_f}\right)$$

$$\text{V.} \quad Z\underset{k_2}{\overset{k_1}{\rightleftharpoons}}Y\underset{k_4}{\overset{k_3}{\rightleftharpoons}}O + ne \rightleftharpoons R;\quad K_1 = k_1/k_2\,;\quad K_2 = k_3/k_4$$

TABLE VI—*continued*

Case a: Dc process is controlled solely by diffusion;

$$F_{dc}(t) = (1 + e^j)/(w + e^j)$$

where

$$w = K_1 K_2/(1 + K_1 + K_1 K_2)$$

Case b: Dc process is controlled kinetically by diffusion and charge transfer:

$$F_{dc}(t) = \frac{(1 + e^j)}{(w + e^j)} \{1 + (\alpha w e^{-j} - \beta) e^{\lambda_{2p}^2 t} \operatorname{erfc}(\lambda_{2p} t^{\frac{1}{2}})\}$$

where

$$\lambda_{2p} = (k_s/D^{\frac{1}{2}})(w e^{-\alpha j} + e^{\beta j})$$

$$\text{VI. } O + ne \rightleftharpoons R \overset{k_4}{\underset{k_3}{\rightleftharpoons}} Y \overset{k_2}{\underset{k_1}{\rightleftharpoons}} Z; \quad K_1 = k_1/k_2; \quad K_2 = k_3/k_4$$

Case a: Dc process is controlled solely by diffusion:

$$F_{dc}(t) = [(1 + e^{-j})y]/(e^{-j} + y)$$

where

$$y = K_1 K_2/(1 + K_1 + K_1 K_2)$$

Case b: Dc process is controlled kinetically by diffusion and charge transfer:

$$F_{dc}(t) = \frac{(1 + e^{-j})}{(e^{-j} + y)} \{y + (\alpha e^{-j} - \beta y) e^{\lambda_{2F}^2 t} \operatorname{erfc}(\lambda_{2F} t^{\frac{1}{2}})\}$$

where

$$\lambda_{2F} = (k_s/D^{\frac{1}{2}})(e^{-\alpha j} + y e^{\beta j})$$

$$\text{VII. } Z \overset{k_1}{\underset{k_2}{\rightleftharpoons}} O + ne \rightleftharpoons R; \quad K_1 = k_1/k_2; \quad K_2 = k_3/k_4$$

$$\underset{Y}{\overset{k_3 \downarrow \uparrow k_4}{}}$$

Case a: Dc process is controlled solely by diffusion

$$F_{dc}(t) = (1 + e^j)/(w + e^j)$$

where

$$w = K_1 K_2/(K_1 K_2 + K_1 + K_2)$$

Case b: Dc process is controlled kinetically by diffusion and charge transfer:

$$F_{dc}(t) = \frac{(1 + e^j)}{(w + e^j)} \{1 + (\alpha w e^{-j} - \beta) e^{\lambda_{3p}^2 t} \operatorname{erfc}(\lambda_{3p} t^{\frac{1}{2}})\}$$

TABLE VI—continued

where

$$\lambda_{3p} = (k_s/D^{\ddagger})(we^{-\alpha j} + e^{\beta j})$$

VIII. $O + ne \rightleftharpoons R \overset{k_2}{\underset{k_1}{\rightleftharpoons}} Z; \quad K_1 = k_1/k_2; \quad K_2 = k_3/k_4$

$$k_4 \updownarrow k_3$$
$$Y$$

Case a: Dc process is controlled by diffusion:

$$F_{dc}(t) = [(1 + e^{-j})y]/(e^{-j} + y)$$

where

$$y = K_1 K_2/(K_1 K_2 + K_2 + K_1)$$

Case b: Dc process is controlled kinetically by diffusion and charge transfer:

$$F_{dc}(t) = \frac{(1 + e^{-j})}{(e^{-j} + y)}\{y + (\alpha e^{-j} - \beta y)e^{\lambda_{3F}^2 t}\,\mathrm{erfc}(\lambda_{3F}t^{\ddagger})\}$$

where

$$\lambda_{3F} = (k_s/D^{\ddagger})(e^{-\alpha j} + ye^{\beta j})$$

IX. $Y \overset{k_1}{\underset{k_2}{\rightleftharpoons}} O + ne \rightleftharpoons R; \quad K = k_1/k_2$

$$\uparrow k_{CO} \downarrow$$
$$k_{CY}$$

Case a: Dc process is controlled kinetically by the catalytic steps (reoxidation) and/or diffusion:

$$F_{dc}(t) = (1 + e^j)(1 + K)/M$$

where

$$M = K + (1 + K)e^j$$

Case b: Dc process is controlled kinetically by the catalytic steps, charge transfer, and diffusion:

$$F_{dc}(t) = \frac{(1 + e^j)(1 + K)}{M}\left\{1 + \left(\frac{\alpha K e^{-j}}{1 + K} - \beta\right)\left(\frac{1}{k_c - \lambda_k^2}\right)\right.$$

$$\left. \times [k_c - \lambda_K k_c^{\ddagger}\,\mathrm{erf}(k_c t)^{\ddagger} - \lambda_K^2 \exp[(\lambda_K^2 - k_c)t]\,\mathrm{erfc}(\lambda_K t^{\ddagger})]\right\}$$

$$k_c = k_{CO} + k_{CY}$$

$$\lambda_K = \frac{k_s}{D^{\ddagger}}\left(\frac{Ke^{-\alpha j}}{1 + K} + e^{\beta j}\right)$$

X. $O + ne \rightleftharpoons R \overset{k_2}{\underset{k_1}{\rightleftharpoons}} Y; \quad K = k_1/k_2$

$$\uparrow k_{CR} \downarrow$$
$$k_{CY}$$

TABLE VI—*continued*

Case a: Dc process is controlled kinetically by the catalytic steps and/or diffusion:

$$F_{dc}(t) = \frac{(1 + e^{-J})(1 + K)}{M_f}$$

where

$$M_f = (1 + K)e^{-J} + K$$

Case b: Dc process is controlled kinetically by the catalytic steps, charge transfer, and diffusion:

$$F_{dc}(t) = \frac{(1 + e^{-J})(1 + K)}{M_f} \left\{ \frac{K}{1 + K} + \left(\alpha e^{-J} - \frac{K\beta}{1 + K} \right) \left(\frac{1}{k_c - \lambda_f^2} \right) \right.$$

$$\left. \times [k_c - \lambda_f k_c^{\frac{1}{2}} \operatorname{erf}(k_c t)^{\frac{1}{2}} - \lambda_f^2 \exp[(\lambda_f^2 - k_c)t] \operatorname{erfc}(\lambda_f t^{\frac{1}{2}})] \right\}$$

$$k_c = k_{CY} + k_{CR}$$

$$\lambda_f = \frac{k_s}{D^{\frac{1}{2}}} \left(e^{-\alpha J} + \frac{K e^{\beta J}}{1 + K} \right)$$

[a] Only those cases for which a specific expression is available for $\psi_0(t)$ are included.

Equations for second-harmonic currents with coupled first-order chemical reactions have been derived by solution of the integral equation for $p = 2$ *(130)*. The derivation contains no unusual features, and a general expression can be written for the second harmonic applicable to any mechanism involving first-order reactions coupled with a single charge-transfer step.

As one would expect, the equations are extremely cumbersome, and only one example will be given here. For the single-step preceding reaction mechanism, the second-harmonic current is given by *(130)*

$$I(2\omega t) = \frac{n^3 F^3 A C_O^*(\omega D_O)^{\frac{1}{2}} \Delta E^2}{R^2 T^2} \left[\frac{P^2 + L^2}{V_2^2 + S_2^2} \right]^{\frac{1}{2}} \sin\left(2\omega t + \cot^{-1} \frac{LV_2 + PS_2}{LS_2 - PV_2} \right)$$

$$(121)$$

$$P = \frac{1}{(1 + e^J)} \left\{ \left[\frac{1}{V^2 + S^2} \left(\alpha V \frac{(2\omega)^{\frac{1}{2}}}{\lambda} + \frac{V + S}{(1 + e^{-J})} \right) - \frac{1}{2} \right] e^J \int_0^t \frac{\psi_0(t - u)\, du}{(\pi u)^{\frac{1}{2}}} \right.$$

$$\left. + \left[\frac{1}{V^2 + S^2} \left(\alpha V \frac{(2\omega)^{\frac{1}{2}}}{\lambda} + \frac{V + S}{(1 + e^{-J})} \right) - \frac{\alpha}{2} \right] \frac{\alpha D^{\frac{1}{2}} \psi_0(t)}{k_s e^{-\alpha J}} \right\}$$

$$(122)$$

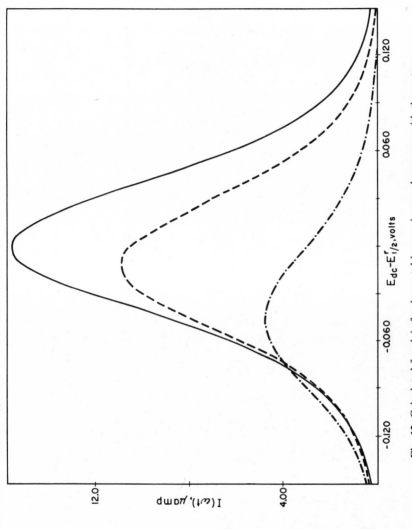

Fig. 12. Calculated faradaic fundamental-harmonic ac polarograms with the system

$$Y \underset{k_2}{\overset{k_1}{\rightleftharpoons}} O + ne \rightleftharpoons R$$

Parameters: $\omega = 2500 \text{ sec}^{-1}$, $\alpha = 0.500$, $D_O = D_R = 1.0 \times 10^{-5} \text{ cm}^2 \text{ sec}^{-1}$, $A = 0.035 \text{cm}^2$, $C_O^* + C_Y^* = 1.0 \times 10^{-3} M$, $T = 298°K$, $\Delta E = 5.00 \text{ mV}$, $n = 1$; —, $K = \infty$, $- -$, $K = 1$, $k_1 = 1.00 \times 10^3 \text{ sec}^{-1}$, $-\cdot-$, $K = 0.1$, $k_1 = 1.00 \times 10^3 \text{ sec}^{-1}$.

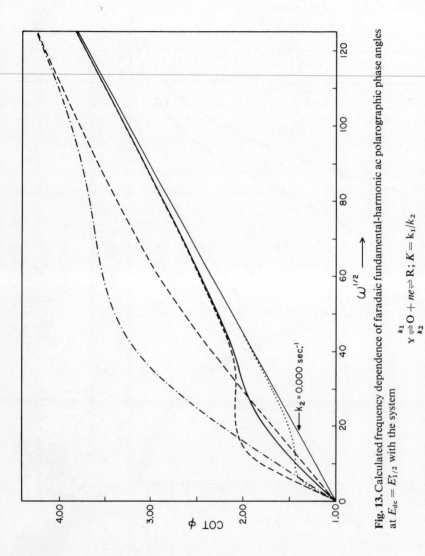

Fig. 13. Calculated frequency dependence of faradaic fundamental-harmonic ac polarographic phase angles at $E_{dc} = E_{1/2}$ with the system

$$Y \underset{k_2}{\overset{k_1}{\rightleftharpoons}} O + ne \rightleftharpoons R; K = k_1/k_2$$

[from (148)]. Parameters: $k_s = 0.100\ cm\ sec^{-1}$, $\alpha = 0.500$, $D = 1.00 \times 10^{-5}\ cm^2\ sec^{-1}$, n = 1, $T = 298°K$, $\Delta E = 5.00\ mV$; $\cdots$, $K = 1.00$; $k_1 = 50.0\ sec^{-1}$, $-\cdot-$, $K = 1.00$; $k_1 = 500\ sec^{-1}$, $---$, $K = 5000$ sec^{-1}, $---$, $K = 0.100$; $k_1 = 50\ sec^{-1}$, $-\cdot\cdot-$, $K = 0.0100$; $k_1 = 50\ sec^{-1}$. (Courtesy of *Analytical Chemistry*.)

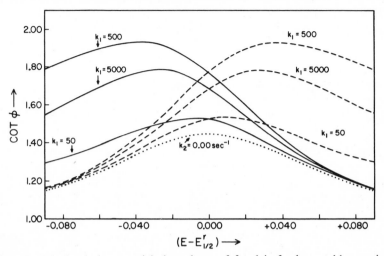

Fig. 14. Calculated dc potential dependence of faradaic fundamental-harmonic ac polarographic phase angle with the system

$$Y \underset{k_2}{\overset{k_1}{\rightleftharpoons}} O + ne \rightleftharpoons R$$

[from (*148*)]. Parameters: $\omega = 400$ sec^{-1}, $K = 1.00$, $k_S = 0.100$ cm sec^{-1}, $\alpha = 0.500$, $D = 1.00 \times 10^{-5}$ cm^2 sec^{-1}, $n = 1$, $T = 298°$K, $E = E_{dc}$, $\Delta E = 5.00$ mV; —, preceding reaction, – –, following reaction, ···, no chemical reaction. (Courtesy of *Analytical Chemistry*.)

$$L = \frac{1}{(1 + e^j)} \left[\frac{1}{V^2 + S^2} \left(\alpha S \frac{(2\omega)^{\frac{1}{2}}}{\lambda} + \frac{S - V}{1 + e^{-j}} \right) \right]$$
$$\times \left[e^j \int_0^t \frac{\psi_0(t - u)\, du}{(\pi u)^{\frac{1}{2}}} + \frac{\alpha D^{\frac{1}{2}} \psi_0(t)}{k_s e^{-\alpha j}} \right] \qquad (123)$$

V and S are defined by Eqs. (107) and (108); V_2 and S_2 are the same as V and S, respectively, except that ω is replaced by 2ω. As with the fundamental harmonic, the problem associated with solving the dc problem limits one to giving explicit solutions for $\psi_0(t)$, P, and L only for the three special cases associated with Eqs. (110) through (114). P and L reduce to the following for these cases:

Case a

$$P = \frac{(1 + K)}{M(1 + e^{-j})} \left[\frac{1}{V^2 + S^2} \left(\alpha V \frac{(2\omega)^{\frac{1}{2}}}{\lambda} + \frac{V + S}{1 + e^{-j}} \right) - \frac{1}{2} \right] \qquad (124)$$

$$L = \frac{1 + K}{M(1 + e^{-j})(V^2 + S^2)} \left(\alpha S \frac{(2\omega)^{\frac{1}{2}}}{\lambda} + \frac{S - V}{1 + e^{-j}} \right) \qquad (125)$$

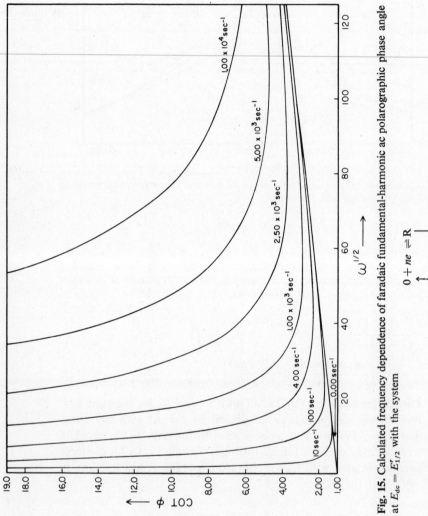

Fig. 15. Calculated frequency dependence of faradaic fundamental-harmonic ac polarographic phase angle at $E_{dc} = E'_{1/2}$ with the system

$$0 + ne \rightleftharpoons R$$

$$\boxed{\downarrow \, k_c}$$

[from (*148*)]. Parameters: $k_s = 0.100$ cm sec^{-1}, $\alpha = 0.500$, $D = 1.00 \times 10^{-5}$ cm^2 sec^{-1}, $n = 1$, $T = 298°$K, $\Delta E = 5.00$ mV, values of k_c indicated on figure. (Courtesy of *Analytical Chemistry*.)

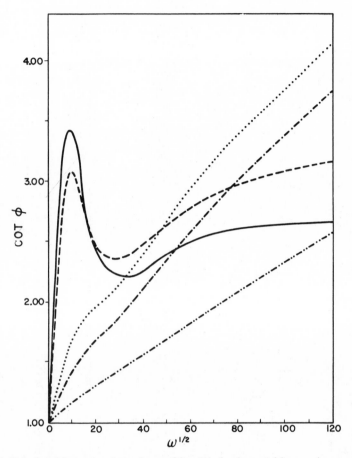

Fig. 16. Calculated frequency dependence of faradaic fundamental-harmonic ac polarographic phase angle at varying dc potentials for the system

$$Z \underset{k_2}{\overset{k_1}{\rightleftharpoons}} Y \underset{k_4}{\overset{k_3}{\rightleftharpoons}} O + ne \rightleftharpoons R$$

[from (*137*)]. Parameters: $k_S = 0.100$ cm sec^{-1}, $\alpha = 0.500$, $D = 1.00 \times 10^{-5}$ cm^2 sec^{-1}, $k_1 = 50.0$ *sec*$^{-1}$, $k_2 = 500$ sec^{-1}, $k_4 = k_3 = 5000$ sec^{-1}, $n = 1$, $T = 298°$K, $\Delta E = 5.00$ mV. The values of $E_{dc} - E_{1/2}^r$ are —, -0.080 volt, – –, -0.060 volt, ···, 0.000 volt, – · –, 0.020 volt, – · · –, 0.060 volt. (Courtesy of *Elsevier.*)

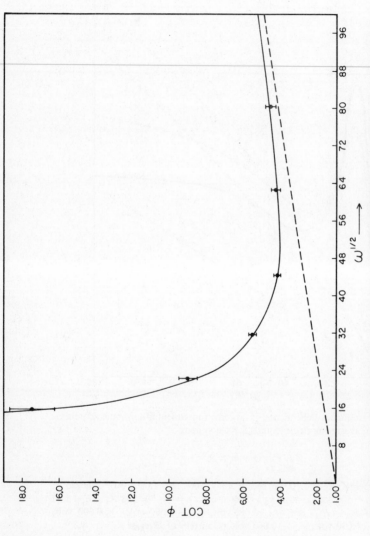

Fig. 17. Frequency dependence of faradaic fundamental-harmonic ac polarographic phase angle observed with 0.840×10^{-3} M Ti^{+4} in 0.200 M $H_2C_2O_4$ and 0.0400 M $KClO_3$ solution [from (33)]. $\Delta E = 5.00$ mV, $E_{dc} = -0.290$ volt vs. SCE. Solid curve corresponds to theory for $k_s = 4.6 \times 10^{-2}$ cm sec^{-1}, $\alpha = 0.35$, $k_C = 1.03 \times 10^3$ sec^{-1}. $D = 0.66 \times 10^{-5}$ cm^2 sec^{-1}. Dashed line corresponds to behavior in absence of chlorate ion. (Courtesy of *Analytical Chemistry*.)

Case b

$$P = \frac{1+K}{M(1+e^{-j})} \left[\frac{1}{V^2+S^2} \left(\alpha V \frac{(2\omega)^{\frac{1}{2}}}{\lambda} + \frac{V+S}{1+e^{-j}} \right) - \frac{1}{2} \right]$$
$$+ \frac{1}{(1+e^j)} \left[\frac{1}{V^2+S^2} \left(\alpha V \frac{(2\omega)^{\frac{1}{2}}}{\lambda} + \frac{V+S}{1+e^{-j}} \right) \left(\alpha - \frac{1+K}{Me^{-j}} \right) \right.$$
$$\left. + \frac{1}{2} \left(\frac{1+K}{Me^{-j}} - \alpha^2 \right) \right] e^{\lambda \kappa^2 t} \operatorname{erfc}(\lambda_K t^{\frac{1}{2}}) \tag{126}$$

$$L = \frac{1}{(1+e^j)(V^2+S^2)} \left(\alpha S \frac{(2\omega)^{\frac{1}{2}}}{\lambda} + \frac{S-V}{1+e^{-j}} \right)$$
$$\times \left[\frac{1+K}{Me^{-j}} + \left(\alpha - \frac{1+K}{Me^{-j}} \right) e^{\lambda \kappa^2 t} \operatorname{erfc}(\lambda_K t^{\frac{1}{2}}) \right] \tag{127}$$

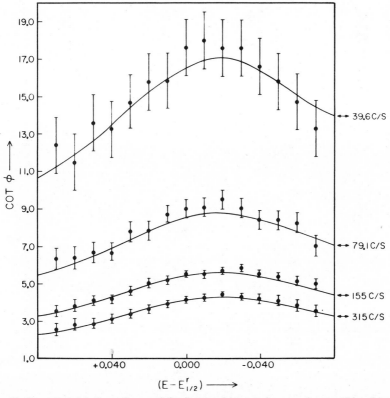

Fig. 18. Dc potential dependence of fundamental-harmonic ac polarographic phase angle observed with 0.840×10^{-3} M Ti^{+4} in 0.200 M H$_2$C$_2$O$_4$ and 0.0400 M KClO$_3$ [from (33)]. Parameters and significance of solid curves same as in Fig. 17 ($E \equiv E_{dc}$). (Courtesy of *Analytical Chemistry*.)

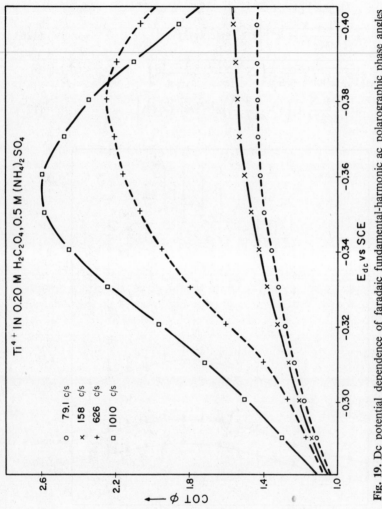

Fig. 19. Dc potential dependence of faradaic fundamental-harmonic ac polarographic phase angles observed with $1.00 \times 10^{-3}\ M\ \mathrm{Ti}^{+4}$ in $0.20\ M\ \mathrm{H_2C_2O_4}$ and $0.50\ M\ \mathrm{(NH_4)_2SO_4}$. $\Delta E = 5.00$ mV, $T = 298°\mathrm{K}$.

Case c

$$P = \frac{[F_{dc}(t)]_c}{4\cosh^2(j/2)} \left[\frac{1}{V^2 + S^2} \left(\alpha V \frac{(2\omega)^{\frac{1}{2}}}{\lambda} + \frac{V + S}{1 + e^{-j}} \right) - \frac{1}{2} \right] \qquad (128)$$

$$L = \frac{[F_{dc}(t)]_c}{4\cosh^2(j/2)(V^2 + S^2)} \left(\alpha S \frac{(2\omega)^{\frac{1}{2}}}{\lambda} + \frac{S - V}{1 + e^{-j}} \right) \qquad (129)$$

where $[F_{dc}(t)]_c$ is the $F_{dc}(t)$ function given by Eq. (114). These equations reduce to expressions for the quasi-reversible case when the chemical reaction influence is negligible ($K \to \infty$ or $k_1, k_2 \to 0$). Figure 20 illustrates the predicted influence of a preceding reaction on the second harmonic. It is apparent that the chemical step significantly alters the shape and magnitude of the second-harmonic current. Whether this effect can be utilized for quantitative assessment of chemical-rate parameters is an open question. The cumbersome nature of the equations may render this sort of application impractical. No second-harmonic experimental work has been reported on systems where coupled chemical reactions could be definitely established, although their influence is suspected for some data (*73*).

2. Second- and Higher-Order Chemical Reactions

Very few reports have been given concerning the application of ac polarography to this type of system. Gerischer's (*95*) theoretical treatment for the faradaic impedance at the equilibrium potential [cf., for example, Delahay (*10*) for discussion] remains the most significant theoretical advance. He showed that the nonlinear partial-differential equations can be linearized for small amplitudes of alternating potential. Although he was concerned with the preceding chemical reaction case, the method is readily applicable to other mechanisms, provided the experiment can be carried out with zero direct current. No extension of Gerischer's approach to ac polarographic conditions (net direct current) has been reported. For systems where one form of the redox couple is unstable (irreversible dimerization or disproportionation of the reduced form), ac polarographic conditions must be employed. Thus equations for such conditions would be extremely useful. A method of treating the ac polarographic problem with coupled higher-order chemical reactions is under investigation in this writer's laboratory (*111*). It applies Gerischer's suggestion that, with sufficiently small ac signals, second-order terms in the *ac* concentrations may be neglected. Preliminary results indicate that useful equations can be obtained for two situations: (a) when a steady state (*132*) exists in the dc concentration profile; (b) when chemical equilibrium exists in the dc process. Our work is too preliminary to discuss at length here.

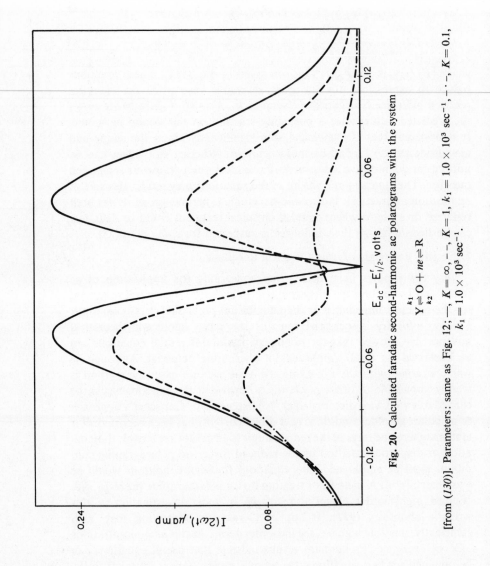

Fig. 20. Calculated faradaic second-harmonic ac polarograms with the system

$$Y \underset{k_2}{\overset{k_1}{\rightleftharpoons}} O + ne \rightleftharpoons R$$

[from (130)]. Parameters: same as Fig. 12; ——, $K = \infty$, - -, $K = 1$, $k_1 = 1.0 \times 10^3$ sec^{-1}, -·-, $K = 0.1$, $k_1 = 1.0 \times 10^3$ sec^{-1}.

Breyer et al. (*154*) studied the ac polarographic wave obtained on reduction of U(VI) which is followed by disproportionation of U(V) (*155–158*). They interpreted their data assuming that electrochemical and chemical (disproportionation) equilibria are maintained in both the ac and dc processes. However, recent studies on the time dependence of ac waves (*122*) and phase-angle data (*122*) with this system indicate that kinetic effects are important, that is, equilibrium conditions do not exist in either the ac or dc process. Also, the disproportionation step appears to be irreversible (*155*).

E. Systems with Multistep Charge Transfer

Multistep charge-transfer reaction schemes are recognized as an important mechanistic category in electrochemistry. For the situation where the electrolysis steps are separated by a wide potential range, theoretical relationships for the single-step mechanism are often applicable. Of primary concern here is the electrode process involving two or more distinct charge-transfer steps occurring in the same potential range. Such cases involving the two-step-reduction mechanism

$$O + n_1 e \rightleftharpoons Y + n_2 e \rightleftharpoons R$$

have been the concern of a number of workers (*159–166*). The two-step reduction with a chemical step interposed between the reduction steps

$$O + n_1 e \rightleftharpoons R_1 \underset{k_1}{\overset{k_2}{\rightleftharpoons}} R_2 + n_2 e \rightleftharpoons R_3$$

[the "ece" mechanism (*167–172*)] is well-established, and the rate constant for the chemical reaction has been evaluated for two systems employing chronopotentiometry (*168*), potentiostatic chronoamperometry (*169*), and linear-scan chronoamperometry (*172*). Other mechanisms in this category such as

$$
\begin{array}{c}
O_1 + ne \\
k_1 \updownarrow k_2 \qquad\quad R \\
O_2 + ne
\end{array}
$$

or

$$
\begin{array}{c}
O + n_1 e \rightleftharpoons R_1 + n_2 e \rightleftharpoons R_2 \\
k_2 \updownarrow k_1 \\
Z
\end{array}
$$

and others are either known or likely to exist [for review, see Ref. (*173*)]. To this author's knowledge, only one theoretical treatment of ac polarographic wave with such mechanisms has appeared (*129*). However, derivation is readily accomplished for many systems of this type with little variation of the approach established in the foregoing discussion. The only serious obstacles to solution prove to be cumbersome algebraic forms and, as in previously discussed systems, the intractability of the dc polarographic problem which often hinders generality of solution. As an example, derivation of the ac wave equation for the two step-reduction mechanism

$$O + n_1 e \rightleftharpoons Y + n_2 e \rightleftharpoons R$$

will be outlined. The system of partial differential equations, initial, and boundary conditions is

$$\frac{\partial C_O}{\partial t} = D_O \frac{\partial^2 C_O}{\partial x^2} \tag{130}$$

$$\frac{\partial C_Y}{\partial t} = D_Y \frac{\partial^2 C_Y}{\partial x^2} \tag{131}$$

$$\frac{\partial C_R}{\partial t} = D_R \frac{\partial^2 C_R}{\partial x^2} \tag{132}$$

For $t = 0$, any x,

$$C_O = C_O^* \tag{133a}$$

$$C_R = C_Y = 0 \tag{133b}$$

For $t > 0$, $x \to \infty$,

$$C_O \to C_O^* \tag{134a}$$

$$C_R \to C_Y \to 0 \tag{134b}$$

For $t > 0$, $x = 0$,

$$D_O \frac{\partial C_O}{\partial x} = \frac{i_1(t)}{n_1 FA} \tag{135}$$

$$D_Y \frac{\partial C_Y}{\partial x} = \frac{i_2(t)}{n_2 FA} - \frac{i_1(t)}{n_1 FA} \tag{136}$$

$$D_R \frac{\partial C_R}{\partial x} = \frac{-i_2(t)}{n_2 FA} \tag{137}$$

(Subscripts 1 and 2 refer to the first- and second-reduction steps.) Solution by the method of Laplace transformation yields, for the surface concentrations (*174*),

$$C_{O_{x=0}} = C_O^* - \int_0^t \frac{i_1(t-u)\, du}{n_1 FA (D_O \pi u)^{\frac{1}{2}}} \tag{138}$$

$$C_{Y_{x=0}} = \int_0^t \frac{i_1(t-u)\, du}{n_1 FA (D_Y \pi u)^{\frac{1}{2}}} - \int_0^t \frac{i_2(t-u)\, du}{n_2 FA (D_Y \pi u)^{\frac{1}{2}}} \tag{139}$$

$$C_{R_{x=0}} = \int_0^t \frac{i_2(t-u)\,du}{n_2 FA(D_R \pi u)^{\frac{1}{2}}} \tag{140}$$

These expressions are then substituted into the absolute rate expressions for the two charge-transfer steps:

$$\frac{i_1(t)}{n_1 FA k_{s,1}} = C_{O_{x=0}} \exp\left\{\frac{-\alpha_1 n_1 F}{RT}[E(t)-E_1^0]\right\}$$

$$- C_{Y_{x=0}} \exp\left\{\frac{(1-\alpha_1)n_1 F}{RT}[E(t)-E_1^0]\right\} \tag{141}$$

$$\frac{i_2(t)}{n_2 FA k_{s,2}} = C_{Y_{x=0}} \exp\left\{\frac{-\alpha_2 n_2 F}{RT}[E(t)-E_2^0]\right\}$$

$$- C_{R_{x=0}} \exp\left\{\frac{(1-\alpha_2)n_2 F}{RT}[E(t)-E_2^0]\right\} \tag{142}$$

Operating on the two resulting equations in the usual manner, one obtains two sets of integral equations in two unknowns (*129*):

$$\psi_{1,p}(t) = \frac{k_{s,1}}{D_1^{\frac{1}{2}}} \left\{ e^{-\alpha_1 j_1} \frac{\alpha_1^p}{p!} (\sin \omega t)^p \right.$$

$$- \sum_{r=0}^{p} \left[(\alpha_1^r e^{-\alpha_1 j_1} + (-1)^r \beta_1^r e^{\beta_1 j_1}) \frac{(\sin \omega t)^r}{r!} \int_0^t \frac{\psi_{1,p-r}(t-u)\,du}{(\pi u)^{\frac{1}{2}}} \right.$$

$$\left. \left. - (-1)^r \beta_1^r e^{\beta_1 j_1} \frac{(\sin \omega t)^r}{r!} \left(\frac{n_2}{n_1}\right)^{p-r} \int_0^t \frac{\psi_{2,p-r}(t-u)\,du}{(\pi u)^{\frac{1}{2}}} \right] \right\} \tag{143}$$

$$\psi_{2,p}(t) = \frac{k_{s,2}}{D_2^{\frac{1}{2}}} \sum_{r=0}^{p} \left[\alpha_2^r e^{-\alpha_2 j_2} \frac{(\sin \omega t)^r}{r!} \left(\frac{n_1}{n_2}\right)^{p-r} \int_0^t \frac{\psi_{1,p-r}(t-u)\,du}{(\pi u)^{\frac{1}{2}}} \right.$$

$$\left. - (\alpha_2^r e^{-\alpha_2 j_2} + (-1)^r \beta_2^r e^{\beta_2 j_2}) \frac{(\sin \omega t)^r}{r!} \int_0^t \frac{\psi_{2,p-r}(t-u)\,du}{(\pi u)^{\frac{1}{2}}} \right] \tag{144}$$

where

$$p = 0,1,2 \ldots. \tag{145}$$

$$\psi_1(t) = \frac{i_1(t)}{n_1 FA C_O^* D_O^{\frac{1}{2}}} = \sum_{p=0}^{\infty} \psi_{1,p}(t) \left(\frac{n_1 F}{RT} \Delta E\right)^p \tag{146}$$

$$\psi_2(t) = \frac{i_2(t)}{n_2 FA C_O^* D_O^{\frac{1}{2}}} = \sum_{p=0}^{\infty} \psi_{2,p}(t) \left(\frac{n_2 F}{RT} \Delta E\right)^p \tag{147}$$

$$D_1 = D_O^{\beta_1} D_Y^{\alpha_1} \tag{148}$$

$$D_2 = D_Y^{\beta_2} D_R^{\alpha_2} \tag{149}$$

$$j_1 = \frac{n_1 F}{RT} (E_{dc} - E_{\frac{r}{2},1}) \tag{150}$$

$$j_2 = \frac{n_2 F}{RT} (E_{dc} - E_{\frac{r}{2},2}) \tag{151}$$

Contrasting this result to that obtained with a single charge-transfer step, we see that an integral equation is generated for each charge-transfer step involved in the faradaic process. In solving a single integral equation related to an ac component, one eventually obtains two linear algebraic equations in two unknowns (the coefficient of the sine and cosine terms). Thus, to obtain ac components from Eqs. (143) and (144), one eventually is confronted with a system of four linear algebraic equations in four unknowns. General solution of Eqs. (143) and (144) has been effected for the fundamental-harmonic and dc components (*129*). The fundamental harmonic is given by

$$I(\omega t) = \frac{2F^2 \, \Delta E \, AC_O^*(2\omega D_O)^{\frac{1}{2}}}{RT\chi} [\delta^2 + \xi^2]^{\frac{1}{2}} \sin\left(\omega t + \cot^{-1}\frac{\delta}{\xi}\right) \tag{152}$$

where

$$\chi = \frac{4e^{\beta_1 j_1}e^{-\alpha_2 j_2}}{(1+e^{-j_1})(1+e^{j_2})} + 4(e^{-\alpha_1 j_1} + e^{\beta_1 j_1})(e^{-\alpha_2 j_2} + e^{\beta_2 j_2}) - 8e^{-\alpha_2 j_2}e^{\beta_1 j_1}$$

$$+ (2\omega)^{\frac{1}{2}} \left[\frac{4D_2^{\frac{1}{2}}(e^{-\alpha_1 j_1} + e^{\beta_1 j_1})}{k_{s,2}} + \frac{4D_1^{\frac{1}{2}}(e^{-\alpha_2 j_2} + e^{\beta_2 j_2})}{k_{s,1}} \right.$$

$$\left. - 4e^{\beta_1 j_1}e^{-\alpha_2 j_2}\left(\frac{\lambda_2 + \lambda_1}{\lambda_1 \lambda_2}\right) \right]$$

$$+ (2\omega)\left[\frac{2D_1^{\frac{1}{2}}(e^{-\alpha_2 j_2} + e^{\beta_2 j_2})}{k_{s,1}\lambda_1} + \frac{2D_2^{\frac{1}{2}}(e^{-\alpha_1 j_1} + e^{\beta_1 j_1})}{k_{s,2}\lambda_2} + \frac{4D_1^{\frac{1}{2}}D_2^{\frac{1}{2}}}{k_{s,1}k_{s,2}} \right]$$

$$+ (2\omega)^{\frac{3}{2}}\left[\frac{2D_1^{\frac{1}{2}}D_2^{\frac{1}{2}}}{k_{s,1}k_{s,2}}\left(\frac{\lambda_1 + \lambda_2}{\lambda_1 \lambda_2}\right) \right] + \frac{(2\omega)^2 D_1^{\frac{1}{2}}D_2^{\frac{1}{2}}}{k_{s,1}k_{s,2}\lambda_1 \lambda_2} \tag{153}$$

$$\delta = n_1^2 F_1(t)\left\{ e^{-\alpha_2 j_2} + e^{\beta_2 j_2} + \left(\frac{n_2}{n_1}\right)e^{-\alpha_2 j_2} - \frac{e^{-\alpha_2 j_2}}{(1+e^{-j_1})} \right.$$

$$\left. - \frac{(n_2/n_1)e^{-\alpha_2 j_2}}{(1+e^{-j_1})(1+e^{j_2})} \right.$$

$$+ (2\omega)^{\frac{1}{2}} \left[\frac{D_2^{\frac{1}{2}}}{k_{s,2}} + \frac{(e^{-\alpha_2 j_2} + e^{\beta_2 j_2})}{\lambda_1} + \left(\frac{n_2}{n_1}\right) e^{-\alpha_2 j_2} \left(\frac{\lambda_1 + \lambda_2}{\lambda_1 \lambda_2}\right) \right]$$

$$+ (2\omega) \left[\frac{D_2^{\frac{1}{2}}}{k_{s,2}} \left(\frac{\lambda_1 + \lambda_2}{\lambda_1 \lambda_2}\right) + \frac{(n_2/n_1) e^{-\alpha_2 j_2}}{2\lambda_1 \lambda_2} \right] + \frac{(2\omega)^{\frac{3}{2}} D_2^{\frac{1}{2}}}{2 k_{s,2} \lambda_1 \lambda_2} \Big\}$$

$$+ n_2^2 F_2(t) \Big\{ e^{-\alpha_1 j_1} + e^{\beta_1 j_1} + \left(\frac{n_1}{n_2}\right) e^{\beta_1 j_1} - \frac{e^{\beta_1 j_1}}{(1 + e^{j_2})}$$

$$- \frac{(n_1/n_2) e^{\beta_1 j_1}}{(1 + e^{-j_1})(1 + e^{j_2})}$$

$$+ (2\omega)^{\frac{1}{2}} \left[\frac{D_1^{\frac{1}{2}}}{k_{s,1}} + \frac{(e^{-\alpha_1 j_1} + e^{\beta_1 j_1})}{\lambda_2} + \left(\frac{n_1}{n_2}\right) e^{\beta_1 j_1} \left(\frac{\lambda_1 + \lambda_2}{\lambda_1 \lambda_2}\right) \right]$$

$$+ (2\omega) \left[\frac{D_1^{\frac{1}{2}}}{k_{s,1}} \left(\frac{\lambda_1 + \lambda_2}{\lambda_1 \lambda_2}\right) + \frac{(n_1/n_2) e^{\beta_1 j_1}}{2\lambda_1 \lambda_2} \right] + \frac{(2\omega)^{\frac{3}{2}} D_1^{\frac{1}{2}}}{2 k_{s,1} \lambda_1 \lambda_2} \Big\} \tag{154}$$

$$\xi = n_1^2 F_1(t) \Big\{ e^{-\alpha_2 j_2} + e^{\beta_2 j_2} + \left(\frac{n_2}{n_1}\right) e^{-\alpha_2 j_2} - \frac{e^{-\alpha_2 j_2}}{(1 + e^{-j_1})}$$

$$- \frac{(n_2/n_1) e^{-\alpha_2 j_2}}{(1 + e^{-j_1})(1 + e^{j_2})}$$

$$+ (2\omega)^{\frac{1}{2}} \left[\frac{D_2^{\frac{1}{2}}}{k_{s,2}} - \frac{e^{-\alpha_2 j_2}}{\lambda_2 (1 + e^{-j_1})} \right] + (2\omega) \left[\frac{D_2^{\frac{1}{2}}}{2 k_{s,2} \lambda_2} - \frac{(n_2/n_1) e^{-\alpha_2 j_2}}{2\lambda_1 \lambda_2} \right] \Big\}$$

$$+ n_2^2 F_2(t) \Big\{ e^{-\alpha_1 j_1} + e^{\beta_1 j_1} + \left(\frac{n_1}{n_2}\right) e^{\beta_1 j_1} - \frac{e^{\beta_1 j_1}}{(1 + e^{j_2})}$$

$$- \frac{(n_1/n_2) e^{\beta_1 j_1}}{(1 + e^{-j_1})(1 + e^{j_2})}$$

$$+ (2\omega)^{\frac{1}{2}} \left[\frac{D_1^{\frac{1}{2}}}{k_{s,1}} - \frac{e^{\beta_1 j_1}}{\lambda_1 (1 + e^{j_2})} \right] + (2\omega) \left[\frac{D_1^{\frac{1}{2}}}{2 k_{s,1} \lambda_1} - \frac{(n_1/n_2) e^{\beta_1 j_1}}{2\lambda_1 \lambda_2} \right] \tag{155}$$

$$F_1(t) = \frac{e^{-\alpha_1 j_1}}{(a - b)(1 + e^{-j_2} + e^{j_1})} \Big\{ \left[\frac{k_{s,1} e^{-j_2} e^{\beta_1 j_1}}{D_1^{\frac{1}{2}}} \right.$$

$$- (\alpha_1 + \alpha_1 e^{-j_2} - \beta_1 e^{j_1})(b - \lambda_2) \Big] e^{b^2 t} \, \text{erfc}(b t^{\frac{1}{2}})$$

$$+ \left[(a - \lambda_2)(\alpha_1 + \alpha_1 e^{-j_2} - \beta_1 e^{j_1}) - \frac{k_{s,1} e^{-j_2} e^{\beta_1 j_1}}{D_1^{\frac{1}{2}}} \right] e^{a^2 t} \, \text{erfc}(a t^{\frac{1}{2}}) \Big\}$$

$$+ \frac{e^{\beta_1 j_1}}{(1 + e^{-j_2} + e^{j_1})} \tag{156}$$

$$F_2(t) = \frac{e^{-\alpha_2 j_2}}{(1 + e^{-j_2} + e^{j_1})} + \frac{e^{-\alpha_2 j_2}}{(a - b)(1 + e^{-j_2} + e^{j_1})}$$

$$\times \left\{ \left[\frac{k_{s,1} e^{-\alpha_1 j_1}}{D_1^{\frac{1}{2}}} (\beta_2 + \beta_2 e^{j_1} - \alpha_2 e^{-j_2}) - a + \lambda_2 \right] e^{a^2 t} \, \text{erfc}(at^{\frac{1}{2}}) \right.$$

$$\left. + \left[b - \lambda_2 - \frac{k_{s,1} e^{-\alpha_1 j_1}}{D_1^{\frac{1}{2}}} (\beta_2 + \beta_2 e^{j_1} - \alpha_2 e^{-j_2}) \right] e^{b^2 t} \, \text{erfc}(bt^{\frac{1}{2}}) \right\} \quad (157)$$

$$a = \frac{\lambda_1 + \lambda_2 - [(\lambda_1 + \lambda_2)^2 - 4K]^{\frac{1}{2}}}{2} \quad (158)$$

$$b = \frac{\lambda_1 + \lambda_2 + [(\lambda_1 + \lambda_2)^2 - 4K]^{\frac{1}{2}}}{2} \quad (159)$$

$$K = \lambda_1 \lambda_2 - \frac{k_{s,1} k_{s,2} e^{-\alpha_2 j_2} e^{\beta_1 j_1}}{D_1^{\frac{1}{2}} D_2^{\frac{1}{2}}} \quad (160)$$

$$\lambda_1 = \frac{k_{s,1}}{D_1^{\frac{1}{2}}} (e^{-\alpha_1 j_1} + e^{\beta_1 j_1}) \quad (161)$$

$$\lambda_2 = \frac{k_{s,2}}{D_2^{\frac{1}{2}}} (e^{-\alpha_2 j_2} + e^{\beta_2 j_2}) \quad (162)$$

The obvious complexities of these equations given above make difficult unambiguous assessment of rate parameters when both charge-transfer steps contribute kinetically to the over-all process. However, with the aid of high-speed digital computers, detailed predictions of these equations can be studied conveniently. Such information gives the electrochemist some feeling for the expected ac polarographic behavior with this type of system which should prove useful in mechanistic diagnosis. Even if the latter proves to be the only practical application of these equations, this author believes the effort in derivation and study is justified. This view is held partly because of the relation of this work to the important unanswered question regarding how many multielectron processes actually involve simultaneous transfer of more than one electron as opposed to two-step mechanisms, such as the one under consideration or the "ece" mechanism.

An interesting property of the theoretical equations for the two-step mechanism is the dependence of phase angle on $F_1(t)$ and $F_2(t)$. These terms are responsive to conditions in the dc process. When the dc process is kinetically affected by the charge-transfer steps, the exponent-error function complement product is significant, and therefore the phase angle is time dependent. Such a prediction is expected for any mechanism involving more than one charge-transfer step operative at the same dc potential. This behavior contrasts to mechanisms involving single-step charge-transfer processes with or without coupled first-order chemical reactions, where time-dependance is predicted only for current amplitude (when $\Delta E \leq 8$ mV).

Equation (152) simplifies considerably for a number of special cases. For the trivial case where E_2^0 is much more negative than E_1^0 (two resolved ac waves), the equation for each wave reduces to the expression for the single-step quasi-reversible system [Eqs. (59) through (62)] with electrochemical rate and thermodynamic parameters for the first step applying in the vicinity of E_1^0, and those for the second step around E_2^0. When the over-all process is diffusion controlled ($\omega^{\frac{1}{2}}/\lambda_1$ and $\omega^{\frac{1}{2}}/\lambda_2$ approach zero) one obtains for the ac wave equation (*129*)

$$I(\omega t) = \frac{F^2 A C_O^*(\omega D_O)^{\frac{1}{2}} \Delta E}{RT(1 + e^{-j_2} + e^{j_1})^2} \left\{ n_1^2 \left[e^{j_1} + \left(\frac{n_1 + n_2}{n_1} \right) e^{j_1} e^{-j_2} \right] \right.$$

$$\left. + n_2^2 \left[e^{-j_2} + \left(\frac{n_1 + n_2}{n_2} \right) e^{j_1} e^{-j_2} \right] \right\} \sin\left(\omega t + \frac{\pi}{4} \right) \qquad (163)$$

It is seen that the phase angle is predicted to be 45°, just as with the single-step diffusion-controlled mechanism. Considerable simplification is also obtained for the cases in which the rates of the two steps differ markedly and when E_1^0 is much more anodic than E_2^0. Some phase-angle and current-amplitude behavior predicted for the two-step mechanism is shown in Figs. 21, 22, and 23.

The same theoretical approach is applicable routinely to other multistep charge-transfer mechanisms such as the "ece" mechanism (*111*). In no case is the theorist rewarded with the luxury of simple equations, conveniently applicable to experimental data.

F. Adsorption

Quantitative theory for ac polarographic waves influenced in any way by adsorption processes is virtually nonexistent. Some work applicable to the tensammetric wave has been published, as have some interesting calculations concerned with ac measurements at the equilibrium potential. However, little has been accomplished for the more general conditions comprising simultaneous flow of direct and alternating current. The lack

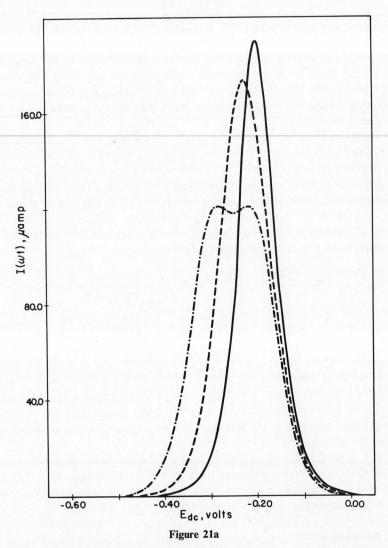

Figure 21a

of extensive progress in this area is primarily because satisfactory understanding of adsorption processes at electrodes has not been achieved for
many situations, particularly when heterogeneous charge transfer and
adsorption occur simultaneously. Few theories characterizing adsorption
processes at electrodes in the absence of alternating currents have achieved
wide acceptance. Thus, the stimulus for extending these ideas to ac methods

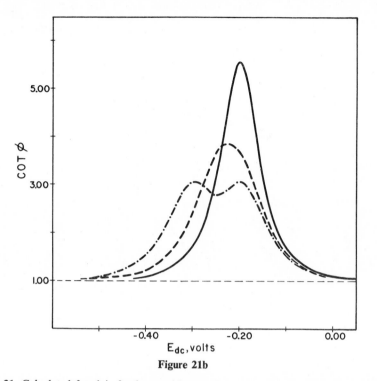

Figure 21b

Fig. 21. Calculated faradaic fundamental-harmonic ac polarograms and phase angles with a two-step reduction [from (129)]. Parameters: $D_O = D_Y = D_R = 1.00 \times 10^{-5}\,cm^2$ sec^{-1}, $A = 0.035\,cm^2$, $c_O^* = 5.0 \times 10^{-3}\,M$, $T = 298°K$, $n_1 = n_2 = 1.00$, $\Delta E = 5.00\,mV$, $\alpha_1 = \alpha_2 = 0.500$, $\omega = 8100\,sec^{-1}$, $k_{S,1} = k_{S,2} = 0.100$; —, $E_1^0 = E_2^0 = -0.200$ volt, $--$, $E_1^0 = -0.200$ volt, $E_2^0 = -0.250$ volt, $- \cdot -$, $E_1^0 = -0.200$ volt, $E_2^0 = -0.300$ volt. (a) current amplitudes; (b) phase angles.

is often lacking. However, with regard to our understanding of adsorption processes at electrodes, a rapidly improving situation is apparent from perusal of the most recent literature on this subject (the past five years). Promise for future progress is evident. Work by Delahay and co-workers $(10,67,68,175–178)$, Lorenz et al. $(179–184)$, Koutecky et al. $(185–188)$, Grahame (106), Parsons $(114,189)$, Mohilner (162), Frumkin and Melik-Gaïkazyan $(190,191)$, and Reinmuth (192) is indicative of this progress. Owing to the currency of these developments, they have not been applied to the theory of ac polarography (or many other techniques). However, because the ac method has been long recognized as one of the most suitable techniques for the study of adsorption processes, ac polarography should not be long in benefiting from advances in this area. Although a reasonable

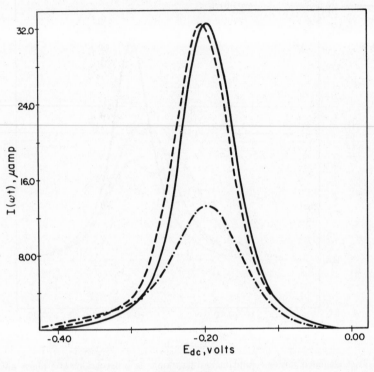

Figure 22a

quantitative discussion of the influence of adsorption processes on the ac polarographic wave is not possible at this time, the importance of this phenomena warrants at least discussion of some qualitative concepts.

Adsorption processes may influence ac polarographic data in the following ways: (a) Adsorption of electroinactive surfactants has a profound effect on the double layer capacity leading to considerable depression of the double-layer charging current in the region of the electrocapillary maximum (ecm) and often a sharp peak at potentials (both positive and negative) well removed from the ecm, where the coulombic forces between the electrode and ions of the electrolyte lead to desorption (9); (b) Adsorbed electroinactive surfactants can have a profound influence on kinetics of charge transfer and other steps associated with the electrode process (193); (c) Adsorption of electroactive species may markedly influence the characteristics of the faradaic alternating current through alteration of the mass-transfer process or by chemical reactions occurring only on the electrode

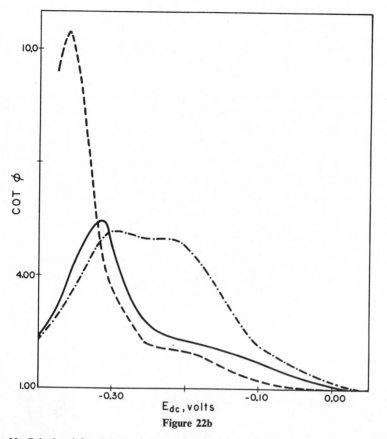

Figure 22b

Fig. 22. Calculated faradaic fundamental-harmonic ac polarograms and phase angles with a two-step reduction [from (129)]. Parameters: $D_O = D_Y = D_R = 1.00 \times 10^{-5}$ cm^2 sec^{-1}, $A = 0.035$ cm^2, $C_O^* = 5.0 \times 10^{-3}$ M, $T = 298°$K, $n_1 = n_2 = 1.00$, $\Delta E = 5.00$ mV, $\alpha_1 = \alpha_2 = 0.500$, $\omega = 100$ sec^{-1}, $k_{S,2} = 1.00 \times 10^{-3}$ cm sec^{-1}, $k_{S,1} = 0.100$ cm sec^{-1}, $\tau = 9$ sec^{-1}; —, $E_1^0 = -0.200$ volt, $E_2^0 = -0.220$ volt, – –, $E_1^0 = -0.200$ volt, $E_2^0 = -0.300$ volt, – · –, $E_1^0 = -0.200$ volt, $E_2^0 = -0.220$ volt, $k_{S,1} = 10^{-2}$ cm^2 sec^{-1}, $k_{S,2} = 10^{-3}$ cm^2 sec^{-1}; (a) current amplitudes, (b) phase angles.

surface [for example, hydrogen-atom combination on platinum (106)] and, as with electroinactive surfactants, the double-layer capacity can be altered.

The effects to be considered in developing a quantitative theory for ac polarography with any of the above phenomena are numerous. To compute the extent of surface coverage, one must have some knowledge of the adsorption isotherm and the rate of adsorption. Both quantities depend on the electrode potential and, thus, will vary over the ac polarographic wave.

81

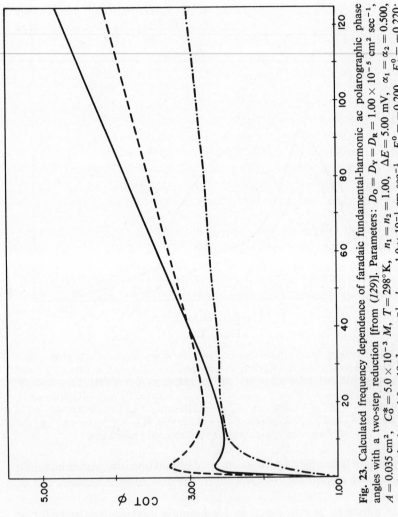

Fig. 23. Calculated frequency dependence of faradaic fundamental-harmonic ac polarographic phase angles with a two-step reduction [from (129)]. Parameters: $D_O = D_Y = D_R = 1.00 \times 10^{-5}$ cm^2 sec^{-1}, $A = 0.035$ cm^2, $C_O^* = 5.0 \times 10^{-3}$ M, $T = 298°$K, $n_1 = n_2 = 1.00$, $\Delta E = 5.00$ mV, $\alpha_1 = \alpha_2 = 0.500$, $\tau = 9$ sec^{-1}, $k_{s,1} = 1.0 \times 10^{-3}$ cm sec^{-1}, $k_{s,2} = 1.0 \times 10^{-1}$ cm sec^{-1}, $E_1^0 = -0.200$, $E_2^0 = -0.220$; ——, $E_{dc} = -0.200$ volt, - -, $E_{dc} = -0.175$ volt, - · -, $E_{dc} = -0.320$ volt.

82

The development of a theory properly describing these phenomena has been a "rate-determining step" in evolving a theory accounting for adsorption in electrochemical measurements. However, recent work by Delahay and Mohilner (*87,175*) is noteworthy and may prove to be the keynote to an era of considerable progress along these lines. In addition to these factors, the extent of surface coverage is dependent also on the rate of the mass-transfer process. When near-equilibrium conditions for adsorption can be assumed, treating the mass-transfer problem is not difficult (*114*). This would be the case when a stationary electrode is immersed in a solution for a time sufficient to achieve adsorption equilibrium, and then a *small* electrical perturbation is applied (for example, a small amplitude alternating potential). Unfortunately, with the commonly employed dropping mercury electrode, adsorption equilibrium is often not achieved and the complexity of the mass-transfer problem is enhanced.

An accurate model for adsorption kinetics and the isotherm, together with proper treatment of the mass-transfer process, is sufficient to develop a theory for the tensammetric process in ac polarography (that is, the influence of adsorbed species on the differential capacity of the electrical double layer). Lorenz (*179*) presented a rigorous theory for this *nonfaradaic impedance* with adsorption of a neutral surfactant on an electrode at near-equilibrium conditions. His equations are expressed without assuming a specific dependance of the rate and isotherm for the adsorption process on experimental parameters. For example, the phase-angle expression is given by

$$\tan \phi = \frac{1 + (2D/\omega)^{\frac{1}{2}}/(\partial\Gamma/\partial_\infty C)_E}{1 - (2D\omega)^{\frac{1}{2}}/(\partial\Gamma/\partial_\infty C)_E(\partial v/\partial\Gamma)_{E_0,C}} \tag{164}$$

(see Sec. VI for notation definitions). A more specific set of equations was developed by Lorenz and Möckel (*180*), who assumed adsorption follows a Langmuir model and the net rate of adsorption is given by

$$v = k_a(\Gamma_s - \Gamma)C_{x=0} - k_d\Gamma \tag{165}$$

As pointed out by Delahay and Mohilner (*87*), their recent theory can be readily applied to the equation of Lorenz [Eq. (164)] to give the nonfaradaic impedance for an adsorption process obeying the Temkin isotherm. They assume the applicability of the basic equations of Temkin (*194*):

$$b\Gamma = -\Delta G^0 + RT \ln a_0 \tag{166}$$

$$v_a = k_a a_0 \exp[-(\lambda^* b/RT)\Gamma]\exp[-\rho \, \Delta G^q/RT] \tag{167}$$

$$v_d = k_d \exp\{[(1 - \lambda^*)b/RT]\Gamma\}\exp[(1 - \rho) \, \Delta G^q/RT] \tag{168}$$

and develop the rate law

$$v = v^0\{\exp[-(\lambda^* b/RT)\delta\Gamma]\exp[-(\rho/RT)\delta(\Delta G^q)]$$
$$- \exp[((1 - \lambda^*)b/RT)\delta\Gamma]\exp[((1 - \rho)/RT)\delta(\Delta G^q)]\} \qquad (169)$$

where

$$v^0 = k^0 a_0^{1-\lambda^*} \exp[(\lambda^* - \rho)\,\Delta G^q/RT] \qquad (170a)$$

$$k^0 = k_a \exp(\lambda^*\,\Delta G^n/RT) \qquad (170b)$$

which, interestingly, is analogous to the rate law for charge transfer at electrodes [Eq. (50)]. If the Temkin isotherm is found to be obeyed by most neutral organic substances, the combined theories of Delahay and Mohilner and Lorenz will represent a significant advance in the quantitative treatment of the tensammetric process in ac polarography. The theory will be applicable to the DME only for relatively high concentrations of surface-active material, owing to the assumption of near-equilibrium conditions. A general theory for the DME involves handling a more complex mass-transfer problem such as considered by Ward and Tordai (*195*), Delahay and co-workers (*176,177*), and Reinmuth (*192*). In this regard, Delahay and Trachtenberg have suggested a simple test to determine whether adsorption equilibrium is achieved at the DME (*176*). One examines the dependence of alternating current on mercury-column height. Independence of column height indicates adsorption equilibrium (or negligible adsorption). This test is valid if a faradaic process does not introduce a time dependence (see below).

One must consider the factors just discussed to develop a theory for the influence of electroinactive surfactants on the faradaic process. In addition, a mechanism whereby the faradaic process is altered by surfactant must be postulated. The latter remains a point of conjecture, and several possibilities have been proposed. They include deceleration of electron transfer (*6,21, 185–188,196–200*), inhibited penetration of the film (*201–207*), inhibition of a chemical step (*208–211*), and interactions between the adsorbed substance and the depolarizer (*212–215*). Reilley and Stumm (*193*) recently discussed the rationale behind these proposals. It is probable that each mechanism might be operative in specific cases, and all merit some consideration.

Koutecky and co-workers (*185–188*) have presented quantitative treatments for instantaneous dc polarographic currents for cases where the surfactant inhibits charge transfer. Silvestroni and Rampazzo (*212*) have derived expressions for instantaneous dc polarographic currents for the case in which the adsorbed substance interacts with the depolarizer. It appears possible to extend these treatments to ac polarography. Their applicability remains a matter of discussion. There is little question that inhibition of the faradaic process by surface-active substances can play an important role in ac polarography. Delahay and Trachtenberg (*176,216*) observed continual decrease of the apparent charge-transfer kinetics with time at a hanging mercury drop with the cadmium system in solutions of normal analytical purity. Several authors [for example, (*217,218*)] have indicated the necessity of special treatment of solutions with activated charcoal to prevent contamination of stationary electrodes by surface active impurities.

Adsorption of the electroactive species probably represents the most important adsorption scheme in ac polarography. One can apply some simple qualitative concepts to develop some feeling regarding how such phenomena might influence the faradaic impedance. Perhaps the most convincing arguments can be applied to the phase angle (*108*). A well-known relation in ac circuit theory is [for example, (*219*)]

$$P = EI \cos \phi \tag{171}$$

where P is the average power dissipation in the ac process, E is the amplitude of alternating voltage, I is the amplitude of alternating current, and ϕ is the phase angle. A phase angle of $0°$ represents maximum power dissipation. The more conservative of energy the ac process, the closer the phase angle approaches $90°$. It was seen in the theory developed above that a diffusion-controlled process yields a predicted phase angle of $45°$. This implies that the mass-transfer process leads to some, but not total, energy dissipation. Qualitatively, this is the expected result because, although some oxidized or reduced species generated on one half-cycle can be dissipated through the diffusion process (that is, diffusion into the solution to be lost to the electrode process), some can also be made available through diffusion for the reverse electrode reaction on the following half-cycle. It was also seen that the influence of any slow kinetic step in addition to diffusion led to the prediction [Eqs. (59), (77), (103), (152)] of smaller phase angles. This is in agreement with the concept that additional slow steps lead to a faradaic process which is less efficient in the passage of alternating current, that is, less conservative of energy. Applying these ideas to processes involving adsorption of the electroactive form, one concludes that the over-all mass-transfer process in the presence of adsorption will be more conservative of electrical energy, because adsorption leads to excess surface concentrations and less freedom for these surface concentrations to be dissipated by the diffusion process. In view of Eq. (171), this suggests that adsorption of the electroactive form will increase the phase angle, provided kinetics of adsorption are not too slow. For very rapid electrode processes and/or low frequencies, the possibility is suggested that the phase angle may exceed 45°, a phenomena not predicted by any of the mechanisms considered earlier. Although a quantitative theory for ac polarography (dc and ac flow) has not been forthcoming, several workers have presented theoretical equations for the faradaic impedance at the equilibrium potential. Most noteworthy are contributions of Gerischer (220), Laitinen and Randles (221), Llopis et al. (222,223), Barker (61), and Senda and Delahay (67). Although these calculations differ in generality and rigor, all predict a net increase in phase angle, because of adsorption, and the possibility of phase angles in excess of 45° under appropriate conditions, in agreement with the foregoing intuitive arguments. The analysis of Senda and Delahay is most general in that no particular form for the adsorption isotherm and rate equation is introduced. The theory of Delahay and Mohilner (87) can be incorporated into the latter work to yield the faradaic impedance for adsorption of the electroactive form which obeys the Temkin equations. A number of experimental studies have appeared to confirm qualitatively these theoretical predictions. Randles and Somerton (21) observed phase angles larger than 45° in the reduction of tetramethyl-*p*-phenylenediamine, which is adsorbed at platinum. Their explanation of this " abnormality " in terms of adsorption appears to be the earliest suggestion that adsorption can yield such a result. Laitinen and Randles (221) made similar observations for the reduction of *tris*-ethylenediamine cobaltic ion, as did Randles (127) and Tamamushi and Tanaka (224) with the thallous-thallium amalgam system. Adsorption was presumably demonstrated with the latter system by Frumkin (191). Bauer and co-workers (225) reported abnormally large phase angles for the cadmium ion-cadmium amalgam system. The combination of intuitive arguments, quantitative theory, and these experimental results would seem to suggest that the development of a theory for the faradaic impedance with adsorbed electroactive material is proceeding along a reasonable path. However, some very recent work by Sluyters and co-workers (226–231) has introduced some doubt into the picture. They have presented a method for analysis of ac data based on plotting the readout in the complex impedance (or admittance) plane which enables one to determine the double-layer capacity and ohmic resistance in the presence of the faradaic wave (the method is discussed in Sec. V). They have shown that the abnormally large phase angle (> 45°) observed with the thallium system can be attributed simply to the influence of the electroactive species on the double-layer capacity (231). Their analysis yields an apparent faradaic impedance corresponding to a diffusion-controlled ac wave. It is concluded that earlier workers measuring the double-layer capacity in the electrolyte solution in the absence of the thallium couple were incorrect in assuming it the same in the presence of the thallium system. There seems little doubt that the thallium couple significantly alters the double-layer capacity, and that the double-layer capacity in the presence of this system must be employed to calculate accurately the faradaic impedance. However, the marked influence on the double-layer capacity seems reasonable only if strong adsorption exists, because the thallium species is present in very small concentrations relative to ions of the supporting electrolyte. If so, one would expect that the faradaic impedance should be significantly influenced by the adsorption, contrary to the results of Sluyters et al. This seems to present a paradox unless (a) the adsorbed form is not electro-active [this seems unlikely, although there appears to be at least one precedent (*232*)]; (b) because the method of data analysis employed by Sluyters et al. assumes a specific form of the equivalent circuit representing the cell impedance, they might be including in their correction the additional capacitive component introduced into the *faradaic impedance* by adsorption in addition to the variation in double-layer capacity resulting from thallium; (c) the substantial influence of the thallium system on the double-layer capacity of mercury arises without surface excesses sufficiently large to alter the faradaic impedance by a measurable amount [it is known that trace amounts of heavy metals dissolved in mercury can significantly alter certain bulk properties such as flow characteristics (*233*)]; (d) the intuitive and quantitative theories for adsorption discussed above are incorrect. These possibilities amount to pure conjecture directed toward illustrating that some questions remain unanswered.

G. Double-Layer Effects

It should be mentioned at the outset that adsorption and double-layer effects cannot be considered on a separate basis. Characteristics of adsorption processes on metals in electrolyte solutions are influenced intimately by the electrical double layer and vice versa. However, in the absence of strong interactions between the electrode material and species in solution (specific adsorption), electrode processes may still be perturbed significantly by forces operative within the diffuse double layer. These influences were not considered in the preceding theoretical discussion.

The earliest attempt to interpret the influence of the double-layer structure was presented by Frumkin (*234*) in 1933. He pointed out that the concentration of reacting particles in the double layer is different from the bulk of the solution, and the electrical energy available for assisting or retarding

88

D. E. Smith

the electrode reaction is less than the potential difference between electrode and solution. Introducing both effects into the absolute rate equation [Eq. (50)], we obtain [cf., for example, (10) and (114)]

$$\frac{i(t)}{nFA} = k_s \exp\left[\frac{(\alpha n - z_0)F\,\Delta\Phi_E}{RT}\right]\left\{C_{O_{x=0}}\exp\left[\frac{-\alpha nF}{RT}(E(t) - E^0)\right]\right.$$

$$\left. - C_{R_{x=0}}\exp\left[\frac{(1-\alpha)nF}{RT}(E(t) - E^0)\right]\right\} \tag{172}$$

where $\Delta\Phi_E$ is the potential difference between the plane of closest approach and the bulk of the solution and z_0 is the ionic valence of the oxidized form. Within the framework of this theory, one can say

$$k_{a,s} = k_s \exp\left[\frac{(\alpha n - z_0)F\,\Delta\Phi_E}{RT}\right] \tag{173}$$

where $k_{a,s}$ is the apparent heterogeneous rate constant (uncorrected for double-layer effects). Parsons (114) has pointed out that splitting the double-layer effect into two parts is a device for calculating the effect of changes in double-layer structure on the free energy of the activated state and that Eq. (173) can be deduced on the basis of energy diagrams of the type used by Horiuti and Polanyi (235). It is apparent that the Frumkin correction for the double layer introduces only minor complication into the theory discussed earlier. The functional relationships remain essentially unchanged, and the double-layer correction may be implemented simply by correcting the apparent heterogeneous rate constant in accord with Eq. (173). However, it should be noted that this type of double-layer correction on ac polarographic data involves the implicit assumption that sinusoidal variations in $\Delta\Phi_E$ may be neglected, an assumption which may not be accurate in some cases.

The correction for activity effects has a similar property. In all rigor, activities should be employed in Eqs. (50) and (172). If this is done, one obtains as the apparent rate constant $k_{a,s}^*$:

$$k_{a,s}^* = f_O^\beta f_R^\alpha k_s \exp\left[\frac{(\alpha n - z_0)F\,\Delta\Phi_E}{RT}\right] \tag{174}$$

It has been pointed out that correction for the double layer by the Frumkin theory is valid only when the thickness of the diffuse double layer is negligible in comparison with the diffusion layer (10,236). For cases involving systems with coupled chemical reactions, the additional requirement that the reaction-layer thickness (see below) be much larger than the thickness of the diffuse double layer must also be invoked. If these conditions are not fulfilled, perturbations on the mass-transfer process resulting from the presence of the electrical double layer must be considered. Perturbations may be due to (a) differences between diffusion coefficients in the double layer and bulk of the solution; (b) differences between chemical rate constants in the double layer and bulk of the solution; (c) migration in the diffuse double layer. Rangarajan and Doss (*237*) have treated a perturbation falling under category (a) for the quasi-reversible case at the equilibrium dc potential. Their model assumed a step-function variation in diffusion coefficients at a certain distance from the electrode surface. If the diffuse double layer alters the diffusion coefficient, it is unlikely that the variation approaches the form of a step function. However, the model is useful in that it gives one some feeling for the importance of such an effect on the diffusion coefficient. As expected, their equations indicate a negligible influence on the faradaic impedance when the ac diffusion-layer thickness, which is approximately $(D/\omega)^{\frac{1}{2}}$, is much larger than the region adjacent to the interface in which the diffusion coefficient differs from the bulk value. On the other hand, when these quantities are comparable, the impedance is significantly altered. For the case in which the diffusion coefficient in the perturbing region (double layer) is larger than in the bulk of the solution, and charge transfer is rapid, the phase angle can exceed 45° as with specific adsorption. Because the double-layer thickness is less than about 10^{-7} cm for supporting electrolyte concentrations in excess of 0.1 M, this effect should not be important except at frequencies in the multimegacycle range.

The case in which the double layer perturbs the rate of a chemical reaction has not been treated for either ac polarography or impedance measurements at the equilibrium dc potential. However, results of a calculation by Testa and Reinmuth (*238*) for chronopotentiometry are qualitatively applicable. It was shown that the double-layer effect would be negligible if the reaction-layer thickness is much larger than the diffuse double-layer thickness. Testa and Reinmuth were considering the case of an irreversible decomposition following charge transfer for which the reaction layer thickness μ is defined as

$$\mu = (D/k)^{\frac{1}{2}} \tag{175}$$

where k is the first-order rate constant for the chemical decomposition and D the diffusion coefficient. In the more general case of a reversible first-order chemical reaction, one has

$$\mu = (D/k_b)^{\frac{1}{2}} \tag{176}$$

where k_b is the reverse rate constant (leading to electroinactive form). In view of the usual order of magnitude of D and the double-layer thickness, a negligible double-layer effect of this type is expected for all but the most rapid chemical reactions ($k > 10^6 \text{ sec}^{-1}$). For such fast reactions, there is little question that the high electric-field gradient within the diffuse double layer can alter rates of chemical reactions involving ionic reactants as pointed out by Delahay and Vielstich (*239*). The field effect on ionic reactions was calculated by Onsager (*240*) and is the basis for a popular chemical-relaxation method for studying rapid homogeneous ionic reactions (*241*).

Migration within the diffuse double layer has received the most attention of the possible corrections beyond the Frumkin theory. The effect arises as a result of the high electric field in the double layer. Levich (*242*), Matsuda and Delahay (*236,243*), Rangarajan and co-workers (*237,244–246*), and Gierst and Hurwitz (*247,248*) have examined this problem, including the quasi-reversible case and the system with a preceding homogeneous first-order chemical reaction. For the quasi-reversible case, incorporation of the influence of migration in the diffuse double layer leads to the mass-transfer boundary-value problem expressed by the relations (*236,244*) [in place of Eq. (1) to (7)]

$$\frac{\partial C_i}{\partial t} = D_i \left[\frac{\partial}{\partial x} \left(\frac{\partial C_i}{\partial x} + \frac{z_i F}{RT} \frac{d\Phi}{dx} C_i \right) \right] \tag{177}$$

For $t = 0$, any x,

$$C_i = C_i^* \exp\left(\frac{-z_i F\Phi}{RT} \right) \tag{178a}$$

For $t > 0$, $x \to \infty$,

$$C_i \to C_i^* \tag{178b}$$

For $t > 0$, $x = 0$,

$$D_i \left(\frac{\partial C_i}{\partial x} + \frac{z_i F}{RT} \frac{d\Phi}{dx} C_i \right) = \frac{\pm i(t)}{nFA} \tag{179}$$

(positive sign for $i = \text{O}$ and negative for $i = \text{R}$) where Φ, the potential-distance profile in the diffuse double layer, is taken from the Gouy-Chapman theory, that is,

$$\frac{d\Phi}{dx} = -\left(\frac{2RT}{|z|F} \right) \kappa \sinh\left(\frac{|z|F\Phi}{2RT} \right) \tag{180}$$

$$\frac{1}{\kappa} = (RT\varepsilon/8\pi z^2 F^2 C_t)^{\frac{1}{2}} \tag{181}$$

Matsuda and Delahay (*236,243*) first treated this problem, and Rangarajan (*244*) later presented a modified treatment. The mathematics of the solution are complicated. The interface concentrations in Laplace transform space involve higher-order transcendental functions [Bessel (*236*) or Gaussian hypergeometric functions (*244*)] which discourage generality in inverse transformation. However, a number of interesting cases were calculated and, among the conclusions reached, were (a) the double-layer effect can be significant, even at relatively low frequencies for electrolysis of ionic species when the potential gradient in the diffuse double layer is sufficiently large; (b) for the case of repulsion by the double layer, a plot of cot ϕ versus $\omega^{\frac{1}{2}}$ is linear, and the apparent heterogeneous rate constant is smaller than the true rate constant—the correction is not overly complicated at frequencies less than 1 mc; (c) for the case of attraction, a plot of cot ϕ versus $\omega^{\frac{1}{2}}$ may be nonlinear and ϕ may exceed 45° with large k_s and low frequencies, similar to the predictions for specific adsorption. Matsuda (*243*), Gierst and Hurwitz (*247,248*), and Rangarajan (*244*) considered this problem for a system with a coupled preceding chemical reaction. Although not applied specifically to ac methods, the general conclusions are applicable to ac polarography. Mathematical complexity again hinders generality, but the range of conditions treated is significant, and the conclusions are interesting. It is shown for most conditions that when the reaction-layer thickness is much larger than the diffuse double-layer thickness, the effect of migration in the double layer on the apparent chemical rate constant is negligible. On the other hand, when the reaction-layer thickness is smaller than the double-layer thickness, the effect on the apparent chemical rate constant can be analyzed in terms of a simple Boltzmann distribution (*243,247*). For intermediate situations, the mathematical relation between the apparent and true chemical rate constants assumes a rather complicated form. Rangarajan (*244*) and Gierst and Hurwitz (*247*) have indicated the possibility that the rate-limiting factor can be mass transfer of the electroactive form through the diffuse double layer (the *dynamic ψ effect*). Conditions required for this situation, in particular $K \ll 1$, may make the dynamic ψ effect of little interest for ac polarography.

Even though the faradaic impedance at the equilibrium potential was treated, none of this theoretical work considering double-layer corrections beyond the Frumkin theory has been applied to simultaneous flow of direct and alternating current which occurs in ac polarography. Thus, much remains to be accomplished.

One characteristic of the theory for double-layer effects related to the applicability of simpler theory neglecting such effects is worth noting.

In many (but not all) cases, the double-layer correction does not significantly alter the qualitative functional form of the simpler theory (for example, linear plots remain linear, etc.) so that mechanistic conclusions are not influenced. In addition, apparent rate constants (chemical or electrochemical) obtained upon application of the simple theory can often be subjected to simple corrections to incorporate the influence of the double-layer. The most important exception is the case of significant attractive migration in the diffuse double layer.

H. Time Dependence of the AC Polarographic Wave

Time-dependent terms appear in a number of the ac wave equations discussed above. It was indicated that such terms represent a response to characteristics of the dc process. Time dependence of the ac polarographic wave, as studied through the mercury-column-height dependence, recently has been a subject of much interest (*121–126*). For this reason a few general remarks on the subject appear in order.

Time dependence of the alternating current, and the resulting response to mercury-column height predicted in the foregoing equations are a manifestation of nonequilibrium conditions in the dc surface concentrations. These concentration components approach equilibrium with time, giving rise to a time-dependent impedance. Senda (*126*) was first to point out the origin of this phenomena. The study of column-height dependence of ac polarographic waves represents a rather sensitive probe into conditions in the dc polarographic process. Ac waves dependent on mercury-column height are predicted when any rate process in addition to diffusion influences this process kinetically (*121,129*). It appears that, through this phenomenon, one can detect kinetic effects in the dc process with relative ease under conditions where the effect might be detected through the dc polarogram only after careful analysis.

Whether the phase angle is time dependent is a point of interest. Theory for the small-amplitude ac wave with single-step charge-transfer mechanisms including any number of coupled first-order chemical reactions predicts that only the amplitude can show a time dependence. As seen above, time-dependent terms appear in phase-angle expressions when multistep charge transfer affects the dc process. In addition, approximate theoretical relationships indicate that the phase angle can be time dependent with only a single charge-transfer step when systems with coupled second-order chemical reactions are operative (*121*). Finally, it is believed that slow adsorption processes can bring about time-dependent phase angles (*121*), although this concept is founded only on an intuitive basis. Thus, the appearance of phase angles responsive to mercury-column height is suggestive of multistep charge-transfer, coupled second- or higher-order chemical reactions, and/or possibly adsorption processes. Such distinctions appear only when sufficiently small amplitudes ($\Delta E \leq 8/n$ mV) are employed. As indicated earlier, time-dependent terms appear in phase-angle equations at larger amplitudes for any mechanism.

Many of the above concepts have received at least qualitative verification in recent work by Aylward and Hayes and others (*122,123,125*) and by Hung and Smith (*121,122*). The earliest report of an ac wave dependent on column height appears to have been given by Florence and Aylward (*124*).

Very recently, a notably different source of time dependence not evident in the foregoing theoretical treatments has been suggested (*248a,248b*). As in the cases already discussed, the effect originates in the dc process. However, it differs in that it is not associated with nonequilibrium conditions induced by effects of slow rate processes other than diffusion. Rather, the time dependence arises because of the influence of spherical diffusion on the dc process and, thus, may occur even with a strictly diffusion-controlled dc process. Unlike planar diffusion, spherical diffusion may induce time-dependent dc surface concentrations of electroactive species which are manifested as a time-dependent impedance. Calculations based on a spherical-diffusion mass-transfer model indicate that time dependence due solely to this source normally will be significant only with systems involving a reduced form in the amalgam state (*248b*). Thus, it appears that this effect may be significant for a somewhat limited category of electrode processes.

This effect of spherical diffusion is discussed further in the next section.

I. Contributions of Electrode Growth and Geometry

All theory discussed above assumes the applicability of the stationary-plane electrode model. Application of these equations to ac polarographic data obtained with a DME amounts to neglecting contributions of curvature and motion relative to the solution of the electrode surface. Most ac polarographic work employs this approximation. The theoretical studies of Gerischer (*95*) and Koutecky (*96*) seemed to support the planar-diffusion approximation. Further support is found in the following intuitive concepts: (a) the relatively slow motion of the electrode surface would not influence the rapidly fluctuating alternating current; (b) the relatively narrow ac concentration profile would not be influenced by electrode curvature. However, Gerischer's work indicating negligible influence of electrode geometry under normal conditions was applicable only at the equilibrium potential—that is, in absence of direct current. Koutecky's equations suggesting negligible influence of drop growth applied to the reversible process. Thus, these theoretical studies did not encompass a wide range of experimental conditions or electrode-reaction mechanisms Matsuda's (97) more recent theoretical treatment of the quasi-reversible process shed additional light on the subject. His approach included rigorous consideration of contributions of drop growth. Ac wave equations were obtained which differed from Eqs. (59) through (62), but only in the form of the $F(t)$ term. Matsuda obtained

$$F(t) = 1 + \left(\frac{7}{3\pi}\right)^{\frac{1}{2}} (\alpha e^{-j} - \beta) \frac{(1.61 + \lambda t^{\frac{1}{2}})}{(1.13 + \lambda t^{\frac{1}{2}})^2} \qquad (182)$$

and the same results for I_{rev}, $G(\omega^{\frac{1}{2}}\lambda^{-1})$, and $\cot \phi$ as with the stationary plane. Equation (182) is consistent with Eq. (60), where $F(t)$ is expressed in terms of $\psi_0(t)$, but differs from what one obtains on combination of Eqs. (60) and (62), owing to a difference in the form of the solution for the direct current. Matsuda employs a dc solution for the expanding plane (249). As with planar diffusion, $F(t)$ for the expanding plane becomes unity if $\lambda t^{\frac{1}{2}} > 50$. The difference in theory based on the two electrode models appears only when the dc process is subject to kinetic influence of charge transfer. Because Koutecky's treatment is applicable only to reversible processes, there is no disparity between the work of Koutecky and Matsuda. The drop-growth contribution predicted by Matsuda appears to arise because the faradaic impedance is markedly dependent on the magnitudes of the dc concentrations of oxidized and reduced forms at the interface. When the dc process is controlled solely by diffusion, the Nernst equation is obeyed and the interface concentrations are independent of electrode growth. However, if nonnernstian conditions exist, deviations of dc concentrations from equilibrium are influenced by the motion of the electrode surface into the solution (which aids mass transfer). This effect is manifested in the differences between equations based on stationary- and expanding-plane models. The foregoing discussion suggests that, if the dc interface-concentration components are in a non-equilibrium state, any factor influencing the dc mass-transfer process can, in turn, affect the magnitude of the ac wave. Electrode geometry would fall in this category. In addition, one is led to conclude that the nonequilibrium state need not be concerned only with the charge-transfer step. Non-equilibrium conditions related to coupled homogeneous chemical reactions and adsorption may introduce electrode geometry and growth contributions.

The hypothesis that electrode geometry can be important is readily subjected to theoretical examination by examining equations for the ac polarographic wave with a quasi-reversible process for diffusion to a stationary sphere. For example, one finds for diffusion to a stationary sphere, both redox forms soluble in solution and equal diffusion coefficients (*250*):

$$F(t) = 1 + \frac{(\alpha e^{-j} - \beta)}{(D^{\frac{1}{2}}r_0^{-1} + \lambda)} \left[\lambda e^{a^2 t} \operatorname{erfc}(at^{\frac{1}{2}}) + D^{\frac{1}{2}}r_0^{-1} \right] \tag{183}$$

where

$$a = D_0^{\frac{1}{2}}r_0^{-1} + \lambda \tag{184}$$

All other terms in the expression for the ac wave are the same as for the planar diffusion model. Interestingly, Eq. (183) also is consistent with Eq. (60). It differs from the other forms of $F(t)$ in the solution to the direct current which, in this case, is for diffusion to a stationary sphere. As in the other examples, $F(t)$ becomes unity if the dc process is nernstian ($\lambda t^{1/2} > 50$). Thus, the difference between Eq. (183) and the stationary-plane expression manifests the contribution of electrode curvature, a contribution which is significant only when the dc process is nonnernstian, at least for the special case represented by Eq. (183). Equations (182) and (183) indicate that an accurate ac polarographic wave equation for the quasi-reversible system, encompassing the possibility of nonnernstian dc behavior, should incorporate the contributions of electrode growth and geometry when the DME is the electrode in question. The fact that Eqs. (182) and (183) predict significant drop growth and geometry contributions only when the dc process is influenced by charge-transfer kinetics is consistent with the foregoing concept that these factors contribute because they alter the deviations of the dc surface concentrations from equilibrium when the latter are in a nonnernstian state. However, to conclude that this is the only mechanism whereby drop growth and geometry can contribute to the ac wave is premature, as a more general treatment of the stationary-sphere electrode model shows (*248b*). For the case of both redox forms soluble in solution, removal of the restriction of equal diffusion coefficients yields an expression for $F(t)$ of considerably greater complexity than Eq. (183) which *does not* reduce to unity for charge-transfer rates corresponding to nernstian conditions ($\lambda t^{1/2} > 50$). Instead, one obtains (*248b*)

$$F(t) = 1 + \frac{D_0^{\frac{1}{2}} - D_R^{\frac{1}{2}}}{e^j D_0^{\frac{1}{2}} + D_R^{\frac{1}{2}}} \left[1 - \exp(a^2 t) \operatorname{erfc}(at^{\frac{1}{2}}) \right] \tag{185a}$$

where

$$a = \frac{e^j D_0^{\frac{1}{2}} + D_R^{\frac{1}{2}}}{r_0(1 + e^j)} \tag{185b}$$

Similarly, for the case of the reduced form soluble in the electrode phase (amalgam formation), an expression for $F(t)$ is obtained which does not reduce to unity for $\lambda t^{1/2} > 50$, but yields (248b)

$$F(t) = 1 + \frac{D_0^{\frac{1}{2}} + D_R^{\frac{1}{2}}}{e^j D_0^{\frac{1}{2}} - D_R^{\frac{1}{2}}} [1 - \exp(b^2 t)\, \mathrm{erfc}(bt^{\frac{1}{2}})] \qquad (186a)$$

where

$$b = \frac{e^j D_0^{\frac{1}{2}} - D_R^{\frac{1}{2}}}{r_0 (1 + e^j)} \qquad (186b)$$

The fact that expressions for $F(t)$ in Eqs. (185a) and (186a) incorporate effects of electrode geometry indicates that spherical diffusion, by itself, serves as a source of this influence. In other words, although coupling of spherical diffusion with another slow rate process (charge transfer) appears to be a sufficient condition for a significant contribution of spherical diffusion in ac polarography, it is not, in general, a necessary condition. The spherical diffusion-induced time dependence discussed in the previous section also is apparent in Eqs. (185a) and (186a). One observes in Eq. (185a) that, when both forms of the redox couple are soluble in the solution phase and exhibit equal diffusion coefficients, $F(t)$ equals unity and the spherical diffusion contribution disappears; a result consistent with the prediction of Eq. (183) for these conditions. The requirement that diffusion coefficients differ suppresses considerably the contribution of spherical diffusion for the case represented by Eq. (185a). Indeed, calculations indicate that, except for unusually divergent values of D_0 and D_R, one may assume that the effect of spherical diffusion is negligible for systems involving both redox forms soluble in solution and nernstian dc processes (248b). Experimental studies cited early in this chapter seem to support this conclusion (21,27,30,81,111). Such is not the case when amalgam formation is involved. Calculations employing Eq. (186a) predict a significant contribution of spherical diffusion, regardless of values of diffusion coefficients (248b), so that this effect must be expected under most conditions for systems involving the amalgam state. Recent experimental studies on such systems support these ideas (248a).

The foregoing discussion is concerned only with the simple quasi-reversible case. For systems involving more complex mechanisms, the problem of assessing the nature of contributions of drop growth and geometry becomes more difficult. However, a theoretical demonstration has been given (250) that coupled chemical reactions can introduce drop growth and geometry effects when nonequilibrium conditions exist with respect to the chemical reactions.

As a result of the investigations discussed above, one may conclude that accurate ac polarographic work with the DME will require incorporation of contributions of drop growth and geometry into theoretical expressions whenever investigations are concerned with systems exhibiting amalgam formation and/or nonequilibrium conditions in the dc process with respect to either the electrochemical charge-transfer step or coupled chemical reactions. Apparently the only type of system immune to these effects is one in which both forms of the redox couple are soluble in the solution phase and the dc process is controlled solely by diffusion, provided diffusion coefficients of the two redox forms do not differ markedly. These conclusions represent a considerable departure from the original concept that the ac polarographic wave would not be significantly influenced by electrode growth and geometry. However, the latter thinking neglects subtle aspects of the " coupling " between the dc and ac processes in the ac polarographic experiment.

Although the importance of drop growth and geometry contributions has been demonstrated, the need for ac wave equations which incorporate both of these effects has not been satisfied. Ac polarographic wave expressions employing Eqs. (182), (183), (185a), or (186a) account for only one of these effects. They are useful in demonstrating the importance of drop growth or geometry and most likely represent more accurate expressions than those based on the planar-diffusion model. However, something better is needed. To date only one attempt to suggest an equation simultaneously accounting for drop growth and geometry has been reported. This equation is applicable to the quasi-reversible case with both redox forms characterized by solubility in the solution and approximately equal diffusion coefficients. It is obtained by noting that, for the case in question, Eq. (60) seems to apply for all electrode models considered. That is, as mentioned above, the expressions for $F(t)$ accounting for drop growth [Eq. (182)] and drop geometry [Eq. (183)] differ from each other and the planar diffusion model only in the form of the expression for $\psi_0(t)$ (i.e., the expression for the dc polarographic wave) which is substituted in Eq. (60). One then concludes that the completely rigorous expression for $F(t)$ which accounts simultaneously for drop growth and geometry can be obtained simply by substituting in Eq. (60) an expression for $\psi_0(t)$ of corresponding rigor. An expression for $\psi_0(t)$ with the desired properties has been presented by Koutecky and Cizek (*251*). Substitution of this relation in Eq. (60) yields

$$F(t) = 1 + (7/3\pi)^{\frac{1}{2}}[(\alpha e^{-j} - \beta)/\lambda]\{1 + 0.7868\mathscr{E}_1\}$$

$$\times \{F(\chi) - \mathscr{E}_1[0.7868F(\chi) - G_A(\chi)]\} \tag{187a}$$

where

$$\chi = (12/7)^{\frac{1}{2}}\lambda t^{\frac{1}{2}} \tag{187b}$$

and $F(\chi)$, $G_A(\chi)$, and $\mathscr{E}_1$ are functions given by Koutecky and Cizek (251). The validity of Eq. (187a) is a matter of speculation since its origin does not reside in a straightforward derivation.

One significant prediction of the theory lies in the fact that the phase angle is independent of $F(t)$ and, thus, is not influenced by drop growth and geometry for any mechanism involving rate control by a *single* charge-transfer step and/or any number and variety of coupled *first-order* homogeneous chemical reactions. Preliminary theoretical studies indicate that the phase angle will be subjected to these perturbations in situations involving coupled second- (or higher-) order chemical reactions, multistep charge transfer, or adsorption.

Figure 24 shows ac polarographic waves calculated on the basis of various electrode models for a quasi-reversible system in which both redox forms are characterized by solubility in the solution phase and equal diffusion coefficients. The magnitude of k_s is sufficiently small that the dc process is nonnernstian. The importance of drop growth and geometry contributions under such conditions is apparent.

J. Stationary Electrodes

Nearly all ac polarographic theory assumes a constant dc potential (for example, all above equations), despite the fact that many experiments employ a linear-scanning direct potential. The origin of this assumption lies in the fact that, when employing normal scan rates with a DME, the potential change is negligible over the life of a mercury drop. Apparently because the vast majority of experimental work employs the DME, little attention has been given to theory for stationary electrodes for which this assumption is not strictly applicable. However, one is led to believe intuitively that equations for the ac polarographic wave assuming constant dc potential might also be applied to ac polarography with stationary electrodes under conditions of linear scan if the scan rate is small compared to the rate of change of alternating potential. Underkofler and Shain (27) examined this problem quantitatively for the reversible ac polarographic wave. Their equations supported these intuitive expectations as did experimental results. Good agreement between theory and experiment was observed for scan rates ranging from 2 to 150 mv/sec with ferric ion in oxalate media. The equation tested was Eq. (36) with E_{dc} given by

$$E_{dc} = E_{init} - vt \tag{188}$$

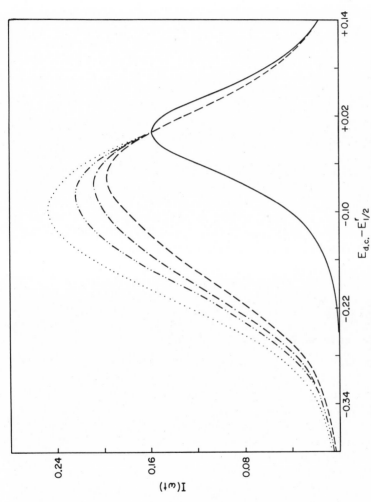

Fig. 24. Calculated faradaic fundamental-harmonic ac polarograms for the quasi-reversible system with various electrode models [from (250)]. Parameters: $k_s = 5.00 \times 10^{-4}$ cm sec^{-1}, $\alpha = 0.500$, drop time $= 6.0$ sec, $n = 1$, $T = 298°$K, $\Delta E = 5.00$ mV, $\omega = 125.66$, $C_O^* = 1.00 \times 10^{-3}$ M, mercury flow rate $= 1.40$ mgsec^{-1}, $D_O = D_R = 5.00 \times 10^{-6}$ cm^{2}sec^{-1}; ——, Breyer-Bauer-Kambara model (assumes nernstian dc behavior), – –, planar diffusion, – · · –, spherical diffusion, – · · –, expanding plane, · · ·, expanding sphere. (Courtesy of *Elsevier*.)

where v is the scan rate and E_{init} is the initial potential. One can extend this work to include all electrode-reaction mechanisms discussed above. It is readily shown that, with sufficiently slow scan rates (presumably within the limit found by Underkofler and Shain), the ac wave equations given here apply to stationary electrodes with linear scan with one modification. The terms responsive to the dc process must employ expressions for the direct current in linear-scan chronoamperometry (for the appropriate electrode geometry), rather than expressions for constant-potential chronoamperometry. This is essentially the same approach used by Okamoto (*252*) for oscillographic square-wave polarography. For example, in the equations for the quasi-reversible case, one substitutes for $\psi_0(t)$ in Eq. (60) the solution of Matsuda and Ayabe (*253*) to obtain an expression for the ac wave applicable to a stationary planar electrode with linear potential scan. The E_{dc} term is always given by Eq. (188). This distinction between linear-scan and constant dc potential theory suggests that, for appropriate systems, notable differences may be observed between ac polarographic behavior at a DME and at stationary electrodes. Among other things, ac polarograms at stationary electrodes should prove noticeably dependent on scan rate for many systems because of the corresponding dependence of the $\psi_0(t)$ function on this parameter. A dependence on scan rate is not expected only for cases in which the ac wave expression is independent of $\psi_0(t)$, as in any case discussed above where $F(t)$ is equal to unity. Presumably, the latter situation prevailed in the study reported by Underkofler and Shain (*27*).

K. Theory of Some Related AC Techniques

Four experimental methods are possible for combining control of direct current or potential with control of sinusoidal alternating current or potential to study the small-amplitude faradaic impedance. Ac polarography represents one combination. Two others have been suggested in the literature. Ac chronopotentiometry employs control of the direct and alternating components of *current* while measuring the resulting alternating component of potential as a function of time (or, alternatively, as a function of direct potential). This method was suggested by Takemori and coworkers (*254*). A technique proposed by Bauer (*255*) combines control of the dc potential and alternating current while measuring the alternating-potential component as a function of dc potential. A fourth possibility, not suggested in the literature, is to control direct current and alternating potential while measuring alternating current as a function of time (or dc potential). These techniques are related closely, both in principle and theory. In other words, the impedance of a polarographic cell can be studied as a function of dc potential by controlling either the alternating current or potential and measuring the uncontrolled ac parameter. The dc potential can be scanned either by controlling this potential or controlling direct current, permitting the normal chronopotentiometric process to proceed. The theoretical relations discussed above for fundamental-harmonic ac polarography are directly applicable to the other three techniques if small-amplitude alternating potentials (controlled or uncontrolled) are employed and rate of dc potential scan is not overly large (*149*). One expresses the ac polarographic theory in terms of an impedance [for example, Eq. (45)], expressing the functions dependent on direct current [$F(t)$, $F_{dc}(t)$, etc.] in the general form, that is, in terms of $\psi_0(t)$ as in Eqs. (60) and (104). The theoretical equation for the measured alternating current or potential is then obtained for the technique of interest from the generalized form of Ohm's law ($E = iZ$) and by substituting the appropriate form of direct current for the $\psi_0(t)$ term. Several conclusions can be drawn regarding the relationship between the four techniques: (a) The faradaic impedance is independent of the mode of control of the ac variable; (b) to observe a difference in the faradaic impedance between any two experimental techniques with the same electrode process, dc potential, and frequency, the techniques must employ different modes of control of the *dc* parameters; (*c*) to observe a difference in phase angle between the experimental techniques, conditions just mentioned under item (b) must apply and the electrode process must involve higher-order coupled chemical reactions, multistep charge transfer, and/or adsorption processes which kinetically influence the dc process.

It is seen that the status of the dc polarographic process is of major concern here, just as in the discussion of time dependence and contributions of electrode geometry and growth. This re-emphasizes the point that ac polarography serves not only an as approach to examining rapid electrode processes, but also as an effective " probe " for investigating the dc process.

L. Faradaic Rectification and Intermodulation Polarography

For reasons mentioned earlier, theory for the faradaic rectification method with normally employed experimental conditions ($I_{dc} = 0$ or $E_{dc} = E_e$, where E_e is the equilibrium potential) will not be discussed at length. However, some brief remarks are in order.

The theory for faradaic rectification with $E_{dc} = E_e$ and alternating-voltage (av) control is obtained directly from the theoretical approach described above by considering the more general case where $C_R^* \neq 0$, deriving the dc faradaic rectification component in the manner shown and centering ones attention on the result for $E_{dc} = E_e$ (E_e is given by the Nernst equation). Such equations are converted readily to the case of $E_{dc} = E_e$ and ac control. For $I_{dc} = 0$ and either av or ac control, derivation of the rectification voltage requires additional modification of the above approach, but equations remain conveniently accessible. The theory for faradaic rectification is well developed, primarily because of extensive work of Delahay and co-workers (10,57,65,67–72) and Barker (61). Although most effort has been concerned with the quasi-reversible case, derivation of equations for systems with coupled chemical reactions has been outlined.

Derivation of equations for intermodulation polarography amounts to modifying Eq. (15) to

$$E(t) = E_{dc} - \Delta E_1 \sin \omega_1 t - \Delta E_2 \sin \omega_2 t \qquad (189)$$

and proceeding in the usual manner. Solution of the resulting integral equation for $p = 1$ yields fundamental-harmonic components corresponding to the frequencies ω_1 and ω_2. Solution for $p = 2$, in addition to giving terms in $\sin^2 \omega_1 t$ and $\sin^2 \omega_2 t$ corresponding to dc rectification and second-harmonic components, yields the cross term $\sin \omega_1 t \sin \omega_2 t$ which corresponds to current components with sum and difference frequencies. Of interest in intermodulation polarography is the difference frequency (8,73). Theory and experimental results for intermodulation polarography have been discussed by Paynter and Reinmuth (8,73). Among other things, it was shown that the second-harmonic and intermodulation polarograms are identical in form for a reversible process. Also, when $\omega_1 - \omega_2$ becomes small, the form of the intermodulation wave approaches that of the dc rectification component. A comparison of the relative merits of second-harmonic and intermodulation polarography was given.

IV. INSTRUMENTATION

A. General

The utility of a technique is intimately dependent on the quality of the tools available for its implementation. Thus, instrumental design represents a continually active area of interest in ac polarography, as with most other electrochemical techniques. The importance of ac polarography as an electroanalytical method has stimulated the development of a variety of commercial instruments ranging from conventional ac polarographs to more sophisticated devices such as square-wave polarographs or phase-selective ac polarographs. Of course, instrumentation required for bridge measurements was available commercially long before the advent of faradaic-impedance measurements. A detailed review of all commercial and/or published instrumental approaches would be rather lengthy, and in view of recent excellent treatments of this subject by Breyer and Bauer (*9*) and by Gerischer (*256*), this will not be attempted. Instead, the discussion will be confined to a single instrumental concept which was not mentioned in these reviews: the operational-amplifier method.

The ac polarograph constructed from operational amplifiers provides a number of advantages. Foremost among these is drastic reduction of ohmic resistance. With a three-electrode configuration, the operational-amplifier potential-control loop provides for *automatic* compensation of much of the series resistance normally associated with the electrolytic cell and external circuitry. This is particularly important in ac polarography, where the low impedance of the electrode-solution interface enhances such problems. With typical operational-amplifier circuitry, the only uncompensated resistance is the ohmic drop between the tip of the reference electrode probe and the summing point of the current-measuring amplifier (see below). When employing a reference electrode with a Luggin probe, the uncompensated resistance encompasses the ohmic losses in a few millimeters of solution, plus the resistance of the working electrode. The latter predominates with a normal DME and may be of the order of 50 to 100 ohms. However, specially designed DMEs are readily constructed which reduce this resistance to a few ohms (*21*). Resistance in many stationary electrodes (hanging mercury drops, platinum wires, etc.) will be much less than 1 ohm. With proper placing of electrodes, the effective ohmic resistance in aqueous solutions may be approximately 1 ohm (*257*). Excellent discussions of the subject of uncompensated ohmic resistance with operational-amplifier circuitry have been given by Booman and Holbrook (*257*) and Schaap and McKinney (*40*). At least equally important are advantages associated with the extreme versatility of the operational-amplifier approach. Virtually any modern electrochemical technique is accessible, including the myriad of ac methods discussed earlier. The operational amplifier is found useful not only in the heart of the instrument, that is, the potential- or current-control loop, but also for construction of a variety of supporting units such as precision rectifiers, auxiliary amplifiers, voltage ramp generators, and many others. With a reasonable supply of amplifiers suitably modified to perform the desired functions, the operator may change from one technique to another simply by switching a few wires and/or a plug-in module. One can bring the leads from the amplifiers to a patchboard and "program" the technique of interest in the analog-computer style. Versatility of the operational amplifier goes beyond providing an instrument capable of performing, one at a time, a variety of electrochemical techniques. Within the framework of the ac polarographic method, one can simultaneously perform a variety of ac experiments in addition to dc polarography. Because the operational amplifier has a low-output impedance, simultaneous loading of the output with a variety of signal-conditioning devices (low-pass filters, tuned amplifiers, phase-sensitive detectors) will not degrade performance. Thus, the ac polarographic signal can be treated to provide simultaneously dc, total fundamental-harmonic, resistive, and/or capacitive components of the fundamental harmonic, second harmonics, the intermodulation component (if two sinusoidal potentials are applied), etc. One is limited only by the number of recorders or other suitable readout devices available. If sufficiently small amplitudes are employed, the dc rectification current will be negligible, and the direct current may be taken as the conventional dc polarographic current. One can carry this idea to unrealistic extremes, but something is to be said for simultaneous recording of the dc polarogram and/or polarograms of two components of the fundamental harmonic. Kinetic studies in ac polarography usually require this data, and savings in operator time realized through simultaneous readout will soon pay the cost of an additional recorder.

An unfortunate disadvantage of operational amplifiers employed in most existing electroanalytical instrumentation is limited-frequency response. This has restricted ac polarographic investigation with operational-amplifier instrumentation to frequencies of ∼1 kc or less. Noise levels in conventional vacuum-tube operational amplifiers represent a restriction in low-level signal measurement (trace analysis, second-harmonic studies, etc.). However, recent advances in the design of solid-state operational amplifiers promise to eliminate or greatly reduce both disadvantages. High-speed solid-state operational amplifiers nearly two orders of magnitude faster than earlier models have recently become available (*258*). Noise levels in these devices appear to be at least an order of magnitude less than the vacuum-tube devices [in our hands (*259*)]. This development makes the operational-amplifier concept competitive with most other approaches to ac polarographic instrumentation with regard to accessible frequency and noise levels.

B. The Operational Amplifier

A variety of excellent discussions on basic principles of operational amplifiers can be found in the literature. Of particular interest are articles by Reilley (*260*), Booman and Holbrook (*257*), and Schwarz and Shain (*261*).

Only a few basic points will be discussed here to provide the reader some frame of reference. The remainder of this section will be devoted to operational-amplifier circuits useful in ac polarography.

An operational amplifier is a high-gain wide-band amplifier constructed to exhibit high stability and low-input currents. In a general sense, two inputs and two outputs are available, as indicated in Fig. 25(a). Each input is inverting with respect to the corresponding output, and the amplifier responds to the potential difference between the inputs. An operational amplifier with all four terminals truly floating with respect to ground can be constructed from only a few commercially available models, such as the Philbrick P-2. In these cases, individual floating power supplies are required, since the negative output [Fig. 25 (a)] is usually the power-supply common. This requirement is not a significant disadvantage with solid-state operational amplifiers where batteries can furnish the dc power. If the power-supply common is grounded, one has the three-terminal amplifier illustrated in Fig. 25(*b*). One input is inverting, and one noninverting. Actually, most commercially available chopper-stabilized operational amplifiers are not designed to operate with both inputs active. As a result, most circuitry employed in electroanalytical work utilizes a two-terminal network designated by Fig. 25(*c*). Both the + input and − output are grounded, and the amplifier inverts the polarity of the output with respect to the input. The two-terminal amplifier responds to a potential difference between the − input and ground and, with the appropriate feedback configurations, maintains the − input at a virtual ground. This is achieved by virtue of the high-open-loop gain of the amplifier ($\sim 10^5$ to 10^9 at dc). Normal configurations with the two-terminal amplifier utilize an input and feedback inpedance, as illustrated in Fig. 26. If the open-loop gain of the amplifier at the frequency of interest is sufficiently large and if the input current is negligible, the relationship between the input and output voltages is given by (*260*)

$$e_0 = -\frac{Z_f}{Z_i}\, e_i \tag{190}$$

When purely resistive impedances are employed, calculation with Eq. (190) is simple. However, with reactive impedances, the term Z_f/Z_i is complex, and some mathematical sophistication is required to effect calculation, particularly when both amplitude and phase relations are desired. The reader is referred to a review by Reilley (*260*), which discusses application of the Laplace-transform method in calculation of the response characteristics for various impedance configurations. For the present

(A) FOUR-TERMINAL SYMBOL FOR THE
     OPERATIONAL AMPLIFIER

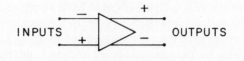

(B) THREE-TERMINAL SYMBOL FOR THE
     OPERATIONAL AMPLIFIER (COMMON
     GROUNDED)

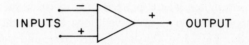

(C) TWO-TERMINAL SYMBOL FOR THE
     OPERATIONAL AMPLIFIER (COMMON
     AND + INPUT GROUNDED)

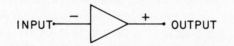

Fig. 25. Operational-amplifier designation. (*a*) Four-terminal symbol (all terminals floating); (*b*) three-terminal symbol (common grounded); (*c*) two-terminal symbol (common and + input grounded).

purposes, Eq. (190) suffices to give a qualitative understanding of the behavior of circuits discussed. It is important, for applications involving ac signals, to keep in mind that all operational-amplifier configurations discussed assume negligibility of amplifier input currents and that the amplifier open-loop gain is essentially infinite (terms in the network equations involving open-loop gain are negligible). As frequency is increased, input currents increase, and open-loop gain decreases. This trend places an upper-frequency limit on the successful operation of all circuits discussed which is dependent on the characteristics of the input and feedback configurations. In this regard, amplifier characteristics such as open-loop-gain–bandwidth product, input resistive and capacitive

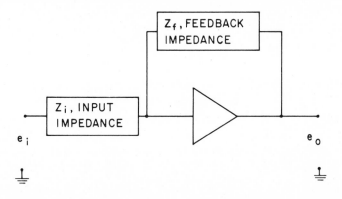

Fig. 26. Two-terminal operational amplifier in normal operating configuration.

impedances, etc., are of particular interest to the worker concerned with ac applications. In addition, output capability (voltage and current) is often important. For example, with high resistance solvents, the C amplifier in the potential-control loop (Fig. 28) will have to furnish voltages much larger than the potential applied between the reference and working electrodes to overcome solution iR drop. In some applications, the amplifier output may be significantly loaded with respect to current demands. Most operational amplifiers furnish 1 to 2 mamp, but current boosters can raise this limit considerably.

C. The Follower Amplifier

For successful application of an operational-amplifier potential-control loop with a three-electrode configuration, an impedance-matching device is desirable. This unit senses the potential of the reference electrode relative to the working electrode and furnishes an output signal equal to this potential difference at low-output impedance, without drawing significant current from the reference electrode. Such a device is normally called a *follower amplifier*, or a *follower*. Followers can be constructed with a very high input impedance ($\sim 10^{12}$ ohms), high-current-output capabilities (~ 20 mamp), and gains of precisely unity (± 0.1 per cent). A follower may be constructed with the rather simple circuit shown in Fig. 27 (*a*). This circuit represents an application of the three-terminal amplifier shown in Fig. 25(*b*) and can be utilized with chopper-stabilized amplifiers with full differential inputs, such as the Philbrick P-2, or with a variety of unstabilized models. Disadvantages regarding drift inherent in the unstabilized variety

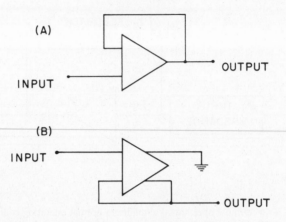

(A)

INPUT

(B)

INPUT

Fig. 27. The follower amplifier. (*a*) + input active configuration; (*b*) − input active configuration (*Booman follower*).

and unavailability of stabilized amplifiers with full differential inputs in early development of operational-amplifier electroanalytical instrumentation led to the development of alternative follower circuits by De Ford (*37*) and Booman and Holbrook (*257*), which utilize the more conventional chopper-stabilized amplifier. The DeFord follower circuit employs a floating chopper-amplifier section enabling use without a separate power supply. The Booman circuit has advantages regarding common-mode error, but requires a separate floating power supply. The latter requirement is an inconvenience with vacuum-tube amplifiers but, as mentioned earlier, is little problem with the solid-state models. The Booman configuration is shown in Fig. 27(*b*). In our hands, this circuit with a Philbrick SP456 operational amplifier yields a follower which is accurate with respect to gain and phase shift up to 100 kc (*259*). It should be noted that the followers shown in Fig. 27 are designed to give an output equal to the potential at the input relative to ground potential.

D. The Controlled-Potential Configuration

A three-electrode controlled-potential configuration commonly employed in electroanalytical work is shown in Fig. 28 (*34,37,261*). Amplifier *I* serves to maintain the potential of the working electrode at virtual ground (the potential of the input of amplifier *I*), less the iR drop in the electrode and its connecting leads. It also provides a signal proportional to the current flowing between the working and auxiliary electrodes ($e_0 = iR_m$). The follower, which responds to the potential of its input relative to ground

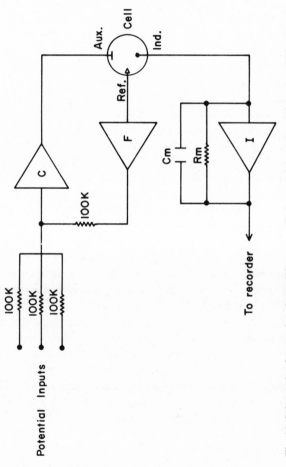

Fig. 28. Schematic of operational-amplifier controlled-potential configuration. Amplifiers *I* and *C*: operational amplifiers in normal two-terminal inverting configuration; Amplifier *F*: follower amplifier, *Cm*: damping condensor. (Courtesy of *Analytical Chemistry*.)

(the input of amplifier I), thus yields an output equal to the potential of the reference electrode relative to the working electrode, less the iR drop between these points. The stable state for the loop comprising amplifier C and the follower corresponds to ground potential at the input of amplifier C. If all resistors connected to this input are identical, the stable state is achieved when the output of the follower is equal and opposite in polarity to the sum of the potentials applied at the input resistors. This, in turn, implies that the potential of the working electrode relative to the reference will be equal to the sum of the potential inputs (same polarity) less the iR drop between the input of amplifier I and input of the follower. Because essentially no current flows through the reference electrode, iR drop in the reference electrode compartment is negligible, even with low-conducting solutions and high-resistance contacts between reference-electrode solution and external solution. Ohmic losses in the solution between the tip of the reference electrode and the working electrode are significant because current is flowing in the external solution. Also, iR drop in the working electrode and leads connecting it to amplifier I must be considered. As mentioned earlier, the magnitude of these ohmic losses can be made extremely small compared to the situation with the conventional two-electrode configuration. The three-electrode configuration was employed to reduce ohmic losses before the advent of operational-amplifier circuitry. Point-by-point measurement was required. The important contribution of the operational-amplifier-control loop is that it permits adaptation of the three-electrode cell to precision automatic-recording instrumentation.

The circuit illustrated in Fig. 28 represents the "heart" of the instrument for controlled-potential techniques. Adaptation to ac polarography simply requires suitable signal sources for the potential inputs and signal-conditioning devices (tuned amplifiers, rectifiers, phase-sensitive detectors, etc.) to operate on the current signal available at the output of amplifier I. The operational amplifier serves also as a convenient and desirable basis for constructing many of these latter devices. The construction of many of the signal-source and signal-conditioning devices to be discussed can be accomplished more economically without the use of operational amplifiers. However, the version constructed from operational amplifiers almost invariably provides considerably better stability, accuracy, and reliability than the less-expensive alternatives.

E. Signal Sources

Three types of signal sources are required in ac polarography: (a) a dc potential source whose output can be varied conveniently to serve as source of the initial potential; (b) a source of a linear-scanning dc potential—a ramp generator whose scan rate and direction can be altered conveniently; (c) a sinusoidal oscillator. High-quality-signal sources of these types can be constructed from operational amplifiers. A frequently utilized precision dc voltage source is shown in Fig. 29(*a*). The first stage of this circuit, discussed in several sources [for example, (*260*)], yields a voltage equal to the potential of the standard cell and furnishes substantial currents without draining the standard cell. The reversing switch permits a change in polarity. The second stage is an amplifier of variable gain, permitting a selection of a continuous range of precision voltages. The accuracy of this circuit is limited only by the accuracy of the resistors and the standard-cell potential. The ramp-generator utilized, probably universally, in operational-amplifier instrumentation is an electronic integrator with a constant-input signal (*37,38,260*). The device is illustrated in Fig. 29(*b*). The source of the input can be the same standard-voltage source serving as the first stage in Fig. 29(*a*). Sinusoidal oscillators are readily constructed from a variety of operational-amplifier circuits (often when least expected). A circuit for a variable-frequency sinusoidal oscillator discussed by Reilley (*260*) is shown in Fig. 30(*a*). Figure 30(*b*) shows another oscillator which has been used in ac polarography (*34*). The latter is basically a tuned amplifier (see below) with some positive feedback, through capacitor *C*, to effect oscillation. It provides a very stable frequency source, which is important when highly tuned amplifiers are employed in the ac polarograph. It suffers the disadvantage that frequency variation is not as easily accomplished as with the circuit shown in Fig. 30(*a*).

F. Tuned Amplifiers

Highly selective tuned amplifiers are useful in ac polarography for the measurement of higher harmonic components (*34,78,81,82*) in the presence of the much larger fundamental harmonic. They also have been utilized to eliminate power-frequency-noise components, and the small, but significant, higher harmonics for precision measurement of the fundamental harmonic (*33,81*). A useful circuit is shown in Fig. 31 (*34,81,262*). It consists of a twin-T network (*263*) as the feedback impedance and a resistor at the input. The twin-T circuit is characterized by an impedance versus frequency behavior showing a sharp maximum centered about a particular frequency. Thus, the gain of the amplifier as a function of frequency exhibits a sharp peak, that is, the amplifier is tuned to a narrow band of frequencies. Q values (angular frequency divided by bandwidth) of the order of 500 may be achieved readily with a single stage. Although this circuit proves very

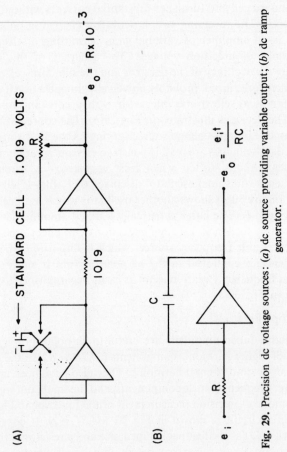

Fig. 29. Precision dc voltage sources: (*a*) dc source providing variable output; (*b*) dc ramp generator.

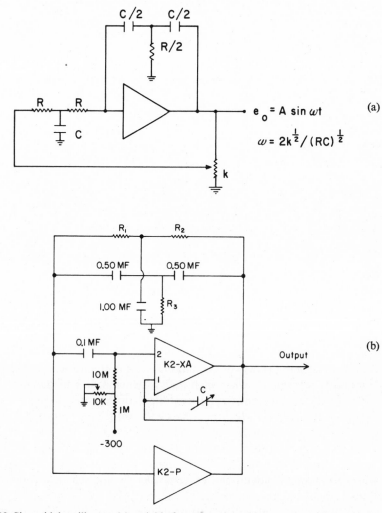

Fig. 30. Sinusoidal oscillators: (*a*) variable-frequency sinusoidal generator; (*b*) "twin-T" oscillator constructed from Philbrick K2-X–K2-P combination. (Courtesy of *Analytical Chemistry.*)

helpful in eliminating undesirable ac signals, its successful use requires care in excess of the normal demands of most operational-amplifier devices. The high-frequency selectivity demands a reasonably stable oscillator (*34*). Slight drift in oscillator frequency which normally may represent no problem can result in a substantial drift in both amplitude and phase response of the ac polarograph employing tuned amplifiers. An oscillator with a frequency

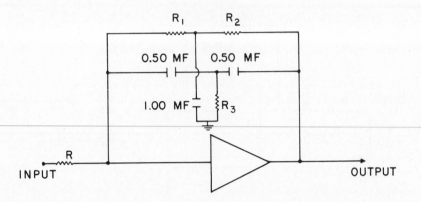

Fig. 31. Tuned amplifier: $R_1 = R_2 = 2R_3 = 3.18 \times 10^5/f_0$. ($\pm 1\%$ components, $f_0 =$ resonant frequency.)

stability of the order of 1 part in 10^4 or better is desirable. Because the response of the circuit is critically dependent on the matching of the resistors and condensors in the twin-T network, slight drift of these components relative to one another as a result of ambient temperature changes can introduce detectable drift in tuned-amplifier response. Thus, low-temperature-coefficient components should be used. Another source of drift resulting from ambient temperature change may arise because the open-loop-gain term in the network equations is not insignificant at the resonant frequency, particularly when high gain is demanded. This makes the response of the circuit somewhat susceptible to drift in open-loop gain of the operational amplifier, which may be significant (the operational amplifier is designed to have high, but not necessarily constant, open-loop gain). In any case, this writer has observed that forced ventilation is sometimes required, in addition to a stable oscillator, to achieve high stability in an ac polarograph employing the circuit in Fig. 31 with vacuum-tube operational amplifiers. Investigations of the performance of the same circuit with a number of solid-state operational amplifiers (*259*) indicates decidedly better performance with regard to stability (at least one order of magnitude improvement), indicating that the low heat dissipation of the solid-state amplifier is a decided advantage.

G. Phase-Sensitive Detectors

Phase-sensitive detectors can be constructed from a variety of circuits. Most devices with this purpose, such as the circuit employed in the Univector Polarograph (*41,42,45*), are full-wave synchronous rectifiers in which a switch (electromechanical or electronic) alternates the polarity of the output signal relative to the polarity of the input at a frequency equal to that of the input signal. The dc component of the output signal is measured. In terms of a Fourier analysis, this operation amounts to multiplying the signal of interest by a square wave of the same frequency whose voltage level varies between plus and minus unity. If the input signal is represented by

$$e_i = A \sin \omega t + B \cos \omega t \tag{191}$$

the output signal may be written (for perfect switching) (*264*) as

$$e_0 = (A \sin \omega t + B \cos \omega t) \sum_{n=0}^{\infty} C_n \sin(2n + 1)\omega t \tag{192}$$

where

$$C_n = 4/(2n + 1)\pi \tag{193a}$$

$$n = 0, 1, 2, 3, \ldots \tag{193b}$$

If we consider. the trigonometric identities (*265*),

$$\sin^2 \omega t = \tfrac{1}{2}(1 - \cos 2\omega t) \tag{194}$$

$$\cos^2 \omega t = \tfrac{1}{2}(1 + \cos 2\omega t) \tag{195}$$

$$\sin \omega t \cos \omega t = \tfrac{1}{2} \sin 2\omega t \tag{196}$$

$$\sin \omega_1 t \sin \omega_2 t = \tfrac{1}{2} \cos(\omega_1 - \omega_2)t - \tfrac{1}{2} \cos(\omega_1 + \omega_2)t \tag{197}$$

$$\cos \omega_1 t \cos \omega_2 t = \tfrac{1}{2} \cos(\omega_1 - \omega_2)t + \tfrac{1}{2} \cos(\omega_1 + \omega_2)t \tag{198}$$

$$\sin \omega_1 t \cos \omega_2 t = \tfrac{1}{2} \sin(\omega_1 + \omega_2)t + \tfrac{1}{2} \sin(\omega_1 - \omega_2)t \tag{199}$$

it is seen that the measured dc component is derived from the cross product [Eq. (192)]

$$A \sin \omega t \cdot \frac{4}{\pi} \sin \omega t$$

The circuit is phase-sensitive because the input term $B \cos \omega t$ generates only ac components. An important characteristic of the phase-sensitive detector, not usually noted in the electrochemical literature, is its inherent frequency selectivity. It is apparent from Eqs. (194) through (199) that, if the input signal contains noise components with frequencies other than the frequency of interest (ω), they will not generate a dc component unless their frequencies are odd harmonics of ω. Thus, with a low-pass filter of reasonably long time constant, most noise components will be suppressed. It should be possible to effect phase-sensitive detection of second harmonics without the aid of tuned amplifiers simply by making the switching frequency of the synchronous rectifier twice the fundamental harmonic frequency. One also concludes that, in phase-sensitive detection of the fundamental harmonic through synchronous rectification, an excessively nonlinear faradaic process will yield a dc output with significant contributions from odd harmonics. The phase characteristics of the harmonics may differ from the fundamental in such a way that the higher harmonics are favored with appropriate settings of the reference-signal phase. The unusual results obtained in the study of tensammetric waves of cyclohexanol with the Cambridge Univector Polarograph (45) might be attributed to these effects. Tuned amplification prior to phase-sensitive detection is preferable because it eliminates confusion from unusually large higher harmonics when the type of detector under consideration is employed.

Figure 32 shows a full-wave synchronous rectifier which can be constructed from operational amplifiers (266). Amplifier A and its associated diode bridge comprise a half-wave synchronous rectifier. A reference square wave of amplitude much larger than the input signal and of the same frequency is applied across the diode bridge at points b and c. The diode bridge shorts point a to ground on the half-cycle of the reference signal in which the diodes are conducting. On the alternate half-cycle, the diodes are blocked and the bridge has negligible effect on point a. Thus the diode bridge acts as a switch making the amplifier responsive to the input signal during alternate half-cycles. The individual half-wave synchronous rectifier represents a reasonably good phase-sensitive detector. Its operation is equivalent to multiplication of the input signal by a square wave whose voltage level switches between unity and zero. Thus, the square-wave multiplier has a dc component, and the output signal is given by

$$e_0 = \tfrac{1}{2}(A \sin \omega t + B \cos \omega t)\left[1 + \sum_{n=0}^{\infty} C_n \sin(2n + 1)\omega t \right] \qquad (200)$$

Disadvantages arise from the facts that a dc component on the input signal can yield a dc output leading to possible error, and fundamental-harmonic ac components are found on the output signal placing greater demands on the low-pass filter, that is, a larger time constant is required to reduce ac "ripple" to an acceptable level. Thus, the full-wave synchronous rectifier is preferred and can be constructed by combining two half-wave synchronous rectification operations as shown in Fig. 32. The polarity of the diodes in the bridges associated with amplifiers A and B is reversed, so that they respond to the input during opposite half-cycles of the reference signal.

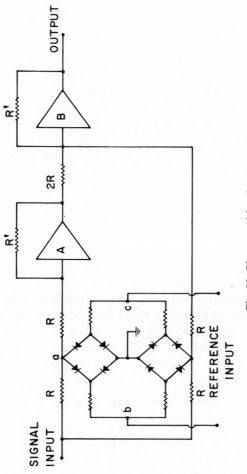

Fig. 32. Phase-sensitive detector.

117

Amplifier *B* combines the operations of half-wave synchronous rectification and addition to yield full-wave synchronous rectification. A variety of other configurations employing operational amplifiers for synchronous rectification are possible.

One aspect of phase-sensitive detection through synchronous rectification which can prove undesirable in some applications is the fact that odd harmonics on the input signal contribute to the dc output. Full advantage is not taken of the frequency selectivity inherent in the multiplication operation. The "ideal" phase-sensitive detector is a true electronic multiplier in which one can multiply the signal to be analyzed by a pure sinusoidal reference signal, rather than a square wave (*34*). Only the input component with the same frequency and phase as the sinusoidal reference yields a dc component at the output, as seen in Eqs. (194) through (199). Electronic multipliers with bandwidths up to 100 kc are available commercially from analog-computer manufacturers. The ubiquitous operational amplifier is one of the main components in these circuits. Logarithmic conversion employing operational-amplifier–transistor circuits has recently been reported (*267*) and appears to represent a convenient approach to construction of electronic multipliers suitable for phase-sensitive detection.

As a means of obtaining highly selective frequency response, phase-sensitive detection has an important advantage over tuned amplification because it can provide extremely selective response without difficulties associated with oscillator-frequency drift. The "resonant" frequency of the phase-sensitive detector is the frequency of the reference signal. If the reference signal is derived from the oscillator providing the alternating potential for the ac polarograph (the usual procedure), drift in the oscillator is of no consequence. The phase-sensitive detector center frequency is "locked" to the frequency of interest, regardless of the bandwidth. The effective bandwidth is determined by the low-pass filter and is given by (*268*)

$$\text{bandwidth} = 1/2RC \quad \text{cps} \tag{201}$$

where RC is the time constant of the low-pass filter. In ac polarographic studies with a DME, directed to kinetic applications, one is normally interested in the instantaneous current. For such work, a time constant of about 0.5 sec is realistic (for drop lives of 5 sec or longer), corresponding to a bandwidth of 1 cps or a Q value of ω. When preservation of drop oscillations is not important, as in analytical applications, or when stationary electrodes are employed, a time constant of the order of 10 sec would be reasonable. This corresponds to the extremely narrow bandwidth of 0.05 cps and a Q value of 20ω.

H. Conventional Rectifiers

When one is interested in detecting the total alternating component of interest, rather than a vectorial component, a conventional rectifier is required, since most recorders accept only dc signals. A precision full-wave rectifier is shown in Fig. 33 (*269*). Taking the half-wave rectified signal from point *a* completely eliminates error from finite forward resistance of the diodes. If the resistor values are much smaller than the back resistance of the diodes, which is easily accomplished, this circuit should be accurate to millivolt levels.

I. Phase Shifters

To vary the phase angle between two signals in a controllable fashion is an operation occasionally required in ac polarography, particularly in phase-sensitive detection. Operational-amplifier circuits can be employed for this purpose. Such circuits may be found advantageous relative to their simpler less-expensive counterparts, because phase shift can be controlled with precision and variable amplification is provided. Figure 34 illustrates a convenient circuit which will generate phase shifts through 360° with a gain of precisely unity for all values of phase shift.

J. Trigger Circuits

A trigger circuit (*34,270*) is often required to convert a sine wave to a square wave or, in timing circuits, to indicate when a voltage has reached or exceeded a predetermined level. Such a circuit obeys the mathematical relationship

$$e_0 = e_1 \quad \text{(for } e_i < e^*) \tag{202}$$

$$e_0 = e_2 \quad \text{(for } e_i \geq e^*) \tag{203}$$

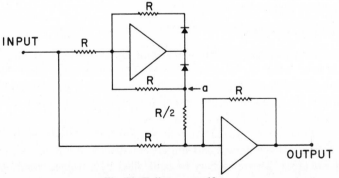

Fig. 33. Full-wave rectifier.

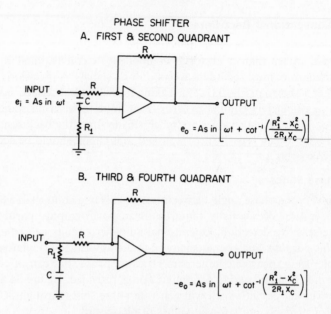

Fig. 34. Phase-shifting network.

where e_0 is the output and e_i the input voltage. Circuits of this variety have been used in recording ac polarographic phase angles (*34*). A generalized version is shown in Fig. 35. The voltage e^* at which the trigger changes state is controlled by the voltage applied at point a, e_a, and the ratio of the magnitudes of resistors R_1 and R_2. For converting an unbiased sine wave to a square wave, e^* should be zero, so e_a is set at 0 volts. The voltage outputs in the two stable states are controlled by the voltage levels at points b and c, which are determined by voltages e_b and e_c and the values of resistors in the voltage divider networks.

K. Holding Circuits

A circuit which will respond to an input signal and store, or "hold," this signal level when the source is removed is required to effect "strobe" readout with the DME (measurement of current at a particular instant in drop life). An excellent holding circuit may be constructed from a follower, as shown in Fig. 36 (*271,272*). If the resistance of the relay at open circuit and the input impedance of the follower are sufficiently large, this circuit will hold a voltage almost indefinitely, relative to the times required for strobe readout. The relay may be controlled by a trigger circuit which is activated when a ramp voltage (the "clock") reaches a certain level.

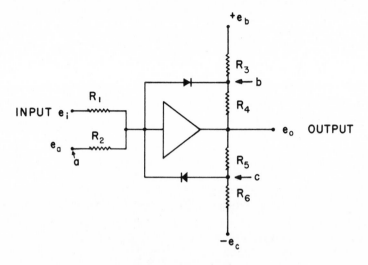

Fig. 35. Trigger circuit.

L. Fundamental-Harmonic AC Polarography

The above circuits represent most of the basic units required to perform ac polarography or any of its modifications. A conventional (fundamental-harmonic) ac polarograph is constructed from the potential-control loop, potential sources, tuned amplifier, and rectifier described in Secs. IV. D, E, F, and H, respectively, together with a suitable recorder. A schematic of the polarograph is shown in Fig. 37. The tuned amplifier is recommended to remove power-frequency noise which can be sufficiently large in vacuum-tube instrumentation to introduce significant error if not suppressed (*34*). Error from abnormally large higher harmonics is also avoided. However, noise components are greatly reduced with solid-state instrumentation and one may often forego the use of tuned amplifiers in fundamental-harmonic measurements.

M. Phase-Selective AC Polarography

Phase-selective detection of the fundamental harmonic requires a phase-sensitive rectifier in place of the conventional rectifier indicated in Fig. 37. The reference signal for the phase-sensitive detector is derived from the applied alternating potential signal which is available at the output of the follower or from the oscillator. This signal should be amplified to ensure that the reference signal is sufficiently large. As indicated earlier, phase-sensitive detection with a DME will benefit from strobe readout. Because

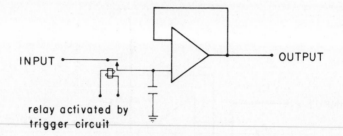

INPUT •

OUTPUT

relay activated by
trigger circuit

Fig. 36. Holding circuit.

the iR voltage loss varies over the life of a mercury drop, the applied potential across the interface varies correspondingly in amplitude and phase. Thus, the reference signal can be in phase with the potential across the interface (optimum conditions) only at a particular point during the mercury drop life. As a result, significant drop oscillations still exist where only double-layer charging current is flowing (the base line), which is the primary reason why phase-sensitive detection with a DME is limited in sensitivity to concentrations of the order of 10^{-6} M or larger. However, strobe readout enables measurement of the current at the instant in drop life corresponding to optimum conditions. In addition, charging current and, therefore, iR vary also with dc potential, so that adjusting the instrument for zero charging current at one potential does not normally ensure a flat, zero signal baseline. However, this problem is much less serious than that associated with iR variation over the life of a mercury drop.

Figure 38 shows one version of a phase-sensitive ac polarograph employing an electronic multiplier as the phase-sensitive detector. The timing circuit employed to control the holding circuit relay also controls a hammer which dislodges the mercury drop at a specified time in the usual manner (54,56). If one uses a phase-sensitive detector of the type illustrated in Fig. 32, a trigger must be introduced to convert the reference signal to a square wave, and provision for floating the reference square wave is also required.

N. Second-Harmonic and Intermodulation AC Polarography

Measurement of higher harmonics is effected through the circuit in Fig. 37 simply by tuning the tuned amplifier to the harmonic of interest. Usually, two tuned-amplifier stages are required to suppress completely the much larger fundamental harmonic. If necessary, tuned amplification of the oscillator signal should be carried out to eliminate all harmonics from the applied alternating potential.

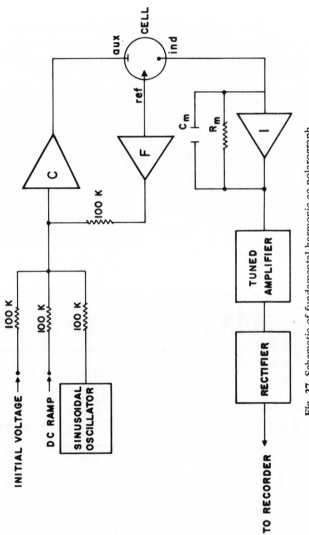

Fig. 37. Schematic of fundamental-harmonic ac polarograph.

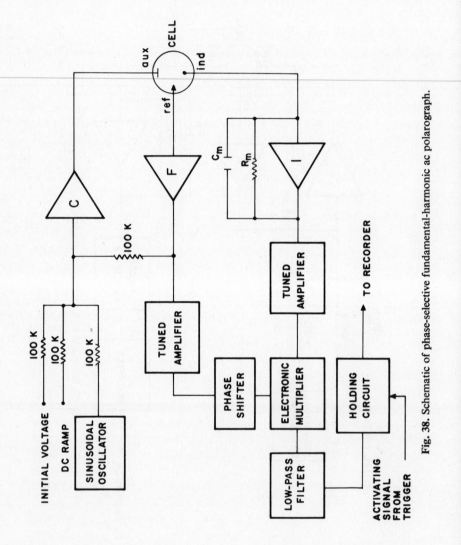

Fig. 38. Schematic of phase-selective fundamental-harmonic ac polarograph.

An intermodulation ac polarograph may be obtained similarly. One applies two sinusoidal inputs of differing frequencies, ω_1 and ω_2, where $\omega_1 - \omega_2$ is precisely fixed. This may be accomplished by mixing the outputs of two oscillators and filtering out undesired harmonics. The tuned amplifier is set to respond to the difference frequency $\omega_1 - \omega_2$.

O. Phase-Selective Second-Harmonic and Intermodulation AC Polarography

Phase-selective detection of second and higher harmonics can be accomplished through slight modification of the circuit shown in Fig. 38. Additional circuitry necessary to generate a second- (or higher-) harmonic-reference signal must be introduced. This is achieved by passing the fundamental-harmonic reference signal through any nonlinear circuit element and then to an amplifier tuned to the harmonic of interest (*34*). The circuit is illustrated in Fig. 39. Essentially the same circuit suffices for phase-sensitive detection of the intermodulation component. One replaces tuned amplifier A by a conventional amplifier to enhance the level of the signal from the follower, which contains both frequency components ω_1 and ω_2. When passed through the nonlinear element, some intermodulation signal is generated. Filtering by amplifier B, which is tuned to the intermodulation frequency, yields the desired reference signal.

P. Phase-Angle Measurement

The circuit illustrated in Fig. 40 was apparently the earliest employed to effect measurement of ac polarographic phase angles with operational-amplifier instrumentation (*32–34*). It provides only point-by-point measurement capability. The resistor and condensor R_m and C_m are precision elements (at least ± 1 per cent), at least one being a decade component. With $C_m = 0$, a purely resistive dummy cell is inserted in place of the polarographic cell, and phase relations between potential and current signals are varied by means of the phase shifter until the Lissajous pattern observed on the oscilloscope is a straight line. The polarographic cell is then inserted and the Lissajous pattern normally "opens" to form an ellipse. Then R_m and C_m are adjusted until the pattern is again a straight line. This operation amounts to compensating for the phase shift introduced by the polarographic cell. With a DME, a straight line can be obtained only at one point in drop life, because of the varying iR term. The end of drop life serves as the most convenient reference at which one can attempt to obtain the linear pattern. The phase shift in the polarographic cell is given by

$$\tan \phi = R_m^* C_m^* \omega \tag{204}$$

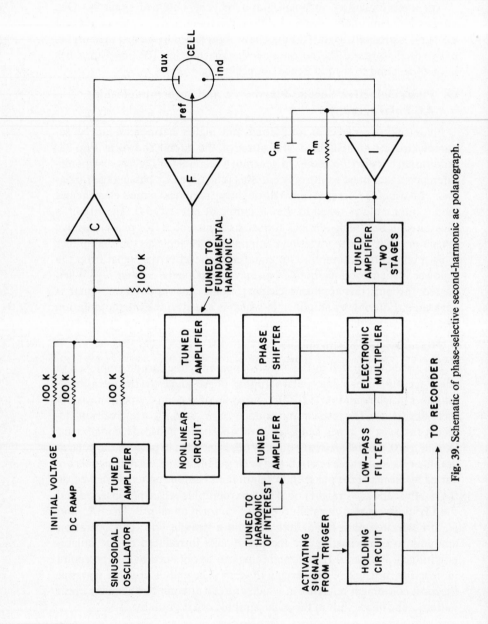

Fig. 39. Schematic of phase-selective second-harmonic ac polarograph.

126

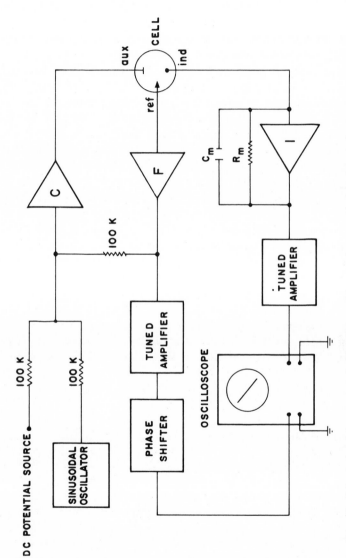

Fig. 40. Schematic of instrument for fundamental-harmonic ac polarographic phase-angle measurement.

where R_m^* and C_m^* are the values required to obtain the linear pattern with the polarographic cell. This method has some disadvantages. It does not provide for automatic recording and is susceptible to error in operator judgment with respect to matching the point of mercury-drop fall with the point at which a linear Lissajous pattern is observed. An automatic-recording approach is preferable.

The schematic shown in Fig. 41 illustrates a circuit which has been applied to automatic recording of phase angles (34). Both potential and current signals are converted to square waves with trigger circuits. The trigger output changes state each time the sine-wave input crosses zero. The amplitude of the resulting square wave is virtually independent of the amplitude of the sinusoidal input. Two triggers in series will effect significant improvement regarding the latter characteristic and square-wave rise time. The square waves derived from the current and potential signals are added (amplifier S) and rectified (full-wave). The magnitude of the signal resulting from these operations is determined by the phase relations between the square waves and is suitable for recording. The dc output of the rectifier is a maximum when the square waves are in phase and zero when they are 180° out of phase. The system can be calibrated by generating known phase shifts with the aid of a dummy cell. This approach to obtaining a signal dependent solely on phase relations is based on conventional methods (273,274). The circuit can be adapted to recording phase angles of second or higher harmonics (34) or intermodulation components. Adaptation requires generation of a reference signal of the frequency of interest as was done for phase-selective detection of these higher-order components.

Q. Possible Future Trends

Future trends normally are perceivable only when some future need can be envisioned. Because much of the theory and instrumentation discussed here is relatively recent in origin, utilization of what is available may prove to be the greatest need. There is much to be said in favor of more extensive quantitative studies with ac polarographic tools (techniques and instrumentation) at hand within their useful range of application, particularly because detailed *quantitative* data on faradaic alternating currents as a function of dc potential and frequency are scarce, even for fundamental harmonics. Quantitative data on second harmonics, etc., are practically nonexistent. Of course, all this suggests no major stimulus for innovations related to ac polarographic instrumentation. However, some needs related to data collection and sensitivity in analytical work are apparent and steps

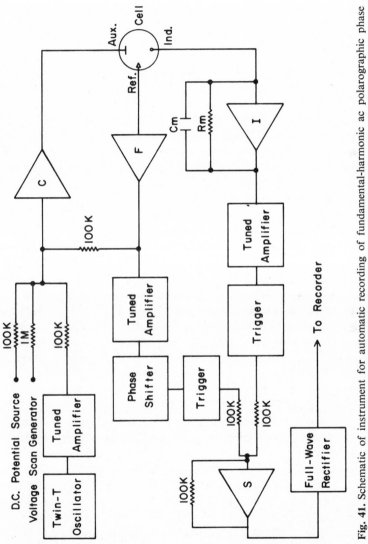

Fig. 41. Schematic of instrument for automatic recording of fundamental-harmonic ac polarographic phase angles [from (34)]. (Courtesy of *Analytical Chemistry*.)

to satisfy them probably will be forthcoming simply because they can be effected with available tools.

1. Multireadout Instrumentation

As mentioned earlier, the ac polarographic signal contains a wealth of information in the form of the dc polarographic current, the fundamental-harmonic amplitude and phase angle, the higher harmonic amplitudes and phase angles, and, if two sinusoidal signals are applied, the output also contains intermodulation components. These current components manifest the impedance of the polarographic cell at dc and over a range of ac frequencies. In essentially all experimental work performed to date, most of this information has been discarded in favor of a single piece of information during any one experimental run. The need to obtain more extensive data without undue expenditure of time, at least with some of the more complicated electrode processes, suggests the desirability of measuring many of these components simultaneously. Within the framework of present experimental thought, this might be looked upon as performing several experiments simultaneously. As mentioned above, the operational-amplifier instrument is ideally suited to such operations. Figure 42 illustrates an instrument which can provide simultaneously the direct current, the total fundamental harmonic, a vectorial component of the fundamental, the total second harmonic, and one of its vectorial components. The requirement of simultaneously recording five separate signals is readily handled by modern data-acquisition systems. Systems which digitize and record (printed or punched-tape) data from many channels by rapid scanning of these channels are available at approximately twice the cost of a high quality x-y recorder. If strobe readout is employed, such recording systems are compatible with the polarographic unit, even with the DME.

2. Subtraction of Undesirable Current Components

The ac polarographic method abounds in approaches to obtaining higher sensitivity by suppressing charging current. Phase-sensitive detection, measurement of higher harmonics, square-wave polarography, etc., represent signal-separation methods inherent in the ac technique. However, alternative methods, devised to enhance the sensitivity of dc polarography, are applicable to ac polarography although they have been almost neglected. One of these approaches employs two polarographic cells, one containing the solution to be analyzed and the other containing the supporting electrolyte plus any interfering electroactive components. Use of two polarizing units and synchronized capillaries generates current components from each

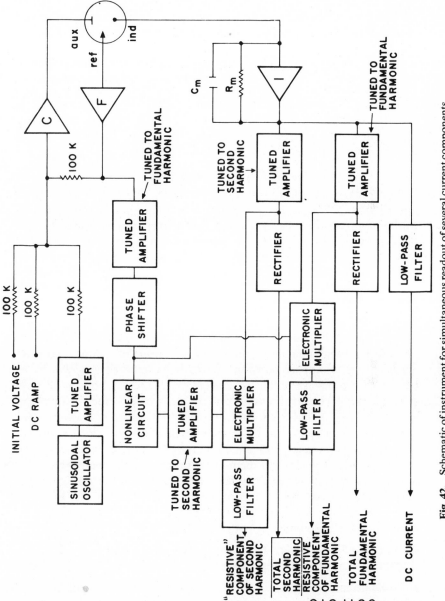

Fig. 42. Schematic of instrument for simultaneous readout of several current components.

131

cell which are subtracted, thus eliminating the undesirable background from the signal of interest. Barker and Faircloth (275) have applied this approach with conventional ac and square-wave polarography. It can be applied to any of the modifications of ac polarography.

3. Differentiation of the AC Polarographic Wave

Another technique used to advantage in dc methods to suppress the base current is differentiation of the readout (38,276). This operation can be performed on the ac polarographic signal, presumably with comparable enhancement of sensitivity. Differentiation with long time constants, as performed by Kelley et al. (38) is preferable.

V. DATA ANALYSIS

Proper data analysis ranks as an important problem to the worker interested in applying ac polarography to the study of kinetics and mechanisms of electrode processes. A significant theoretical framework exists which the experimentalist may apply to the interpretation of ac polarographic data. Extension and strengthening of this framework appears inevitable. Instrumentation is at hand which will permit measurement of ac polarographic currents to accuracies approaching ± 0.1 per cent (relative standard deviation). However, unless instrument readout can be properly analyzed, theoretical and instrumental developments are rendered sterile.

Two aspects of data analysis are apparent. The first entails correcting raw data for the nonfaradaic currents and ohmic losses to obtain the faradaic current at a given applied potential. That this step must be considered nontrivial and can involve pitfalls is suggested by the recent work of Sluyters et al. (231), which was discussed earlier. The second aspect involves examination of faradaic-current data with the ultimate goal of reaching valid conclusions regarding electrode-reaction mechanisms and kinetics. Operationally, the two stages of data analysis may be undertaken separately or simultaneously. The classical method of Randles (6,12) involves the stepwise approach. Data is first corrected for nonfaradaic effects, and the faradaic data is then analyzed. The more recent approach of Sluyters-Rehbach et al. (226–231) in which the data is plotted in the complex plane may be looked upon as encompassing both stages simultaneously.

A. Vectorial Correction for Ohmic Resistance and Double-Layer Charging Current

1. Fundamental Harmonics

Until very recently, nearly all quantitative kinetic studies with ac polarography employed vectorial subtraction of the ohmic losses and double-layer charging current from the applied voltage and total alternating current, respectively, to obtain the faradaic impedance. This represents a simple extension to ac methods of the standard approach employed in dc techniques. The only difference is the necessity of treating the ac components as vectorial quantities because of the phase relations involved. The correction requires knowledge of the ohmic resistance and double-layer charging current. Treating the ohmic resistance and double-layer capacity as series components, one can determine both from experimental data on the cell impedance in absence of the electroactive material. Application of the double-layer capacity measured in this fashion to data with the electroactive substance present assumes no significant influence of the depolarizer on the double-layer capacity. This assumption is normally valid because the effect of the electroactive species is "swamped out" by the much more abundant ions of the supporting electrolyte. However, if the electroactive material is adsorbed, this assumption may not be valid (as with thallium, discussed earlier), and for such situations, the method under consideration is at a disadvantage. In some cases, double-layer capacity measurements can be effected in the presence of the electroactive material by employing very high frequencies at which the faradaic current is negligible.

Correction for the ohmic resistance is indicated from examination of the vectorial relations illustrated in Fig. 43. One is concerned with the triangle comprising the externally applied alternating potential V, the iR drop, and the effective applied potential ΔE. Note that the iR drop introduces a correction regarding both magnitude and phase of the alternating potential. The phase angle ϕ_C must be added to the measured phase angle ϕ_M to obtain the phase angle between the total alternating current and effective alternating potential. From the laws of sines and cosines (277), it can be shown that

$$\sin \phi_C = \frac{i_t R_\Omega \sin \phi_M}{[V^2 + (i_t R_\Omega)^2 - 2V i_t R_\Omega \cos \phi_M]^{\frac{1}{2}}} \tag{205}$$

$$\Delta E = [V^2 + (i_t R_\Omega)^2 - 2V i_t R_\Omega \cos \phi_M]^{\frac{1}{2}} \tag{206}$$

The quantities i_t, ϕ_M, and R_Ω are determined experimentally. ΔE and ϕ_C are then calculated from Eq. (205)and (206). Correction for the double-layer charging current is then accomplished by the vectorial subtraction indicated by the triangle comprising i_t, i_c, and $I(\omega t)$. It can be shown that

$$\cos \phi = \frac{i_t \cos Q}{[i_t^2 + i_c^2 - 2i_c i_t \sin Q]^{\frac{1}{2}}} \tag{207}$$

and

$$I(\omega t) = [i_t^2 + i_c^2 - 2i_c i_t \sin Q]^{\frac{1}{2}} \tag{208}$$

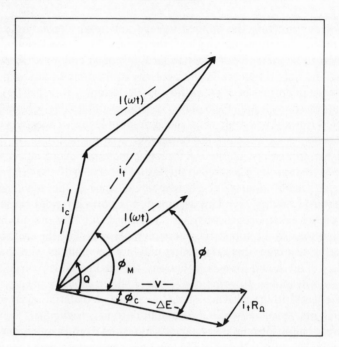

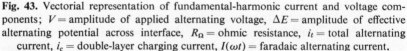

Fig. 43. Vectorial representation of fundamental-harmonic current and voltage components; V = amplitude of applied alternating voltage, ΔE = amplitude of effective alternating potential across interface, R_Ω = ohmic resistance, i_t = total alternating current, i_c = double-layer charging current, $I(\omega t)$ = faradaic alternating current.

where

$$Q = \phi_M + \phi_C \qquad (209)$$

One substitutes in Eqs. (207) and (208) the measured value of i_t and calculated Q. The term i_c represents the charging current at the value of ΔE existing in the presence of the faradaic wave, calculated from Eq. (206). This differs from the i_c measured in absence of the faradaic process because the effective applied potential ΔE is not the same in the two experiments. One employs the relation

$$i_c = (\Delta E/\Delta E')i_c' \qquad (210)$$

where i_c' and $\Delta E'$ are the values corresponding to the experiment carried carried out in absence of the faradaic process. The value of i_c calculated from Eq. (210) is substituted in Eqs. (207) and (208). This yields the phase angle of the faradaic current relative to the potential across the interface, ΔE, and the faradaic ac amplitude corresponding to ΔE.

2. Second Harmonics and Other Nonlinear Components

Because second, higher harmonic, and intermodulation currents are very small, their contribution to the iR term can be neglected (*73*). Correction for ohmic losses is achieved by considering only the losses from the fundamental-harmonic terms. Thus, the effective applied potential ΔE is given by Eq. (206). One sees that correction for iR distortion of these higher-order current components requires measurement of the fundamental harmonic. This consideration adds fuel to arguments in favor of the *multireadout instrument*. Ohmic losses can have a significant influence on the shape of the second-harmonic wave as demonstrated by Paynter (*73*). Maximum iR distortion is encountered at the peak of the fundamental-harmonic wave which normally corresponds to the minimum of the second harmonic. This introduces an enlarged separation of the second-harmonic peaks. Correction for double-layer charging current contributions to the second harmonic, etc., is carried out by vectorial subtraction as for the fundamental harmonic. This requires knowledge of the phase relations of the second harmonic, etc. However, except for studies at very low concentrations or at very high frequencies, charging current contributions to these higher-order currents are negligible and correction is not necessary. Thus, only consideration of ohmic losses normally is required.

B. Analysis of Faradaic-Component Data

1. Fundamental Harmonic

Assuming that the corrections for iR drop and double-layer charging current have been carried out properly, one is then in a position to compare data on the faradaic process with theory. Several aspects of this subject were considered in the theoretical section (for example, Tables II through IV). Because the mechanistic schemes which have been considered are numerous, individual consideration of the manifestations of each is not possible, partly because the space required for discussion would be excessive and partly because equations for all mechanisms discussed have not been examined in sufficient detail. Instead, a procedure for examining data will be outlined and some expectations and problems associated with specific mechanisms will be considered briefly. The procedure given represents a reasonably efficient approach to data analysis, but it is only one of many which might be proposed. Of course, blind adherence to any single data-analysis procedure is not suggested. A specific procedure is outlined primarily to serve as a frame of reference upon which to initiate data analysis. Modification to suit the complexity of the system under investigation is expected to be frequent.

The data-analysis procedure suggested consists of proceeding with an examination of the following data in more or less the order given:

(a) dc polarographic data and/or data from other dc techniques;
(b) column-height dependence of ac parameters;
(c) frequency dependence of cot ϕ;
(d) dc potential dependence of cot ϕ;
(e) frequency dependence of $I(\omega t)$;
(f) dc potential dependence of $I(\omega t)$.

At each stage in the examination, one draws whatever conclusions appear plausible regarding mechanism and quantitative rate and thermodynamic parameters. These conclusions are then subjected to test through the later steps in the procedure.

That the inspection of dc polarographic data is suggested as the first step may seem surprising in a discussion devoted to ac polarography. However, the ac polarographic experiment is not an end unto itself, but is a means to obtaining information not accessible through the more conveniently implemented dc studies (particularly for fast rate processes). To ignore the information provided by dc polarography makes the interpretation of ac data considerably more difficult (and vice versa).

With rapid processes, the dc experiment may provide the standard redox potential, diffusion coefficients, stability constants, and number of electrons transferred. In other situations, detailed information on relatively slow coupled chemical reactions, etc., might be obtained. Because of the numerous systematic investigations employing dc polarography, chances are good that the worker will find much of this type of information already available in the literature.

Examination of the column-height dependence of the ac parameters is suggested as the second step because investigation of the effect of column height may be incorporated conveniently in the initial data-collection effort, and because the presence or lack of a column-height dependence is immediately obvious without resorting to special plots. This data provides, at the outset of the investigation, a sensitive indication whether the dc process is controlled solely by diffusion and whether mercury-drop growth and geometry considerations must be considered in quantitative interpretation of ac data. The analogous experiment for studies employing stationary electrodes is an investigation of the dependence of the ac parameters on scan rate. These matters were discussed in detail in the theoretical section.

Inspection of phase-angle data is suggested as the next stage, because the phase angle is relatively sensitive to many kinetic influences so that qualitative examination of data may lead to immediate mechanistic conclusions. Also, theoretical expressions for the phase angle with most kinetic schemes are simple relative to the current-amplitude expressions, so that quantitative kinetic data follows more easily from phase angles. If $\cot \phi = 1$ at all frequencies, this is indicative of a diffusion-controlled process. A linear plot of $\cot \phi$ versus $\omega^{\frac{1}{2}}$ of positive slope and intercept of unity on the ordinate suggests a simple quasi-reversible system or, possibly, a mechanism involving a coupled preceding or following reaction, where $g^2 \gg 1$ over the frequency range examined [cf., Eq. (120)]. A nonlinear $\cot \phi$ versus $\omega^{\frac{1}{2}}$ plot showing a maximum or shoulder (cf., Figs. 13 to 16) indicates a preceding or following reaction mechanism, or possibly a two-step reduction with the two steps proceeding at substantially different rates (*129*). A double maximum or shoulder indicates the possibility of a two-step coupled chemical reaction (*137*). Values of $\cot \phi$ less than unity suggest adsorption and/or attractive migration in the diffuse double layer. Linear $\cot \phi$ versus $\omega^{\frac{1}{2}}$ plots with intercepts larger than 1.00 may indicate a two-step reduction mechanism where the nonlinear segment occurs at frequencies below those accessible experimentally (cf., Fig. 23). The examination of the dc potential dependence of $\cot \phi$ aids in distinguishing between preceding and following reaction mechanisms (*148*) and between simple catalytic reactions and those complicated by a preceding or following reaction (*137*). Once a tentative mechanism is decided upon, one attempts to calculate kinetic parameters from the appropriate theoretical equations. If, for example, a simple linear $\cot \phi$ versus $\omega^{\frac{1}{2}}$ plot with an intercept at 1.00 on the ordinate is obtained and the maximum on the $\cot \phi$ versus E_{dc} plot does not vary with frequency, chances are excellent that one is dealing with the simple quasi-reversible case. Equations (68) and (69) are then employed to calculate k_s and α (assuming E^0 and $D^{\frac{1}{2}}$ are known from dc measurements). To avoid the possibility that an ambiguity exists and the mechanistic conclusion is incorrect, despite the supporting evidence, one should calculate theoretical curves for $\cot \phi$ versus E_{dc}, $I(\omega t)$ versus $\omega^{\frac{1}{2}}$, and $I(\omega t)$ versus E_{dc} based on these values of k_s and α and compare data with these theoretical curves. In this manner, one employs all available data to make certain that the mechanism and rate parameters concluded on the basis of a limited examination of the data are correct. Some possible ambiguities which can arise from only limited inspection of data are apparent in the mechanisms considered in the discussion of theory. A two-step-reduction mechanism [Eqs. (152) through (162)] may yield a linear $\cot \phi$ versus $\omega^{\frac{1}{2}}$ plot and a peak on the cot ϕ versus E_{dc} curve which is independent of frequency, leading one to the tentative conclusion that the mechanism is the simple single-step quasi-reversible case. Values of k_s and α would then be calculated from Eqs. (68) and (69). However, it would become apparent immediately that the detailed cot ϕ versus E_{dc} behavior and current-amplitude data were not consistent with the tentative mechanism and rate parameters. Hopefully, reexamination of the data would ultimately lead one to the correct mechanistic conclusion. Other such examples can be cited where initial misconceptions will be corrected by detailed comparison of experimental data with theory.

If a coupled chemical reaction is influencing phase-angle data, measurements at high frequencies may minimize sufficiently the chemical kinetic effect so that Eqs. (68) and (69) may be applied to obtain values of k_s and α. Measurements at lower frequencies can then be employed to obtain chemical rate constants. The calculation method has been outlined for the catalytic case (*33*) and is relatively simple. For systems with single-step preceding or following reactions, data on the low-frequency linear segment [for example, Eq. (120)] can be employed to calculate the chemical-rate constants from relatively simple mathematical expressions. The calculation is particularly simple when stability constants are available from dc data. A significant problem in systems showing kinetic influence of coupled chemical reactions, particularly with consecutive equilibria involving complex ions, is determining the species involved in the reaction introducing the kinetic effect. Determination of the species undergoing reduction represents a significant step in solving this problem. Methods based on variation of pH, ligand concentration, etc., have been worked out by Gerischer (*278–280*), Matsuda et al. (*281–284*), Koryta [cf. (*285*)], and others. In any event, a combination of information will lead one to a tentative mechanism and values of charge-transfer and chemical-kinetic rate constants. Again, it is recommended that all available data be employed in an internal consistency check by calculating various theoretical curves from the tentative rate parameters and comparing all data with these predictions.

Situations one may encounter vary considerably, depending on the system complexity and amount of information initially available. Some systems will yield the correct mechanism and rate parameters early in the data-analysis procedure. Other cases may require much time and consideration. A tentative mechanism may be postulated and tentative rate constants calculated only to find the mechanism and/or rate parameters inconsistent with all available data; then one must backtrack to an earlier stage in data analysis, test other mechanisms and/or rate parameters, etc.

It is apparent from the trend of this discussion that the subject of analysis of faradaic-current data is deserving of a chapter in itself. For this reason the discussion will be terminated at this stage with the hope that the reader has obtained some feeling regarding how one can approach data analysis and what problems are involved.

2. Second Harmonics and Other Nonlinear Components

Analysis of higher-order current components has received relatively little attention. Nearly all work has been concerned with the relatively simple quasi-reversible case. Because theory for more complex mechanisms is relatively undeveloped, the subject of mechanistic diagnosis through data on higher-order currents cannot be discussed at this time. Thus, considerations will be confined to possible approaches for calculating k_s and α with the simple quasi-reversible system, neglecting the dc rectification component which has been discussed in detail by Delahay (10).

Both second-harmonic and intermodulation waves are characterized frequently by two peaks (for example, Figs. 2(a), 7 through 9). In extreme cases they will show a single peak, with the possibility of a shoulder appearing on the second harmonic. For normal experimental conditions ($\omega_1 - \omega_2 \ll \omega_1, \omega_2$), the intermodulation current exhibits a null between the two peaks, whereas the minimum observed with the second harmonic does not always correspond to zero current. Paynter and Reinmuth ($8,73$) have suggested the measurement of the dc potential corresponding to zero intermodulation current E_0 as a function of frequency to calculate k_s and α. This technique has all the advantages associated with null methods: for example, it does not require correction for iR drop. Working curves of $E_0 - E_{\frac{1}{2}}^r$ versus $\log[(2\omega D)^{\frac{1}{2}}/k_s]$ are constructed for various values of α from theoretical equations. Experimental data is analyzed by plotting observed $E_0 - E_{\frac{1}{2}}^r$ against $\log f^{\frac{1}{2}}$ (f is the frequency) and comparing this plot with the working curve. The shape of the experimental curve determines α and the separation of the experimental and working curves on the abscissa is employed to calculate k_s. It is desirable to calculate the entire intermodulation wave from values of k_s and α determined in this fashion and compare the calculated wave with the experimental result to ensure one that the proper mechanism is under consideration. Essentially the same method could be applied to second-harmonic data, where the potential of the second-harmonic minimum E_m is the point of interest. In this case, iR corrections are necessary before plotting experimental $E_m - E_{\frac{1}{2}}^r$ versus $\log f^{\frac{1}{2}}$ because the minimum does not correspond to zero current. A null can be obtained in second-harmonic measurements by phase-sensitive detection. Both the "resistive" and "capacitive" components of the second harmonic (that is, the sine and cosine terms) exhibit a null at some dc potential. Because these two components do not necessarily become zero at the same dc potential, the total second harmonic does not always exhibit a null. Measurement of the null potential for the resistive and/or capacitive components as a function of frequency and comparison with working curves, as above, can yield k_s and α. Other possible approaches exist, among which are (a) construction of working curves for the separation of the two peaks (second harmonic or intermodulation) as a function of $\log[(2\omega D)^{\frac{1}{2}}/k_s]$ for varying α values and comparison with experimental data; (b) construction of similar working curves for the magnitude of peak currents. A typical set of working curves for the second harmonic current is shown in Fig. 44.

Analysis of data from harmonics beyond the second can be carried out in the same manner.

C. Plotting AC Polarographic Data in the Complex-Impedance or Admittance Plane

Analysis of ac polarographic data by plotting it in the complex-impedance or admittance plane has been employed recently by Sluyters and co-workers (*226–231*). This approach to examining the response of complex electrical networks is standard in the fields of electronics, electrical engineering, etc. [cf., for example, (*286,287*)], and it is unfortunate that its application to ac polarography was not suggested earlier. It represents a means of combining corrections for iR drop and double-layer charging current and analysis of the faradaic impedance in one operation, eliminating the necessity of additional experiments in absence of the depolarizer. Sluyters et al., have considered the method only within the framework of the simple-reversible and quasi-reversible systems. Extension to more complex kinetic schemes has not been undertaken to this writer's knowledge, although it should be readily accomplished. However, the application to the simple cases shows the merits of the approach, and it is anticipated that the technique will become increasingly popular.

Full details are too extensive to discuss here, but some salient features will be pointed out to illustrate the advantages of plotting ac data in the complex plane. Some advantages become apparent upon consideration of the "base line" in the complex-impedance or admittance plane. These are illustrated in Fig. 45(*a*) and (*b*), respectively. The base line in the complex-impedance plane is a simple vertical line which intersects the abscissa

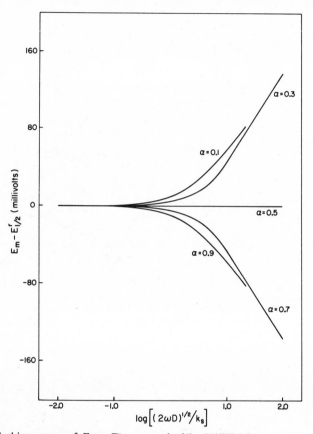

Fig. 44. Working curves of $E_m - E_{1/2}^r$ versus $\log[(2\omega D)^{1/2}/k_s]$ for second harmonic ac polarograms with various values of α.

(resistive-component coordinate) at the value of the ohmic resistance. Variations of the capacitive impedance ($1/\omega C$) correspond to movement of the total impedance along the vertical line. The form of the base line in the impedance plane manifests the fact that the ohmic resistance is constant [variation of ohmic resistance with drop life can occur when a DME and Luggin-probe reference electrode are employed, in which case the base line in Fig. 45(a) applies at a particular instant in drop life]. In the admittance plane, the base line is a semicircle with a diameter of $1/R_\Omega$, as shown in Fig. 45(b). It is only necessary to obtain data over a range of ωC sufficient to define the circle. Graphical extrapolation then yields $1/R_\Omega$.

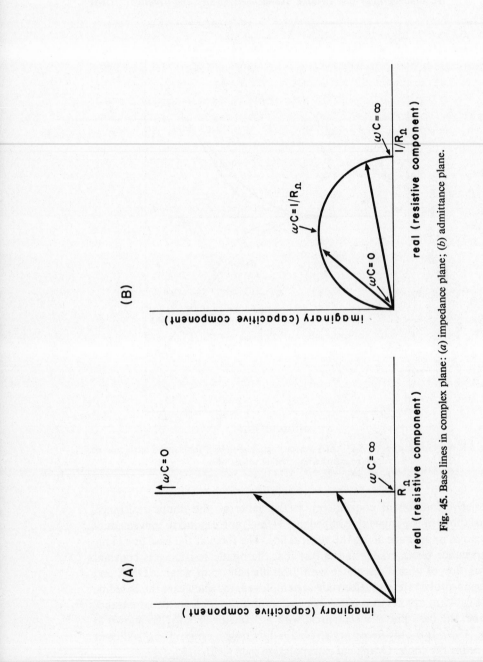

Fig. 45. Base lines in complex plane: (*a*) impedance plane; (*b*) admittance plane.

Sluyters et al. discussed the effects of the faradaic impedance on plots in the impedance and admittance plane with emphasis on the former. For brevity, remarks here will be confined to data plotted in the impedance plane. Any operation performed on data in the impedance plane has an analogous operation in the admittance plane. It is useful to consider first the influence of varying frequency at a fixed dc potential along the faradaic wave (constant interface concentrations of 0 and R). Sluyters (*226*) showed that the impedance over the entire frequency range with a quasi-reversible system could be represented by the solid curve in Fig. 46. The circular portion corresponds to rate control primarily by charge transfer $[(2\omega)^{\frac{1}{2}}/\lambda \gg 1]$, and the linear portion corresponds to mass-transfer control $[(2\omega)^{\frac{1}{2}}/\lambda \ll 1]$. If data can be obtained over a range of frequencies sufficient to define the linear and circular region, one can obtain R_Ω by circular extrapolation

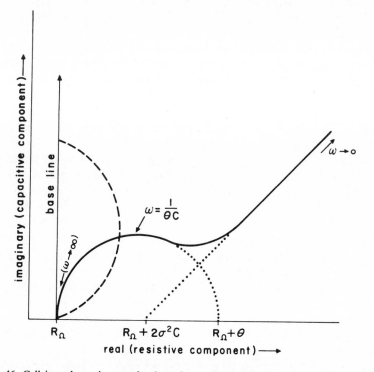

Fig. 46. Cell impedance in complex impedance plane with a quasi-reversible process; —, variation in impedance with frequency at constant dc potential; – –, variation in impedance with bulk concentration (or dc component of interface concentration) at constant frequency.

to the abscissa, θ (see below) from the radius of the semicircle, $2\sigma^2 C$ (see below) by extrapolation of the linear portion to the abscissa, and $1/\theta C$ from the frequency corresponding to the top of the semicircle. C is the double-layer capacity, θ the activation polarization resistance (226), which in the notation of this chapter is given by

$$\theta = \frac{4RT \cosh^2(j/2)}{n^2 F^2 A C_O^*(2\omega D_O)^{\frac{1}{2}}} \left(\frac{(2\omega)^{\frac{1}{2}}}{\lambda} \right) \tag{211}$$

and σ is

$$\sigma = \frac{4RT \cosh^2(j/2)}{n^2 F^2 A C_O^*(2D_O)^{\frac{1}{2}}} \tag{212}$$

It is apparent that one can determine the ohmic resistance, double-layer capacity, diffusion coefficient, and heterogeneous rate parameters, k_s and α. The latter are obtained from data at several values of dc potential. Practically speaking, the curve illustrated in Fig. 46 cannot always be obtained experimentally. If charge transfer is very rapid (θ small), the semicircular region is too small to be clearly defined, and only the linear segment is seen. If the faradaic wave is too small compared to the double-layer charging current, the entire curve is ill-defined. In the extreme case only the vertical base line is obtained.

It was shown that, if the bulk concentration of depolarizer is varied at constant dc potential and frequency, the variation in cell impedance describes a portion of a circle (dashed curve in Fig. 46). The center of this circle may lie on the base line for pure charge-transfer control or to the left of the base line for mixed control by mass- and charge-transfer or pure mass-transfer control. Varying dc potential has an effect similar to that obtained by varying bulk concentration (interface concentrations are varied). However, because double-layer capacity varies over the dc potential comprising the ac polarogram and because the phase of the faradaic component varies with dc potential (except for the reversible case), an ac polarogram plotted in the complex-impedance plane is an open curve as shown in Fig. 47. If the faradaic process were strictly reversible and the double-layer capacity were constant, the polarogram would describe a closed circular region. Rehbach and Sluyters (229,230) devised a method in which the polarograms are obtained with at least three different concentrations of depolarizer. The variation in the impedance with concentration at any dc potential describes a circle (Fig. 47, dashed circles) which intersects the abscissa at R_Ω and the vertical base line at $1/\omega C$. Each point on the polarogram belongs to a second circle (dotted circle, Fig. 47) intersecting the abscissa at R_Ω, with its center on the abscissa. The coordinates of the center

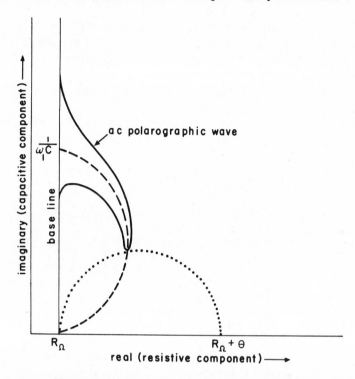

Fig. 47. Ac polarogram in complex-impedance plane.

of the first circle and the diameter of the second, obtained at different dc potentials, suffice to permit calculation of k_s, α, D_O and D_R.

Sluyters et al. used the equivalent-circuit representation for the cell impedance as a basis for their calculations. Reliance on the equivalent-circuit concept presents no problems with the simple quasi-reversible case, but difficulties may arise with more complicated kinetic schemes, particularly those involving adsorption. A general theoretical method which does not employ the equivalent-circuit concept would be rederivation of ac polarorgraphic equations by modifying the potential and current terms to account for the ohmic loss and double-layer charging current. One replaces Eq. (15) by

$$E(t) = E_{dc} - \Delta E \sin \omega t - iR_{\Omega} \qquad (213)$$

and the faradaic current $i(t)$ in surface concentration expressions [Eqs. (10), (11), (98), (99), (138), (139), (140)] by (10)

$$i(t) = i_t + C[dE(t)/dt] \qquad (214)$$

where i_t is the total current density (faradaic plus nonfaradaic). The final equations would give the total alternating current and/or impedance in terms of C, R_Ω and the faradaic parameters. The theoretical predictions could then be plotted in the complex plane.

Higher-order current components could also be plotted in the complex plane, presumably with some advantages.

A polarograph suitable for automatic recording in the complex admittance or impedance plane could be constructed readily from the operational-amplifier circuitry discussed above.

NOTATION DEFINITIONS

A	electrode area
C_i	concentration of species i
C_i^*	initial concentration of species i
$C_{i_{x=0}}$	surface concentration of species i
C_T	total concentration of supporting electrolyte
f_i	activity coefficient of species i
a_i	activity of species i
D_i	diffusion coefficient of species i
i_t	total cell current
$i(t)$	total faradaic current (cathode current positive)
$i_1(t), i_2(t)$	total faradaic current due to first and second reduction steps, respectively
$i_{dc}(t)$	dc faradaic current
$I_2(dc)$	dc faradaic rectification current
$I(\omega t)$	fundamental-harmonic faradaic alternating current
$I(n\omega t)$	nth-harmonic faradaic alternating current
I_p	peak faradaic fundamental-harmonic current
E^0	standard redox potential in European convention
$E_{\frac{1}{2}}^r$	reversible half-wave potential for planar diffusion
$E_{\frac{1}{2},1}^r, E_{\frac{1}{2},2}^r$	reversible half-wave potential of first- and second reduction steps, respectively
$E(t)$	instantaneous applied potential
ΔE	amplitude of applied alternating potential
E_{dc}	dc component of potential
$[E_{dc}]_{max}$	dc potential of maximum cot ϕ
$[E_{dc}]_{peak}$	dc potential of maximum $I(\omega t)$
E_e	equilibrium dc potential
E_1^0, E_2^0	standard redox potential of first and second reduction steps, respectively
e_i	input voltage to amplifier
e_o	output voltage of amplifier

$\Delta\Phi_E$	difference of potential between the plane of closest approach and the bulk of the solution at applied potential E
Φ	potential in the diffuse double layer referred to the potential in the bulk of the solution
n	number of electrons transferred in heterogeneous charge-transfer step.
F	Faraday's constant
R	ideal-gas constant
T	absolute temperature
n_1, n_2	number of electrons transferred in first and second charge-transfer steps, respectively
α	charge-transfer coefficient
α_1, α_2	charge-transfer coefficients for first and second reduction steps, respectively
k_s	heterogeneous rate constant for charge transfer at E^0
$k_{s,1}\ k_{s,2}$	heterogeneous rate constants for the first and second charge transfer steps, respectively (at E_1^0 and E_2^0)
k, k_1, k_2, k_m, k_p	first-order chemical reaction rate constants or combinations thereof
k_a, k_d	rate constants for adsorption and desorption, respectively
k^0	standard rate constant for adsorption
v_a	rate of adsorption step
v_d	rate of desorption step
v	net rate of adsorption
v^0	adsorption exchange rate
$\lambda*$	coverage parameter
Γ	surface concentration
b	characteristic parameter for Temkin isotherm
ρ	charge parameter
ΔG^0	standard free energy of adsorption
ΔG^n	charge-independent part of standard free energy of adsorption
ΔG^q	charge-dependent part of standard free energy of adsorption
$_\infty C$	infinite-frequency capacity of electrode-solution interface
C_f	equivalent-series-capacitive component of faradaic impedance
R_f	equivalent-series-resistive component of faradaic impedance
Z_f	faradaic impedance (absolute value)
Z_{rev}	faradaic impedance with reversible process
R_{rev}	equivalent series resistive component of faradaic impedance with a reversible process $(= 1/\omega C_{rev})$
t	time

x	distance from electrode surface
r_0	radius of spherical electrode
u	auxiliary variable of integration
ω	angular frequency
v	dc potential scan rate
ϕ	phase angle of fundamental-harmonic faradaic alternating current relative to applied alternating potential
R_Ω	uncompensated ohmic resistance
z_0	charge on species i

REFERENCES

1. B. Breyer and F. Gutmann, *Trans. Faraday Soc.*, **42**, 645, 650 (1946).
2. B. Breyer and F. Gutmann, *Trans. Faraday Soc.*, **43**, 785 (1947).
3. B. Breyer and F. Gutmann, *Discussions Faraday Soc.*, **1**, 19 (1947).
4. D. C. Grahame, *J. Amer. Chem. Soc.*, *63*, 1207 (1941).
5. B. Ershler, *Discussions Faraday Soc.*, **1**, 269 (1947).
6. J. E. B. Randles, *Discussions Faraday Soc.*, **1**, 11, 47 (1947).
7. D. N. Hume, *Anal. Chem.*, **36**, 200R (1964).
8. W. H. Reinmuth, *Anal. Chem.*, **36**, 211R (1964).
9. B. Breyer and H. H. Bauer, "Alternating Current Polarography and Tensammetry," in *Chemical Analysis* (P. J. Elving and I. M. Kolthoff, eds.), Vol. 13, Wiley (Interscience), New York, 1963.
10. P. Delahay, in *Advances in Electrochemistry and Electrochemical Engineering* (P. Delahay and C. W. Tobias, eds.), Vol. 1, Chap. 5, Wiley (Interscience), New York, 1961.
11. D. M. Mohilner, *Advances in Electroanalytical Chemistry* (A. J. Bard, ed.), Vol. 1, Chap. 4, Dekker, New York, 1965.
12. P. Delahay, *New Instrumental Methods in Electrochemistry*, Wiley (Interscience), New York, 1954.
13. H. V. Malmstadt, C. G. Enke, and E. C. Toren, Jr., *Electronics for Scientists*, Benjamin, New York, 1962.
14. B. Breyer and S. Hacobian, *Australian J. Sci. Res.*, **A5**, 500 (1952).
15. B. Breyer and S. Hacobian, *Australian J. Chem.*, **9**, 7 (1956).
16. B. Breyer, *Proc. Intern. Congr. Surface Activity, 2nd, London*, **1957**, Vol. 3, p. 34.
17. A. N. Frumkin and B. B. Damaskin, *J. Electroanal. Chem.*, **3**, 36 (1962).
18. B. Breyer and S. Hacobian, *J. Electroanal. Chem.*, **3**, 45 (1962).
19. H. H. Bauer and P. J. Elving, *Australian J. Chem.*, **12**, 335 (1959).
20. H. Schmidt and M. von Stackelberg, *J. Electroanal. Chem.*, **1**, 133 (1959).
21. J. E. B. Randles and K. W. Somerton, *Trans. Faraday Soc.*, **48**, 937, 951 (1952).
22. J. E. B. Randles and H. A. Laitinen, *Trans. Faraday Soc.*, **51**, 54 (1955).
23. D. C. Grahame, *J. Am. Chem. Soc.*, **71**, 2975 (1949).
24. R. Tamamushi and N. Tanaka, *Z. Physik. Chem.*, **21**, 89 (1959).
25. H. Gerischer, in *Advances in Electrochemistry and Electrochemical Engineering* (P. Delahay and C. W. Tobias, eds.), Vol. 1, Chap. 4, Wiley (Interscience), New York, 1961.

26. D. E. Walker, R. N. Adams, and J. R. Alden, *Anal. Chem.*, **33**, 308 (1961).
27. W. L. Underkofler and I. Shain, *Anal. Chem.*, **37**, 218 (1965).
28. C. N. Reilley and R. W. Murray, in *Treatise on Analytical Chemistry* (I. M. Kolthoff and P. J. Elving, eds.), Part 1, Vol. 4, Chap. 43, Wiley (Interscience), New York, 1963.
29. B. Breyer, F. Gutmann, and S. Hacobian, *Australian J. Sci. Res.*, A3, 558, 567 (1950).
30. B. Breyer and S. Hacobian, *Australian J. Chem.*, **7**, 225 (1954).
31. B. Breyer, H. H. Bauer, and S. Hacobian, *Australian J. Chem.*, **8**, 322 (1955).
32. D. E. Smith, Division of Analytical Chemistry, 140th Meeting, ACS, Chicago, Ill., Sept., 1961; D. E. Smith, Ph.D. Thesis, Columbia Univ., New York, 1961; D. E. Smith and W. H. Reinmuth, unpublished work, 1961.
33. D. E. Smith, *Anal. Chem.*, **35**, 610 (1963).
34. D. E. Smith, *Anal. Chem.*, **35**, 1811 (1963).
35. Electro Scientific Industries Catalog No. A-26, March 1963.
36. G. L. Booman, *Anal. Chem.*, **29**, 213 (1957).
37. D. D. DeFord, Division of Analytical Chemistry, 133rd Meeting, ACS, San Francisco, Calif., April, 1958.
38. M. T. Kelley, D. J. Fisher, and H. C. Jones, *Anal. Chem.*, **31**, 1475 (1959); **32**, 1262 (1960).
39. W. L. Underkofler and I. Shain, *Anal. Chem.*, **35**, 1778 (1963).
40. W. B. Schaap and P. S. McKinney, *Anal. Chem.*, **36**, 29 (1964).
41. G. Jessop, British Pat. 640,768 (1950).
42. G. Jessop, British Pat. 776,543 (1957).
43. D. J. Ferrett, G. W. C. Milner, H. I. Shalgosky, and L. J. Slee, *Analyst*, **81**, 506 (1956).
44. L. A. Balchin and D. I. Williams, *Analyst*, **85,** 503 (1960).
45. J. W. Hayes and H. H. Bauer, *J. Electroanal. Chem.*, **3**, 336 (1962).
46. E. Niki, *Rev. Polarog. (Kyoto)*, **3**, 41 (1955).
47. E. Niki, *J. Electrochem. Soc. Japan (Japan Ed.)*, **23**, 526 (1955).
48. E. Niki, *Kagaku No Ryoiki*, **10**, 203 (1956); *CA*, **51**, 17564e.
49. T. Takahashi and E. Niki, *Talanta*, **1**, 245 (1958).
50. D. E. Smith and W. H. Reinmuth, *Anal. Chem.*, **32**, 1892 (1960).
51. W. Erbelding and W. D. Cooke, Division of Analytical Chemistry, 140th Meeting, ACS, Chicago, Ill., Sept. 1961.
52. G. C. Barker and I. L. Jenkins, *Analyst*, **77**, 685 (1952).
53. G. C. Barker, *Anal. Chim. Acta*, **18**, 118 (1958).
54. *The Mervyn Mark IV Square Wave Polarographic Analyzer*, Applications Bulletin, Matheson Scientific, Inc., 1964.
55. R. E. Hamm, *Anal. Chem.*, **30**, 350 (1958).
56. D. D. DeFord and E. H. Nagel, Division of Analytical Chemistry, 144th Meeting, ACS, Los Angeles, Calif., April 1963.
57. P. Delahay, M. Senda, and C. H. Weis, *J. Phys. Chem.*, **64**, 960 (1960); *J. Am. Chem. Soc.*, **83**, 312 (1961).
58. K. S. G. Doss and H. P. Agarwal, *J. Sci. Ind. Res.* (India), **9B**, 280 (1950).
59. K. S. G. Doss and H. P. Agarwal, *Proc. Indian Acad. Sci.*, **34A**, 263 (1951); **35A**, 45 (1962).
60. K. B. Oldham, *J. Electrochem. Soc.*, **107**, 766 (1960).

61. G. C. Barker, in *Transactions of the Symposium on Electrode Processes* (E. Yeager, ed.), Wiley, New York, 1961.

62. G. C. Barker, R. L. Faircloth, and A. W. Gardner, *Nature*, **181**, 247 (1958).

63. I. A. Vdovin, *Dokl. Akad. Nauk SSSR*, **120**, 554 (1958).

64. K. B. Oldham, *Trans. Faraday Soc.*, **53**, 80 (1957).

65. H. Matsuda and P. Delahay, *J. Am. Chem. Soc.*, **82**, 1547 (1960).

66. M. Senda and I. Tachi, *Bull. Chem. Soc. Japan*, **28**, 632 (1955).

67. M. Senda and P. Delahay, *J. Phys. Chem.*, **65**, 1580 (1961).

68. M. Senda and P. Delahay, *J. Am. Chem. Soc.*, **83**, 3763 (1961).

69. M. Senda, H. Imai, and P. Delahay, *J. Phys. Chem.*, **65**, 1253 (1961).

70. H. Imai, *J. Phys. Chem.*, **66**, 1744 (1962).

71. H. Imai and P. Delahay, *J. Phys. Chem.*, **66**, 1108 (1962).

72. H. Imai and P. Delahay, *J. Phys. Chem.*, **66**, 1683 (1962).

73. J. Paynter, Ph.D. Thesis, Columbia Univ., New York, 1964.

74. R. Neeb, *Z. Anal. Chem.*, **188**, 401 (1962).

75. R. Neeb, *Naturwiss.*, **49**, 447 (1962).

76. H. H. Bauer, *J. Electroanal. Chem.*, **1**, 256 (1960).

77. H. H. Bauer and P. J. Elving, *Anal. Chem.*, **30**, 341 (1958).

78. H. H. Bauer, *Australian J. Chem.*, **17**, 715 (1964); (a) correction to **17**, 715 (1964) available with reprints.

79. J. van Cakenberghe, *Bull. Soc. Chim. Belges*, **60**, 3 (1951).

80. A. C. Aten, Ph.D. Thesis, Free University of Amsterdam, The Netherlands, 1959.

81. D. E. Smith and W. H. Reinmuth, *Anal. Chem.*, **33**, 482 (1961).

82. R. Neeb, *Z. Anal. Chem.*, **186**, 53 (1962).

83. H. Gerischer and M. Krause, *Z. Physik. Chem. (Frankfurt)*, **10**, 264 (1957); **14**, 184 (1958).

84. P. Delahay, *Anal. Chem.*, **34**, 1161 (1962).

85. P. Delahay, *J. Phys. Chem.*, **66**, 2204 (1962).

86. P. Delahay and A. Aramata, *J. Phys. Chem.*, **66**, 2208 (1962).

87. P. Delahay and D. M. Mohilner, *J. Am. Chem. Soc.*, **84**, 4247 (1962).

88. P. Delahay and W. H. Reinmuth, *Anal. Chem.*, **34**, 1344 (1962).

89. W. H. Reinmuth, *Anal. Chem.*, **34**, 1272 (1962).

90. W. H. Reinmuth and C. E. Wilson, *Anal. Chem.*, **34**, 1159 (1962).

91. H. Matsuda, *Z. Elektrochem.*, **61**, 489 (1957).

92. D. C. Grahame, *Chem. Revs.*, **41**, 441 (1947).

93. J. Kuta and I. Smoler, in *Advances in Polarography* (I. S. Longmuir, ed.), Vol. 1, Pergamon, New York, 1960, pp. 350–8.

94. J. Kuta and I. Smoler, in *Progress in Polarography* (P. Zuman, ed., with the collaboration of I. M. Kolthoff), Vol. 1, Chap. 3, Wiley (Interscience), New York, 1962.

95. H. Gerischer, *Z. Physik. Chem. (Leipzig)*, **198**, 286 (1951).

96. J. Koutecky, *Collection Czech. Chem. Commun.*, **21**, 433 (1956).

97. H. Matsuda, *Z. Elektrochem.*, **62**, 977 (1958).

98. H. S. Carslaw and J. C. Jaeger, *Conduction of Heat in Solids*, Oxford, New York, 1947, p. 57, Eq. 9.

99. H. Matsuda, *Z. Elektrochem.*, **59**, 494 (1955).

100. T. Kambara and I. Tachi, *J. Phys. Chem.*, **61**, 1405 (1957).

101. W. H. Reinmuth, *Anal. Chem.*, **34**, 1446 (1962).

102. C. D. Hodgman, ed., *Handbook of Chemistry and Physics*, 41st ed., Chemical Rubber Publishing Co., Cleveland, 1959, p. 275.
103. T. Berzins and P. Delahay, *Z. Elektrochem.*, **59**, 792 (1955).
104. M. Fournier, *Comp. Rend.*, **232**, 1673 (1951).
105. G. S. Buchanan and R. L. Werner, *Australian J. Chem.*, **7**, 239 (1954).
106. D. C. Grahame, *J. Electrochem. Soc.*, **98**, 370C (1952).
107. H. H. Bauer, *J. Electroanal. Chem.*, **1**, 2 (1959).
108. W. H. Reinmuth and D. E. Smith, *Anal. Chem.*, **33**, 964 (1961).
109. H. H. Bauer, W. H. Reinmuth, and D. E. Smith, *Anal. Chem.*, **33**, 1803 (1961).
110. H. H. Bauer, *J. Electroanal. Chem.*, **3**, 150 (1962).
111. D. E. Smith, unpublished work, 1965.
112. J. Paynter and W. H. Reinmuth, *Anal. Chem.*, **34**, 1335 (1962).
113. H. H. Bauer and D. C. S. Foo, *Australian J. Chem.*, **17**, 510 (1964).
114. R. Parsons, in *Advances in Electrochemistry and Electrochemical Engineering* (P. Delahay and C. W. Tobias, eds.), Vol. 1, Chap. 1, Wiley (Interscience), New York, 1961.
115. P. Delahay, in *Progress in Polarography* (P. Zuman, ed., with the collaboration of I. M. Kolthoff), Vol. 1, Chap. 4, Wiley (Interscience), New York, 1962.
116. S. Glasstone, K. J. Laidler, and H. Eyring, *The Theory of Rate Processes*, McGraw-Hill, New York, 1941, pp. 575–77.
117. P. Delahay, *J. Am. Chem. Soc.*, **75**, 1430 (1953).
118. M. Smutek, *Collection Czech. Chem. Commun.*, **18**, 171 (1953).
119. T. Kambara and I. Tachi, *Bull. Chem. Soc. Japan*, **25**, 135 (1952).
120. S. K. Rangarajan and K. S. G. Doss, *J. Electroanal. Chem.*, **3**, 217 (1962).
121. H. L. Hung and D. E. Smith, *Anal. Chem.*, **36**, 922 (1964).
122. G. H. Aylward, J. W. Hayes, H. L. Hung, and D. E. Smith, *Anal. Chem.*, **36**, 2218 (1964).
123. G. H. Aylward, J. W. Hayes, and R. Tamamushi, *Proceedings of the First Australian Conference on Electrochemistry* (J. A. Friend and F. Gutmann, eds.), Pergamon Oxford, 1964.
124. T. M. Florence and G. H. Aylward, *Australian J. Chem.*, **15**, 65 (1962).
125. G. H. Aylward and J. W. Hayes, private communication, 1964.
126. M. Senda, *Kagaku No Ryoiki, Zokan*, **50**, 15 (1962).
127. J. E. B. Randles, in *Transactions of the Symposium on Electrode Processes, Philadelphia, May 1959*, Wiley, New York, 1961.
128. J. Paynter, D. E. Smith, and W. H. Reinmuth, unpublished work, 1961.
129. H. L. Hung and D. E. Smith, *J. Electroanal. Chem.*, **11**, 237, 425 (1966).
130. T. McCord and D. E. Smith, unpublished work, 1965.
131. J. R. Delmastro and D. E. Smith, unpublished work, 1965.
132. R. Brdicka, V. Hanus, and J. Koutecky, in *Progress in Polarography* (P. Zuman, ed., with the collaboration of I. M. Kolthoff), Vol. 1, Chap. 7, Wiley (Interscience), New York, 1962.
133. J. Koryta, in *Progress in Polarography* (P. Zuman, ed., with the collaboration of I. M. Kolthoff), Vol. 1, Chap. 12, Wiley (Interscience), New York, 1962.
134. H. Strehlow, *Technique of Organic Chemistry* (A. Weissberger, ed.), Vol. 8, Part 2, Chap. 15, Wiley (Interscience), New York, 1963.
135. R. Brdicka, in *Advances in Polarography* (I. S. Longmuir, ed.), Vol. 2, Pergamon, New York, 1960, pp. 655–673.

136. R. Brdicka, *Z. Elektrochem.*, **64**, 16 (1960).
137. H. L. Hung, J. R. Delmastro, and D. E. Smith, *J. Electroanal. Chem.*, **7**, 1 (1964).
138. V. Hanus and R. Brdicka, *Chem. Listy*, **44**, 291 (1950); *Khimija*, *1*, 28 (1951).
139. J. Kuta, *Collection Czech. Chem. Commun.*, **24**, 2532 (1959).
140. J. Koutecky, *Collection Czech. Chem. Commun.*, **19**, 1093 (1954).
141. K. Wiesner, *Collection Czech. Chem. Commun.*, **12**, 64 (1947).
142. J. M. Los and K. Wiesner, *J. Am. Chem. Soc.*, **75**, 6346 (1953).
143. J. M. Los, L. B. Simpson, and K. Wiesner, *J. Am. Chem. Soc.*, **78**, 1564 (1956).
144. J. Paldus and J. Koutecky, *Collection Czech. Chem. Commun.*, **23**, 376 (1958).
145. J. Koutecky, V. Hanus, and S. G. Mairanovskii, *Zh. Fiz. Khim.*, **34**, 651 (1960).
146. H. Matsuda, P. Delahay, and M. Kleinerman, *J. Am. Chem. Soc.*, **81**, 6379 (1959).
147. S. Satyanarayana, A. K. N. Reddy, and K. S. G. Doss, *Australian J. Chem.*, **13**, 177 (1960).
148. D. E. Smith, *Anal. Chem.*, **35**, 602 (1963).
149. D. E. Smith, *Anal. Chem.*, **36**, 962 (1964).
150. G. H. Aylward and J. W. Hayes, *Anal. Chem.*, **37**, 195, 197 (1965).
151. W. H. Reinmuth, *Anal. Chem.*, **34**, 1272 (1962).
152. R. P. Buck, *J. Electroanal. Chem.*, **5**, 295 (1963).
153. J. Koutecky and R. Brdicka, *Collection Czech. Chem. Commun.*, **12**, 337 (1947).
154. B. Breyer, J. R. Beevers, and H. H. Bauer, *J. Electroanal. Chem.*, **2**, 60 (1961).
155. J. Koutecky and J. Koryta, *Collection Czech. Chem. Commun.*, **19**, 845 (1954).
156. J. Koryta and J. Koutecky, *Collection Czech. Chem. Commun.*, **20**, 423 (1955).
157. D. M. H. Kern and E. F. Orleman, *J. Am. Chem. Soc.*, **71**, 2102 (1949).
158. E. F. Orleman and D. M. H. Kern, *J. Am. Chem. Soc.*, **75**, 3058 (1953).
159. T. Berzins and P. Delahay, *J. Am. Chem. Soc.*, **75**, 5716 (1953).
160. H. A. Laitinen and S. Wawzonek, *J. Am. Chem. Soc.*, **64**, 1764 (1942).
161. K. J. Vetter and G. Thiemke, *Z. Elektrochem.*, **64**, 805 (1960).
162. D. M. Mohilner, *J. Phys. Chem.*, **68**, 623 (1964).
163. K. J. Vetter, *Z. Naturforsch.*, **7a**, 328 (1952); **8a**, 823 (1953).
164. R. M. Hurd, *J. Electrochem. Soc.*, **109**, 327 (1962).
165. H. Mauser, *Z. Elektrochem.*, **62**, 419 (1958).
166. A. C. Riddiford, *J. Chem. Soc.*, **1960**, 1175.
167. A. C. Testa and W. H. Reinmuth, *Anal. Chem.*, **33**, 1320 (1961).
168. A. C. Testa and W. H. Reinmuth, *J. Am. Chem. Soc.*, **83**, 784 (1961).
169. G. S. Alberts and I. Shain, *Anal. Chem.*, **35**, 1859 (1963).
170. L. Holleck and R. Schindler, *Z. Elektrochem.*, **60**, 1138 (1960).
171. M. Suzuki, *Mem. Coll. Agr. Kyoto Univ.*, 67 (1954).
172. R. S. Nicholson and I. Shain, Division of Analytical Chemistry, 148th Meeting, ACS, Chicago, Ill., Sept., 1964; *Anal. Chem.*, **37**, 178, 190 (1965).
173. A. A. Vlcek, in *Progress in Inorganic Chemistry* (F. A. Cotton, ed.), Vol. 5, Wiley (Interscience), New York, 1963, pp. 211–384.
174. H. B. Herman and A. J. Bard, *Anal. Chem.*, **36**, 971 (1964).
175. P. Delahay and D. M. Mohilner, *J. Phys. Chem.*, **66**, 959 (1962).
176. P. Delahay and I. Trachtenberg, *J. Amer. Chem. Soc.*, **79**, 2355 (1957); **80**, 2094 (1958).
177. P. Delahay and C. T. Fike, *J. Amer. Chem. Soc.*, **80**, 2628 (1958).
178. T. Berzins and P. Delahay, *J. Phys. Chem.*, **59**, 906 (1955).
179. W. Lorenz, *Z. Elektrochem.*, **62**, 192 (1958).

180. W. Lorenz and F. Möckel, *Z. Elektrochem.*, **60**, 507 (1956).

181. W. Lorenz and G. Salie, *Z. Physik. Chem. (Leipzig)*, **218**, 259 (1961).

182. W. Lorenz, *Z. Physik. Chem. (Leipzig)*, **218**, 272 (1961).

183. W. Lorenz and G. Kruger, *Z. Physik. Chem. (Leipzig)*, **221**, 231 (1961).

184. G. Kruger, W. Lorenz, and P. Theml, *Z. Physik. Chem. (Leipzig)*, **222**, 81 (1963).

185. J. Koutecky and J. Weber, *Collection Czech. Chem. Commun.*, **25**, 1423 (1960).

186. J. Weber, J. Koutecky, and J. Koryta, *Collection Czech. Chem. Commun.*, **63**, 583 (1959).

187. J. Kuta, J. Weber, and J. Koutecky, *Collection Czech. Chem. Commun.*, **25**, 2376 (1960).

188. J. Weber and J. Koutecky, *Collection Czech. Chem. Commun.*, **25**, 2993 (1960).

189. R. Parsons, *Trans. Faraday Soc.*, **51**, 1518 (1955).

190. A. N. Frumkin and V. I. Melik-Gaikazyan, *Dokl. Akad. Nauk SSSR*, **78**, 855 (1951).

191. A. N. Frumkin, in *Transactions of the Symposium on Electrode Processes, Philadelphia, May 1959* (E. Yeager, ed.), Wiley, New York, 1961.

192. W. H. Reinmuth, *J. Phys. Chem.*, **65**, 473 (1961).

193. C. N. Reilley and W. Stumm, in *Progress in Polarography* (P. Zuman, ed., with the collaboration of I. M. Kolthoff), Vol. 1, Chap. 5, Wiley (Interscience), New York, 1962.

194. M. I. Temkin, *Zh. Fiz. Khim.*, **15**, 296 (1941).

195. A. F. H. Ward and L. Tordai, *J. Chem. Phys.*, **14**, 453 (1946).

196. A. N. Frumkin, *Zh. Fiz. Khim.*, **24**, 244 (1950).

197. A. N. Frumkin, *Dokl. Akad. Nauk SSSR*, **85**, 373 (1952).

198. A. N. Frumkin, *Z. Elektrochem.*, **59**, 807 (1955).

199. A. P. Martirosyan and T. A. Kryukova, *Zh. Fiz. Khim.*, **27**, 851 (1953).

200. R. Tamamushi and T. Yamanaka, *Bull. Chem. Soc. Japan*, **28**, 673 (1955).

201. H. Imai and S. Chaki, *Bull. Chem. Soc. Japan*, **29**, 498 (1956).

202. I. M. Kolthoff and V. Okinaka, *J. Am. Chem. Soc.*, **81**, 2296 (1959).

203. V. V. Losev, *Dokl. Akad. Nauk SSSR*, **107**, 432 (1956).

204. M. A. Loshkarev and A. A. Kryukova, *Dokl. Akad. Nauk SSSR*, **62**, 97 (1948).

205. M. A. Loshkarev and A. A. Kryukova, *Zh. Anal. Khim.*, **6**, 166 (1951).

206. A. G. Stromberg and M. S. Guterman, *Zh. Fiz. Khim.*, **27**, 993 (1953).

207. A. G. Stromberg and L. S. Zagainova, *Dokl. Akad. Nauk SSSR*, **97**, 101 (1954).

208. J. Heyrovsky, *Discussions Faraday Soc.*, **1**, 212 (1947).

209. J. Heyrovsky and M. Matyas, *Collection Czech. Chem. Commun.*, **16**, 455 (1951).

210. J. Heyrovsky, F. Šorm, and J. Forejt, *Collection Czech. Chem. Commun.*, **12**, 11 (1947).

211. M. Matyas, *Collection Czech. Chem. Commun.*, **16**, 496 (1951).

212. P. Silvestroni and L. Rampazoo, *J. Electroanal. Chem.*, **7**, 73 (1964).

213. P. Silvestroni, *Ric. Sci. Suppl.*, **26**, 166 (1956); **28**, 2341 (1958).

214. L. Holleck and B. Kastening, *Z. Elektrochem.*, **63**, 177 (1959).

215. H. Gerischer, *Ann. Rev. Phys. Chem.*, **1961**, 231.

216. P. Delahay, *J. Chim. Phys.*, **54**, 369 (1957).

217. G. C. Barker, *Square Wave Polarography*, Part 3, Atomic Energy Research Establishment, Harwell, 1957.

218. K. M. Joshi, W. Mehl, and R. Parsons, in *Transactions of the Symposium on*

Electrode Processes, Philadelphia, May 1959 (E. Yeager, ed.), Wiley, New York, 1961.

219. G. P. Harnwell, *Principles of Electricity and Electromagnetism*, McGraw-Hill, New York, 1949, p. 161.
220. H. Gerischer, *Z. Physik. Chem.*, **201**, 55 (1952).
221. H. A. Laitinen and J. E. B. Randles, *Trans. Faraday Soc.*, **51**, 54 (1955).
222. J. Llopis, J. Fernandez-Biarge, and M. Perez Fernandez, *Electrochim. Acta*, **1**, 130 (1959).
223. J. Llopis, J. Fernandez-Biarge, and M. Perez Fernandez, in *Transactions of the Symposium on Electrode Processes, Philadelphia, May, 1959* (E. Yeager, ed.), Wiley, New York, 1961.
224. R. Tamamushi and N. Tanaka, *Z. Physik. Chem. N. F.*, **28**, 158 (1961).
225. H. H. Bauer, D. L. Smith, and P. J. Elving, *J. Am. Chem. Soc.*, **82**, 2094 (1960).
226. J. H. Sluyters, *Rec. trav. Chim.*, **79**, 1092 (1960).
227. J. H. Sluyters and J. J. C. Oomen, *Rec. trav. Chim.*, **79**, 1101 (1960).
228. M. Rehbach and J. H. Sluyters, *Rec. trav. Chim.*, **80**, 469 (1961).
229. M. Rehbach and J. H. Sluyters, *Rec. trav. Chim.*, **81**, 301 (1962).
230. M. Sluyters-Rehbach and J. H. Sluyters, *Rec. trav. Chim.*, **82**, 525 (1963); **82**, 536 (1963).
231. M. Sluyters-Rehbach, B. Timmer, and J. H. Sluyters, **82**, 553 (1963).
232. R. A. Osteryoung and F. C. Anson, *Anal. Chem.*, **36**, 975 (1964).
233. J. B. Lawrence, *Instruments*, **25**, March (1952).
234. A. N. Frumkin, *Z. Physik. Chem.*, **164**, 121 (1933).
235. J. Horiuti and M. Polanyi, *Acta Physicochim. SSSR*, **2**, 505 (1935).
236. H. Matsuda and P. Delahay, *J. Phys. Chem.*, **64**, 332 (1960).
237. S. K. Rangarajan and K. S. G. Doss, *J. Electroanal. Chem.*, **5**, 114 (1963).
238. A. C. Testa and W. H. Reinmuth, *Anal. Chem.*, **32**, 1518 (1960).
239. P. Delahay and W. Vielstich, *J. Am. Chem. Soc.*, **77**, 4955 (1955).
240. L. Onsager, *J. Chem. Phys.*, **2**, 599 (1943).
241. M. Eigen and L. de Maeyer, in *Technique of Organic Chemistry* (A. Weissberger, ed.), Vol. 8, Part 2, Chap. 18, Wiley (Interscience), New York, 1963.
242. V. G. Levich, *Dokl. Akad. Nauk SSSR*, **67**, 309 (1949).
243. H. Matsuda, *J. Phys. Chem.*, **64**, 336 (1960); **64**, 339 (1960).
244. S. K. Rangarajan, *Can. J. Chem.*, **41**, 983 (1963); **41**, 1007 (1963); **41**, 1469 (1963).
245. K. Narayanan and S. K. Rangarajan, *Australian J. Chem.*, **16**, 565 (1963).
246. S. K. Rangarajan, *J. Electroanal. Chem.*, **5**, 253 (1963); **5**, 350 (1963).
247. L. Gierst and H. Hurwitz, *Z. Elektrochem.*, **64**, 36 (1960).
248. H. Hurwitz, *Z. Elektrochem.*, **65**, 178 (1961).
248a. T. Biegler and H. A. Laitinen, *Anal. Chem.*, **37**, 572 (1965).
248b. J. R. Delmastro and D. E. Smith, *Anal. Chem.*, **38**, 169 (1966).
249. H. Matsuda and Y. Ayabe, *Bull. Chem. Soc. Japan*, **28**, 422 (1955).
250. J. R. Delmastro and D. E. Smith, *J. Electroanal. Chem.*, **9**, 192 (1965).
251. J. Koutecky and J. Cizek, *Collection Czech. Chem. Commun.*, **21**, 836 (1956).
252. K. Okamoto, *Bull. Chem. Soc. Japan*, **36**, 1381 (1963).
253. H. Matsuda and Y. Ayabe, *Z. Elektrochem.*, **59**, 494 (1955).
254. Y. Takemori, T. Kambara, M. Senda, and I. Tachi, *J. Phys. Chem.*, **61**, 968 (1957).
255. H. H. Bauer, *Rev. Polarog. (Kyoto)*, **11**, 58 (1963).
256. H. Gerischer, *Z. Elektrochem.*, **58**, 9 (1954).

257. G. L. Booman and W. B. Holbrook, *Anal. Chem.*, **35**, 1793 (1963).
258. G. A. Philbrick Researches, Inc., Technical Bulletin, SP456, 1965.
259. E. R. Brown and D. E. Smith, unpublished work, 1965.
260. C. N. Reilley, *J. Chem. Educ.*, **39**, A853, A933 (1962).
261. W. M. Schwarz and I. Shain, *Anal. Chem.*, **35**, 1770 (1963).
262. J. M. Reece, *Electronics*, **32**, 73 (Nov. 6, 1959).
263. H. Fleisher, in *M.I.T. Radiation Lab Series* (G. E. Valley, Jr. and H. Wallman, eds.), Vol. 18, McGraw-Hill, New York, 1948, pp. 387–8.
264. R. D. Stuart, *Fourier Analysis*, Wiley, New York, 1961, pp. 25–27.
265. R. S. Burington, *Handbook of Mathematical Tables and Formulas*, Handbook Publishers, Inc., Sandusky, Ohio, 1948; p. 18.
266. D. D. DeFord, D. E. Smith and E. R. Brown, unpublished work, 1965
267. W. L. Paterson, *Rev. Sci. Instn.*, **34**, 1311 (1963).
268. R. D. Moore and O. C. Chaykowsky, Princeton Applied Research, Technical Bulletin 109, 1963.
269. R. Howe, *Instr. Control Systems*, **34**, 1482 (1961).
270. A. E. Rogers and T. W. Connolly, *Analog Computation in Engineering Design*, McGraw-Hill, New York, 1960, p. 429.
271. E. P. Parry and R. A. Osteryoung, Division of Analytical Chemistry, 148th Meeting, ACS, Chicago, Ill., Sept., 1964.
272. Burr-Brown Research Corporation, *Handbook of Operational Amplifier Applications*, Tucson, Arizona, 1963.
273. J. R. Woodbury, *Electronics*, **34**, 56 (1961).
274. Y. P. Yu, *Electronics*, **31**, 99 (1958).
275. G. C. Barker and R. L. Faircloth, *J. Polarog. Soc.*, **1958**, 11.
276. S. P. Perone, Division of Analytical Chemistry, 148th Meeting, ACS, Chicago, Ill., Sept., 1964.
277. R. S. Burington, *Handbook of Mathematical Tables and Formulas*, Handbook Publishers, Inc., Sandusky, Ohio, 1948, p. 20.
278. H. Gerischer, *Angew. Chem.*, **68**, 20 (1956).
279. H. Gerischer, *Z. Elektrochem.*, **57**, 604 (1953).
280. H. Gerischer, *Z. Physik. Chem. (Leipzig)*, **202**, 302 (1953).
281. H. Matsuda and Y. Ayabe, *Bull. Chem. Soc. Japan*, **29**, 134 (1956).
282. H. Matsuda and Y. Ayabe, *Z. Elektrochem.*, **63**, 1164 (1959).
283. H. Matsuda and Y. Ayabe, *Z. Elektrochem.*, **66**, 469 (1962).
284. H. Matsuda, Y. Ayabe, and K. Adachi, *Z. Elektrochem.*, **67**, 593 (1963).
285. J. Koryta, in *Progress in Polarography* (P. Zuman, ed.), with the collaboration of I. M. Kolthoff, Vol. 1, Chap. 12, Wiley (Interscience), New York, 1962.
286. G. P. Harnwell, *Principles of Electricity and Electromagnetism*, McGraw-Hill, New York, 1949, p. 464.
287. T. S. Gray, *Applied Electronics*, Wiley, New York, 1954, pp. 528, 578, 579.

APPLICATIONS OF CHRONOPOTENTIOMETRY TO PROBLEMS IN ANALYTICAL CHEMISTRY

Donald G. Davis

DEPARTMENT OF CHEMISTRY
LOUISIANA STATE UNIVERSITY IN NEW ORLEANS
NEW ORLEANS, LOUISIANA

I.	Introduction	157
II.	Theory	161
	A. Reversible Processes	161
	B. Irreversible Processes	162
	C. Consecutive and Stepwise Processes	163
	D. Coupled Chemical Reactions	164
	E. Current-Reversal and Cyclic Chronopotentiometry	168
	F. Programmed Current Chronopotentiometry	172
	G. Charging of the Double Layer and Roughness of the Electrode Surface	173
III.	Experimental Methods	175
	A. Apparatus	175
	B. Cells and Electrodes	177
	C. Techniques	180
IV.	Applications	184
	A. Concentration Measurements	184
	B. Electrode Kinetics	187
	C. Chemical Kinetics	188
	D. Adsorption	189
	E. Other Applications	192
	References	193

I. INTRODUCTION

Chronopotentiometry is an electrochemical method characterized by the application of a constant current to an electrochemical cell and the subsequent measurement of the working-electrode potential as a function of time. The resulting chronopotentiogram showing the variation of potential with time can be used for a variety of analytical purposes, including the measurement of concentration of electroactive species, as well as electrode or solution kinetics.

A typical chronopotentiogram, or potential-time curve, is shown in Fig. 1, obtained by the electrolysis of a 0.01 M solution of potassium ferricyanide in $1M$ potassium chloride. A block diagram of the basic chronopotentiometric apparatus is shown in Fig. 2. The constant-current supply maintains the current between the platinum-working and auxiliary electrodes at a preselected constant value. The potential of the working electrode versus the reference electrode is recorded by the *time-Y* recorder. If, as is approximately the case, the ferri-ferrocyanide couple is reversible, the potential of the working electrode (cathode) may be expressed by

$$E = E^{0\prime} + 0.0591 \log(C_0/C_R) \tag{1}$$

at 25°C, where C_O and C_R are the concentrations of the ferricyanide and ferrocyanide ions at the electrode surface. $E^{0\prime}$ is the formal potential of the ferri-ferrocyanide couple in $1M$ potassium chloride.

The original solution, containing only ferricyanide, has a relatively oxidizing potential initially. As soon as the electrolysis is initiated, some of the ferricyanide at the electrode surface becomes reduced to ferrocyanide,

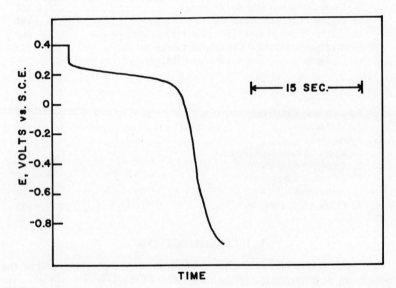

Fig. 1. Chronopotentiogram of 0.01 M potassium ferricyanide in 1 M potassium chloride.

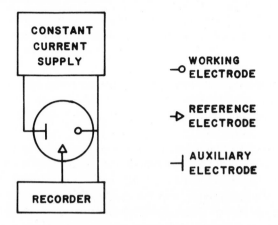

Fig. 2. Chronopotentiometric apparatus.

and the working electrode potential becomes more reducing as may be seen from Eq. (1) (that is, the ratio of C_O/C_R becomes progressively smaller). The potential changes only slowly with time during the period that the ratio C_O/C_R is not too different from 1. As the electrolysis continues, the ferricyanide concentration at the electrode surface decreases to the point where there is no longer sufficient ferricyanide to allow all the constant current to be consumed by ferricyanide. More ferricyanide continually arrives at the electrode surface by diffusion (the solution is unstirred and an excess of electrolyte is provided to prevent convection and migration, respectively), but the magnitude of the current is selected such that the diffusion process cannot long supply the necessary amount of oxidant. Thus the concentration of ferricyanide at the electrode surface drops to zero, and the now "excessive" current forces the electrode to a potential at which a different reaction can occur. A rapid change in potential is noticed. The time from the start of the electrolysis until the rapid potential change is designated the transition time τ. (Fig. 1).

The existance of the transition time was first reported by Sand (*1*) in 1901. He showed that, in an unstirred solution and under the conditions of linear diffusion, the transition time τ (in seconds) is related to concentration by

$$\tau^{\frac{1}{2}} = \frac{\pi^{\frac{1}{2}} n F A D^{\frac{1}{2}} C^0}{2i} \tag{2}$$

where n is the number of faradays per molar unit of reaction, i is the constant current (in amperes), A is the electrode area (in square centimeters), F is the faraday (96,493 coulombs), D is the diffusion coefficient (in square centimeters per second), and C^0 is the bulk concentration in moles per cubic centimeter.

After the work of Sand (*1*), Butler and Armstrong (*2,3*) accomplished intermediary work in the method, and in fact, coined the term *transition time*.

The basic relation expressed in Eq. (2), that is, that the product $i\tau^{\frac{1}{2}}$ is proportional to the concentration of substance reacting at the electrode, has been verified by a number of workers (*4–7*). The credit for pointing up the usefulness of chronopotentiometry to analytical chemists probably belongs to Gierst and Juliard (*7*). The main stimulus for further work in the United States, however, was the monograph of Delahay (*8*) and the back-to-back papers of Delahay and Mamantov (*9*) and Reilley et al. (*10*).

For analytical chemists, the most interesting feature of the Sand equation [Eq. (2)] is, perhaps, that the *square root* of the transition time is proportional to the bulk concentration of the electroactive substance. This property has great analytical utility, but is confusing at first for those used to a linear relationship between the measured variable and concentration. Obviously, the transition time can be greatly changed by variations of the current (or current density). The limits between which the transition time can be adjusted are determined by the conditions that first, convection should not interfere with diffusion, and second, a negligible portion of the applied constant current be used to charge the double layer and alleviate surface effects (that is, electrode oxidation and adsorption). In practice, the extreme range of useful transition times is probably between 1 msec and 2 min. Owing to the state of the art, early workers (*1,4*) attempted to measure transition times on the order of hours, a practice which undoubtedly led to significant errors as a result of convective mass transfer.

Not only was the basic relation between transition time and concentration [Eq. (2)] established early in the twentieth century, but also the appropriate equation for calculating the complete potential time curve for reversible cases (*11*). Although this work preceded that of Heyrovsky by about 30 years, chronopotentiometry is only now being used by electroanalytical chemists. A significant number of important applications have been reported, as well as the complete theory of most important cases. Although chronopotentiometry will probably not replace classical polarography for the reduction of metal ions in the potential range appropriate to the dropping mercury electrode, analysis involving oxidations can be carried out chronopotentiometrically with much better results than polarographically with a rotating platinum electrode, especially in the millimolar-concentration range. More interesting, perhaps, are the recent applications of chronopotentiometry to kinetic measurements, both electron transfer and "chemical," to studies of adsorption on the electrode surface, and as a general electrochemical tool.

II. THEORY

A. Reversible Processes

The chronopotentiogram shown in Fig. 1 may be considered to result from a single electrochemical reaction, without kinetic or catalytic complications. The experimental conditions are selected such that diffusion is the sole means of mass transfer. Additionally, the electrode may be considered a plane, with both oxidized and reduced species soluble in the solution. In a similar case, one with a mercury electrode, solubility in the electrode would replace the latter condition. The Sand (*1*) equation [Eq. (2)] applies to all such cases, whether reversible or irreversible; but the original potential-time relation of Karaoglanoff (*11*) applies only to those cases for which the magnitude of the forward and reverse electrochemical rate constants is large enough that the Nernst equation is obeyed, that is, the electrochemical process may be considered reversible. Such a process is usually symbolized by

$$O + ne \rightleftharpoons R \tag{3}$$

The equation of Karaoglanoff (*11*) for both O and R soluble is

$$E = E_{\tau/4} - \frac{RT}{nF} \ln \frac{t^{\frac{1}{2}}}{\tau^{\frac{1}{2}} - t^{\frac{1}{2}}} \tag{4}$$

where RT/nF has its usual significance, t is the time in seconds, and τ is the transition time. This is analogous to similar equations for polarographic waves, and it may be shown that $E_{\tau/4} = E_{\frac{1}{2}}$ (the polarographic half-wave potential) provided the chronopotentiometry is done with a mercury electrode. $E_{\tau/4}$ is related to the standard potential of the electroactive couple by

$$E_{\tau/4} = E^0 - \frac{RT}{nF} \ln \frac{f_R D_O^{\frac{1}{2}}}{f_O D_R^{\frac{1}{2}}} \tag{5}$$

D being the diffusion coefficient (in square centimeters per second) of the subscripted species and f the activity coefficient. Note that $E_{\tau/4}$ is independent of both current and concentration.

Equation (4) indicates that a plot of $\log \left[(\tau^{\frac{1}{2}} - t^{\frac{1}{2}})/t^{\frac{1}{2}} \right]$ versus E should yield a straight line whose reciprocal slope is 2.303 (RT/nF) or

0.0591/n at 25°C. This has been verified in a number of cases (*10,12*). In the example of Fig. 1, the slope of the straight-line plot was found to be 0.068, indicating a "slight degree of irreversibility."

B. Irreversible Processes

In the derivation of the Sand equation, no assumptions as to the reversibility of the electrode process are introduced. Thus for irreversible electrode reactions, as represented by

$$O + ne \rightarrow R \tag{6}$$

the relationship previously mentioned may be applied to relate transition time to concentration, just as the Ilkovic equation may be applied to both reversible and irreversible polarographic processes.

The electrode process must not, however, be so irreversible that little or no reaction occurs over the available range of potentials. The potential of the working electrode during a "totally" irreversible electrode reaction has been related to the current (*13*); thus

$$i = nAFCk_f^0 \exp\left(\frac{-\alpha n_a FE}{RT}\right) \tag{7}$$

where α is the transfer coefficient of the electrode process, C is the concentration of the electroactive species at the electrode surface, n_a is the number of electrons involved in the rate-determining step, and k_f^0 is the rate constant for the irreversible process in centimeters per second at $E = 0$ (versus the normal hydrogen electrode). This equation is used with the assumptions that the rate of the reverse process is negligible and that the kinetics of the electron-transfer reaction is controlled by a single rate-determining step. By introducing the transition time from Eq. (2), the following potential-time relationship is obtained (*13*):

$$E = \frac{RT}{\alpha n_a F} \ln(\tau^{\frac{1}{2}} - t^{\frac{1}{2}}) + \frac{RT}{\alpha n_a F} \ln \frac{2k_f^0}{(\pi D)^{\frac{1}{2}}} \tag{8}$$

It is apparent from Eq. (8) that a plot of $\ln(\tau^{\frac{1}{2}} - t^{\frac{1}{2}})$ versus E should yield a straight line of slope $RT/\alpha n_a F$. In addition, the $E_{\tau/4}$ for the irreversible case shifts linearly with $\ln C^0$ and with the current density I. If the basic assumption that the reverse process is negligible is not true, no linear $\ln$ plot is found. $E_{\tau/4}$ does vary with current and concentrations but not necessarily in a regular or predictable fashion. This intermediate case (as well as many others) has been discussed by Reinmuth (*14*), who presents a convenient compilation of criteria for various kinetic schemes.

The diagnostic criteria considered above are not sufficient conditions to establish the exact kinetic scheme but can be very useful, especially coupled with current-reversal chronopotentiometry (Sec. II.E) and chemical knowledge.

Values of αn_a and k_f^0 can be determined through the use of Eq. (8) or one of its variations (8) such as

$$E = \frac{RT}{\alpha n_a F} \ln \frac{nFC^0 k_f^0}{I} + \frac{RT}{\alpha n_a F} \ln[1 - (t/\tau)^{\frac{1}{2}}] \qquad (9)$$

which is convenient to work with, since all the unknown quantities can be measured experimentally except αn_a and k_f^0, the kinetic parameters sought. Several cases of this type have been treated experimentally (12). Chrono- potentiometry has certain advantages over polarography for work of this type, since the diffusion process involved is much simpler, electrodes other than mercury can be used, and the reaction can be studied in both directions by reversing the current. Theoretically, chronopotentionmetry should be capable of measuring larger k_f^0's than polarography; but experimentally difficulties, especially at solid electrodes, tend to reduce this advantage [see, for example, (15)].

C. Consecutive and Stepwise Processes

When the solution under investigation contains two or more different electroactive substances with sufficiently different reduction potentials, two or more inflections are observed, provided there exists a difference of at least 0.1 volt in the reduction (or oxidation) potentials of the substances. Equation (2) applies only to the first substance undergoing reaction at the electrode. After the first inflection, when the reaction of the second substance is proceeding, the first substance continues to diffuse to the electrode surface and to react. The current efficiency for the second reaction is there by reduced below 100 per cent, and a longer time is needed to reach the second transition time.

The equation for the relation of the transition time of the second sub- stance to its concentration is (16)

$$(\tau_1 + \tau_2)^{\frac{1}{2}} - \tau_1^{\frac{1}{2}} = \frac{\pi^{\frac{1}{2}} n_2 FAD_2^{\frac{1}{2}} C_2^0}{2i} \qquad (10)$$

τ_2 being measured from τ_1, that is, $\tau_2 + \tau_1$ is the total time to the second inflection point. The equation can be extended to n substances and has been experimentally verified (10,16).

Equation (10) indicates that knowledge of the concentrations or kinds of substances reacting before the substance of interest (substance 2 in this case) is not necessary. Indeed it has been suggested that the τ_2 could be enhanced by the addition of an ion reacting at earlier potentials to increase the lower attainable analytical level (*10*). The upper useful limit of this procedure is reached when the concentration of the added ion is about 2.5 times that of the ion initially present. This procedure, however, suffers from experimental difficulties which essentially negate any analytical advantage, because τ_2 must be measured from a sloping segment of the chronopotentiogram for the added component (*15*).

Similar considerations apply to those substances which react at the electrode in several discrete stages such as vanadium(V) (*17*), copper(II) in hydrochloric acid (*18*), or iodide (oxidation to iodine and iodate) (*19*). In these cases, however, the situation is simplified, since the concentration is identical for each stage, only the various n values effect the relative transition time (*16*).

For a two-stage reduction, we have

$$\tau_2/\tau_1 = (2n_2/n_1) + (n_2/n_1)^2 \tag{11}$$

where n_1 is the number of electrons involved in the first step and n_2 the number in the second step. In the simplest case, when $n_1 = n_2$, Eq. (11) indicates that $3\tau_1 = \tau_2$.

It should be remembered that, for a stepwise reaction, the same relation between the total transition time and concentration holds whether the individual steps are discernable. The over-all transition time can be made the basis for analytical work regardless of the complexity of intermediate steps, provided only that all intermediates result in the same product (*2*).

D. Coupled Chemical Reactions

Frequently, a chemical reaction(s) in solution is closely associated with the electron-transfer reaction. For the simple cases, three situations can be considered:

Preceding chemical reaction

$$Y \underset{k_{-1}}{\overset{k_1}{\rightleftharpoons}} O \overset{ne}{\rightarrow} R \tag{12}$$

Following chemical reaction

$$O \overset{ne}{\rightarrow} R \underset{k_{-1}}{\overset{k_1}{\rightleftharpoons}} Z \tag{13}$$

Catalytic reaction

$$\left[\begin{array}{c} \overset{ne}{\longrightarrow} O \to R \\ R + X \overset{k_1}{\to} O \end{array}\right. \tag{14}$$

Although there are many other combinations, or variations of the above reactions, most work up to now can be roughly broken into these categories.

1. Preceding Chemical Reactions

Gierst and Juliard (7) treated this case by the addition of a kinetic term to Eq. (2), but Delahay and Berzins (13) have given a more rigorous treatment. Assuming that Y is not reduced at the electrode and that standard chronopotentiometric conditions pertain, the rate of reduction of substance O depends on the rate of chemical transformation as well as on the rate of mass transfer of the substances involved. Solution of the resulting boundary value problem gives

$$I\tau_k^{\frac{1}{2}} = \frac{\pi^{\frac{1}{2}}nFC^0D^{\frac{1}{2}}}{2} - \frac{\pi^{\frac{1}{2}}I}{2K(k_1 + k_{-1})^{\frac{1}{2}}}\, \text{erf}[(k_1 + k_{-1})^{\frac{1}{2}}\tau_k^{\frac{1}{2}}] \tag{15}$$

where K is the equilibrium constant for the chemical reaction and τ_k is the measured "kinetic" transition time.

If the argument of the error function is larger than 2, the error function becomes almost equal to unity and Eq. (15) becomes

$$I\tau_k^{\frac{1}{2}} = \frac{\pi^{\frac{1}{2}}nFC^0D^{\frac{1}{2}}}{2} - \frac{\pi^{\frac{1}{2}}I}{2K(k_1 + k_{-1})^{\frac{1}{2}}} \tag{16}$$

This equation predicts that $I\tau^{\frac{1}{2}}$ for the case of a preceding chemical reaction should give a straight line when plotted against I and that $K(k_1 + k_{-1})^{\frac{1}{2}}$ can be measured from the slope (Fig. 3). If K is very large, Eq. (16) reduces to Eq. (2). It must be remembered in this case that the concentration related to τ is that of O actually in solution.

For large values of the current density I, it has been shown (8) that Eq. (15) can be modified to

$$(I\tau_k^{\frac{1}{2}})_{I\to\infty} = \frac{\pi^{\frac{1}{2}}nFC^0D^{\frac{1}{2}}}{2(1 + 1/K)} \tag{17}$$

This shows that in the limit $I\tau^{\frac{1}{2}}$ will again be independent of current density (A in Fig. 3). Whether this is experimentally available depends on the values of the rate constants.

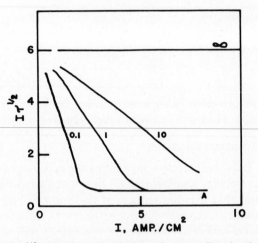

Fig. 3. Variation of $I\tau^{1/2}$ with I for a chemical reaction preceding the electrode reaction. The numbers on each curve are relative values of $K(k_1 + k_{-1})^{1/2}$ as in Eq. (16).

Chronopotentiometry has a real advantage over polarography for studies of this type in that values of $K(k_1 + k_{-1})^{\frac{1}{2}}$ up to 500 $\text{sec}^{-\frac{1}{2}}$ can be measured—perhaps 100 times greater than can be accomplished polarographically *(8,21)*. Naturally, if K is known, the rate constants can be found, since $K = k_1/k_{-1}$. Some of the applications that have been reported are recorded in Sec. IV.C.

More complicated reactions of this general type may also be treated under certain conditions. Delahay and Berzins *(13)* have considered second-order processes such as

$$Y + X \rightleftharpoons O \overset{ne}{\rightarrow} R \tag{18}$$

$$Y \rightleftharpoons O + X$$
$$\tag{19}$$
$$O \overset{ne}{\rightarrow} R$$

Under certain conditions the slope of $I\tau^{\frac{1}{2}}$ versus I plots for the case shown in Eq. (18) decreases when the concentration of X increases and for that of Eq. (19) increases as the concentration of X increases.

Reinmuth *(21)* has considered the cases

$$pY \rightleftharpoons O \overset{ne}{\rightarrow} R \tag{20}$$

$$Y \rightleftharpoons pO \overset{ne}{\rightarrow} R \tag{21}$$

and has shown that in the restricted case in which the equilibrium strongly favors Y, the rate of decrease of $I\tau^{\frac{1}{2}}$ with I is greater than first power for Eq. (21) but less for Eq. (20).

Several-step reactions such as

$$A \xrightarrow{n_1 e} B \xrightarrow{k} C \xrightarrow{n_2 e} D$$

have been considered (*20,22*). This particular case (called the ECE mechanism) is interesting in that different treatments have been developed depending on whether two separate transition times or only one are observable. An example of the latter, *p*-nitrophenol, has been worked out (*23*) and the rate constant for the intervening chemical reaction measured.

It should always be remembered by analytical chemists that, when kinetic complications precede the charge transfer process, $I\tau^{\frac{1}{2}}$ is not directly proportional to concentration. When unwanted kinetic complications appear, it is possible to gain analytical data by taking several chronopotentiograms on each sample and extrapolating a plot of $I\tau^{\frac{1}{2}}$ versus I (or $I^{1/p}$) to zero current density. Analytical results can also be obtained by adjusting I so that τ is constant for all samples (*17*) (if the apparatus used allows this to be done conveniently).

2. Following Chemical Reactions

Reactions of this type may be divided into many subclasses depending on the rates of both the electrode and the chemical reactions. If both are very rapid, the case is that of a regular reversible-electrode reaction, and no information about the following chemical reaction can be obtained. Other cases are of interest, however, since information about the following chemical reaction can be obtained—or at least one case can be distinguished from another. A particularly complete summary of the behavior of the various subclasses of the case under consideration has been given by Reinmuth (*14*).

The situation in which a reversible electrode reaction is followed by a slow reversible chemical reaction has been treated by Delahay (*8*):

$$O \xrightarrow{ne} R \underset{k_{-1}}{\overset{k_1}{\rightleftharpoons}} Y$$

For this case the Sand equation [Eq. (2)] is still valid, since the transition time is governed only by the concentration of O; but the potential-time relationship, which depends on the concentration of R, must be modified to

$$E = E^{0\prime} + \frac{RT}{nF} \ln \frac{\tau^{\frac{1}{2}} - t^{\frac{1}{2}}}{t^{\frac{1}{2}}} - \frac{RT}{nF} \ln \left[\frac{1}{1+K} + \frac{\pi^{\frac{1}{2}}}{2} \frac{K}{1+K} \frac{\mathrm{erf}[(k_1 + k_{-1})^{\frac{1}{2}} t^{\frac{1}{2}}]}{(k_1 + k_{-1})^{\frac{1}{2}} t^{\frac{1}{2}}} \right] \tag{22}$$

From Eq. (22) it is evident that potential-time curves are shifted to less cathodic potentials as the rate constant k_1 for the transformation of R to Y increases. It has been shown (23) that E is a linear function of $\ln I$, and k_1 (under some conditions) can be obtained from the slope of this line.

Cases of the type

$$O \overset{ne}{\rightleftharpoons} R \overset{k}{\to} Y$$

have also been considered (24) and potential-time relations set forth to be used to determine values of k. Of special interest in this connection is the usefulness of current-reversal and cyclic chronopotentiometry for studying cases of this type (Sec. II.E).

3. Catalytic Reactions

If the electroactive species is regenerated by a subsequent chemical reaction [Eq. (14)], the transition time is enhanced and depends on both the concentration of X and the rate constant for its reaction with R (23):

$$\left(\frac{\tau_c}{\tau_d} \right)^{\frac{1}{2}} = \frac{2(kC_x^0 \tau_c)^{\frac{1}{2}}}{\pi^{\frac{1}{2}} \mathrm{erf}[(kC_x^0 \tau_c)^{\frac{1}{2}}]} \tag{23}$$

where τ_c is the measured transition time and τ_d is the transition time for the electrode reaction if no X is present in solution. If τ_c and τ_d are known, the value of the function is then known, and the argument can be read from a table of this function (25); thus k determined.

As in other cases of this type, current-reversal or cyclic chronopotentiometry can often provide useful information about catalytic reactions.

E. Current-Reversal and Cyclic Chronopotentiometry

If the reduction of substance O to R is carried out chronopotentiometrically in the usual way, but the constant current is reversed *at* or *before* the transition time for O, substance R may be reoxidized so that a reverse transition time is recorded, and the potential-time curve thus obtained can be treated quantitatively (16). Current-reversal chronopotentiograms may be seen in Fig. 4 for both a reversible (curve *A*) and an irreversible (curve *B*) case. Provided substance R is soluble in either the solution or the electrode,

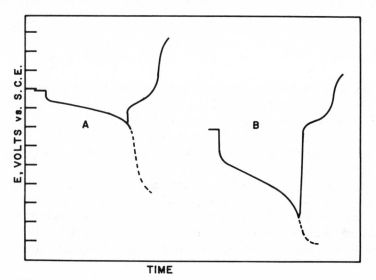

Fig. 4. Current-reversal chronopotentiograms. Curve *A*, reversible case; Curve *B*, irreversible case. The dashed lines indicate the form of the chronopotentiogram if the current had not been reversed.

and the reverse-current density is equal in magnitude to the "forward" current, the reverse-transition time (τ_R) will be equal to one-third of the forward-electrolysis time (t_f). If the current is reversed at exactly the transition time, then

$$\tau_R = \tfrac{1}{3}\tau \tag{24}$$

This decrease in reverse-transition time, as compared to the forward one, is due to the diffusion of a portion of R away from the electrode during electrolysis. If this diffusion is prevented by, for instance, the product R precipitating or plating on the electrode surface, then

$$\tau_R = \tau \tag{25}$$

The reverse potential-time curve for the case in which both O and R are soluble has been derived (*16*). For a reversible-electrode reaction;

$$E = E' + \frac{RT}{nF} \ln \frac{\tau^{\frac{1}{2}} - [(\tau + t')^{\frac{1}{2}} - 2t'^{\frac{1}{2}}]}{(\tau + t')^{\frac{1}{2}} - 2t'^{\frac{1}{2}}} \tag{26}$$

where t' is equal to $t - \tau$, t is the elapsed time since the start of the electrolysis, and τ is the transition time for the forward processes as well as the time of current reversal. The potential analogous to $E_{\tau/4}$ for standard chronopotentiograms is $E_{0.215/\tau'}$ (*26*).

For the case of an irreversible electrode reaction, Eq. (26) becomes

$$E = \frac{RT}{(1-\alpha)n_a F} \ln \frac{\pi^{\frac{1}{2}} D_R^{\frac{1}{2}}}{2k_{b,h}^0} - \frac{RT}{(1-\alpha)n_a F} \ln[(\tau + t')^{\frac{1}{2}} - 2t'^{\frac{1}{2}}] \qquad (27)$$

Since $k_{f,h}^0$ and α can be found from a forward chronopotentiogram [eq. (8)] and $k_{b,h}^0$ and $(1 - \alpha)$ from a current-reversal chronopotentiogram, a complete study of the kinetics of an electrode reaction can thus be made. More important perhaps is the fact that various types of kinetic complications can also be quickly identified by current-reversal chronopotentiometry, as will be seen below.

Current-reversal chronopotentiometry can be used to great advantage in the study of chemical reactions which follow the electrode reaction (27–30).

King and Reilley (29) have considered the application of current-reversal chronopotentiometry to the case of immediate decomposition of the electrochemical products to electroactive species, as generalized:

$$\begin{aligned} A - n_a e &\to Z \xrightarrow{k=\infty} B + C \\ B + n_b e &\to D \\ C + n_c e &\to E \end{aligned} \qquad (28)$$

In such a case, two reverse transition times would be observed, one for B and one for C. For such a case

$$\frac{t_f}{\tau_b} = \left(1 - \frac{n_a}{n_b}\right)^2 - 1 \qquad (29)$$

where t_f is the time of forward electrolysis and τ_b is the transition time for the reduction of species B. Similarly,

$$\frac{t_f}{\tau_b + \tau_c} = \left(1 - \frac{n_a}{n_b + n_c}\right)^2 - 1 \qquad (30)$$

These equations were extended for the reduction of any number of produced electroactive species. Equations (29) and (30) were used to treat the data of Geske for tropilidene (31).

The case in which the rate of the chemical step is not infinitely rapid has been treated by Testa and Reinmuth (30). The general situation may be symbolized by

$$\begin{aligned} A - n_a e &\rightleftharpoons Z \xrightarrow{k} C \\ Z + n_z e &\to D \\ C + n_c e &\to E \end{aligned} \qquad (31)$$

where A, Z, and C are electroactive. The actual case treated was

$$HO\!-\!C_6H_4\!-\!NH_2 - 2e \rightleftharpoons O\!=\!C_6H_4\!=\!NH$$

$$O\!=\!C_6H_4\!=\!NH + 2H_2O \rightarrow O\!=\!C_6H_4\!=\!O + NH_3$$

The following equation was derived and applied (*27,30*):

$$2\,\mathrm{erf}(k\tau_1)^{\frac{1}{2}} = \mathrm{erf}\,k(t_f + \tau_1)^{\frac{1}{2}} \tag{32}$$

where t_f is the time of forward electrolysis, and τ_1 is the first reverse-transition time (for the re-reduction of $O\!=\!C_6H_4\!=\!NH$). Because of the complexity of Eq. (32), results are best calculated by means of a constructed working curve.

Equation (32) has been generalized for variations in forward and reverse *n* values and for catalytic reactions under certain restrictions (*28*).

Two-component systems have also been studied by current-reversal chronopotentiometry (*26,32*) and can be treated quantitatively—at least under the assumption that all species have equal diffusion coefficients.

A variation of current-reversal chronopotentiometry is cyclic chronopotentiometry in which the current is reversed repeatedly at potentials taken at the transition times of various waves. Several cases have been examined, including diffusion-controlled single-component systems (*33*), electron transfer followed by a chemical reaction (*34*), and multicomponent systems and stepwise reactions (*35*). Although these cases can be treated quite exactly, the transition-time relation

$$\tau_n = a_n\tau_1 \tag{33}$$

where τ_n is the *n*th transition time and a_n relates τ_n to τ_1, the complete solution for *n* greater than 2 requires an iterative method for solving the necessary equations, even in the simple single-component case (*33*). Fortunately, computers can be used and FORTRAN programs are included in the original papers, as well as instructive data tables. Kinetic cases particularly can be well treated by this method.

By way of illustration of a simple case, consider the reduction of a single species in the situation that both oxidized and reduced forms are soluble (or reduced form insoluble), currents in both directions are equal, and no reduced form is originally present. The values of a_n [see Eq. (33)] are given for *n* values up to 10 in Table I.

Details of the theory, calculations, and experiments in the application of chronopotentiometric methods to kinetic problems will be considered in a subsequent chapter in this series.

TABLE I

n^a	a_n (R soluble)	a_n (R insoluble)
1	1.000	1.000
2	0.333	1.000
3	0.588	1.174
4	0.355	1.174
5	0.546	1.263
6	0.366	1.263
7	0.525	1.319
8	0.373	1.319
9	0.513	1.359
10	0.378	1.359

[a] All odd n's are oxidations and even n's are reductions.

F. Programmed Current Chronopotentiometry

Although the majority of chronopotentiometric research has involved the use of constant current (often with reversal), the use of step changes in current magnitude (*32*), current which is a square-root function of time (*36–38*), a constant-power function of time (*39,40*), and an exponential function of time (*40*) have been considered and the appropriate equations and apparatus reported.

Simply increasing or decreasing the electrolysis current during the recording of a chronopotentiogram has no analytical advantages and in some cases, especially current decrease, may lead to unnecessary complications (*32*). In the study of kinetics and mechanisms of electrode processes, certain useful applications are to be found, especially by use of current reversal—a subcase of step changes in current, (Sec. II.E). Current cessation also holds some promise for the study of chemical reactions between electrochemically produced reactants (*26,32*).

For the practicing analytical chemist chronopotentiometry with current which is a square-root function of time holds the most interest, since under these conditions the transition time τ is directly proportional to concentration (*36*). The equations which apply for the current-time relation are

$$I = Bt^{\frac{1}{2}} \tag{34}$$

where B is a constant of proportionality; for the transition-time concentration relation,

$$\tau = (2nF/B)C^0(D/\pi)^{\frac{1}{2}} \tag{35}$$

and for the relation between potential and time (reversible reaction),

$$E = E_{\tau/2} + \frac{RT}{nF} \ln \frac{\tau - t}{t} \tag{36}$$

Computations are thus less complex and multicomponent mixtures easier to handle (*38*). This method has not found general usefulness, however, presumably because of the electronic complexity of the apparatus to generate currents which are a square-root-function of time.

Other power functions have also been treated (*39,40*), and it has been shown that for the general current program $I = Bt^r$,

$$B\tau^{r+\frac{1}{2}} = nFD^{\frac{1}{2}}C^0\phi \qquad (r > -\tfrac{1}{2}) \tag{37}$$

where r is the power used and $\phi = \Gamma(r + \tfrac{3}{2})\Gamma(r + 1)$, Γ being the gamma function, numerical values of which are available in tables. Studies completed so far seem to indicate that no appreciable analytical advantage is to be found using complex current functions except that somewhat sharper transition times can be recorded for irreversible systems (*40*) and that adsorption studies (*41*) can be conducted with increased understanding.

G. Charging of the Double Layer and Roughness of the Electrode Surface

Nonfaradaic processes such as charging of the electrical double layer can consume some of the constant current applied during a chronopotentiometric experiment. In addition, oxidation or reduction of the electrode surface and roughness of the surface may distort chronopotentiograms or consume current. Adsorption and surface reactions will be considered subsequently.

1. Charging of the Double Layer

Even when no oxidizable or reducible substance is present in solution, it is found experimentally that current passes when the potential of the electrode is changed. This effect is caused by the formation of an electrical double layer at the electrode surface (*42,43*). The double layer consists of several sublayers. The first (inner Helmholtz layer) consists of solvent molecules and nonhydrated specifically adsorbed, anions (or cations depending on the potential). In the outer Helmholtz layer the ions are hydrated. When the potential of the electrode changes, there occurs a redistribution of the ions near the electrode. This process leads to a flow of current and to some extent may be considered a capacitance (see Chapter 3).

The effect of double-layer charging on chronopotentiometry has been considered by a number of workers (*9,15,44–46*). The quantity of electricity used for the charging process q_c can be related to the quantity used for the faradaic process q_f by

$$\frac{q_c}{q_f} = \frac{ic_d\,\Delta E}{kC^2} \tag{38}$$

where c_d is the differential capacity and k a constant for a given substance and experimental conditions. The effect of the double layer (*44,45*) and consequent distortion of chronopotentiograms (*44*) obviously becomes more important as the current increases and the concentration decreases. The influence of the concentration in Eq. (38) is particularly important. Work at low concentrations can best be accomplished by adjusting the current to as low a value as possible. The limiting factor here is that, at low currents and long transition times, convection may begin to become important. The double-layer distortion limits the lower concentration level that can be measured with reasonable accuracy chronopotentiometrically to about 10^{-5} M.

Bard (*45*) has suggested a method for correcting chronopotentiometric data for double-layer effects, electrode oxidation, or reduction and adsorption. The following equation indicates the method:

$$\frac{I\tau^{\frac{1}{2}}}{C^0} = \frac{nFD^{\frac{1}{2}}\pi^{\frac{1}{2}}}{2} + \frac{B}{C^0\tau^{\frac{1}{2}}} \tag{39}$$

where B is an over-all correction factor. If $I\tau/C$ (observed) is plotted against $\tau^{\frac{1}{2}}$ for a particular solution, a straight line results, the slope of which yields the true chronopotentiometric constant and has an intercept of B/C^0. Each individual value of $I\tau/C^0$ can then be corrected by subtracting B/C^0 and dividing by $\tau^{\frac{1}{2}}$. If one has knowledge of the double-layer capacitance and the amount of electrode oxidation, measurements of the amount of adsorbed material can be made.

2. Surface-Roughness Effects

A certain amount of debate has appeared in the literature (*15,47,48*) as to whether roughness of electrode surface is likely to effect chronopotentiometric results. The general conclusion seems to be that surface roughness will almost always be less important than double-layer effects, even on rough electrodes. At least until definite experimental evidence is brought forth to the contrary, it appears that roughness effects on visually smooth electrodes may be considered quite minor, if they exist at all.

III. EXPERIMENTAL METHODS

A. Apparatus

1. Current Supplies

For most chronopotentiometric work, a constant-current supply (see Fig. 2 for general apparatus diagram) may easily be constructed from several 45-volt batteries and a series of resistors arranged in such a way that currents from a few milliamperes to a few microamperes may be obtained (*10*). The best results are obtained if the resistors are of fairly high wattage, so that heating effects are minimized. The constant current can be measured by inserting a standard resistor (R_s in Fig. 5), in the circuit and measuring the voltage drop across this resistor with a potentiometer during the electrolysis.

If high currents are used continually, it may be convenient to replace the batteries with a variable high-voltage power supply of good quality, such as the General Radio Type 1204-B (Concord, Mass.). In this case it is also advantageous to provide the circuit with a variable resistor to act as a dummy cell (see Fig. 5). This resistor is adjusted to approximate closely the cell resistance, so that current variations will be minimized on connecting

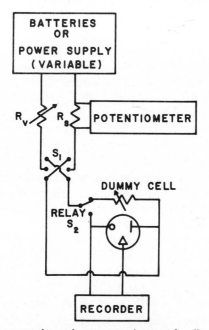

Fig. 5. Chronopotentiometric apparatus (see text for discussion).

the electrolysis cell into the circuit by means of a relay S_2. The total cell resistance should be small compared to R_v, the swamping resistances, so that minor variations in the cell resistance during the electrolysis will not cause noticeable current variations.

In this relatively simple arrangement it is convenient to include in the circuit a current reversing switch S_1, so that current-reversal experiments may be performed (Sec. II.E). For high-speed current-reversal or cyclic chronopotentiometry, the reversal switch S_1 could be replaced with an automatic current-reversing circuit (33).

Constant-current sources of high quality can also be constructed from operational amplifiers (49) by placing the cell in the feedback loop of an amplifier, the input of which is a known potential through a known resistor. If the potential is constant, a constant current will pass through the cell. The main advantage of this type of circuit is not for constant-current chronopotentiometry, but rather for those methods which make use of variable current, such as power of time chronopotentiometry (40).

2. Recorders

A great variety of recorders have been applied to chronopotentiometric studies. The normal criteria of range, expense, etc., can be applied. Standard strip-chart recorders are usually adequate, provided the chart drive is constant with time and the pen speed is at least 1 sec full scale. A preamplifier with a high-input resistance, such as a pH meter, should be used if the current in the recorder would otherwise be great enough to perturb the cell. The author has used a Sargent (Chicago, Ill.) potentiometric variable-range recorder for several years and found it to be satisfactory. The small Varian (Palo Alto, Calif.) G-10 recorders are also satisfactory for many applications.

If an X-Y recorder is available, it can be used with the time-base mode of operation, care being taken to check the accuracy of the time base periodically. These recorders have the advantage of chronopotentiograms which are easier to store than those obtained with strip-chart recorders.

For the measurement of transition times below 1 or 2 sec, an oscilloscope of the appropriate type can be used.

For routine applications it is possible to measure transition times without recording chronopotentiograms (10). This can be accomplished by viewing the scale of a pH meter and manually recording the time from the start of the electrolysis until a certain preselected potential is reached. An automatic instrument has been proposed (50) which will accomplish the same thing by using an electronic-triggering relay to stop a timer when the desired potential is reached.

B. Cells and Electrodes

Ordinarily, cells with three electrodes are used for chronopotentiometry: the working electrode, the auxiliary electrode to complete the electrolysis, and the reference electrode. Typical cells are shown in Figs. 6 and 7. Ideally, the geometry of the cell is arranged in such a way that the current density at the working electrode is as uniform as possible. This is best achieved, in the case of plane electrodes, by orienting both the working and reference electrodes perpendicular to the axis of the cell. For spherical or cylindrical electrodes, this point of design is often violated for the sake of convenience of cell construction.

The reference electrode is usually fitted with a small tip (Luggin-type capillary) which is placed within a few millimeters of the electrode surface to reduce the ohmic drop between the unpolarized reference electrode and the working electrode while relatively substantial currents are passing through the cell.

The material chosen for electrode construction depends on the type of process to be studied. Platinum, gold, or carbon electrodes are usually used for anodic processes or reduction of fairly strong oxidizing agents. Mercury electrodes are generally used for other cathodic processes because of their high hydrogen overvoltage. Mercury passivated with a thin layer of mercurous halide has, however, been used for oxidations at potentials where mercury would normally dissolve (*51,52*).

The most satisfactory type of platinum electrodes are those which are constructed by sealing a disk of platinum foil in a glass tube (*44*), such as

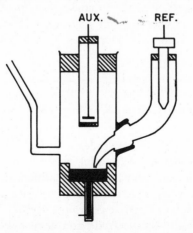

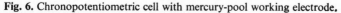

Fig. 6. Chronopotentiometric cell with mercury-pool working electrode.

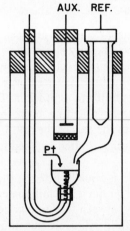

Fig. 7. Chronopotentiometric cell with shielded platinum working electrode.

shown in Fig. 7. The tube extends several millimeters beyond the disk on the solution side to help assure the conditions of semiinfinite linear diffusion and prevent convection. It is convenient to be able to orient this electrode facing either up or down, so that convection is prevented when the electrolysis products are either lighter or heavier than the original solution. Bard (*44*) has shown the extent of this effect, and that, with a properly oriented electrode, transition times up to 300 sec can be measured without convection problems. The design of a particularly convenient cell with an orientable electrode has been given by Mark (*53*).

Perhaps the easiest electrode to construct is made by simply sealing a piece of noble metal wire in a soft glass tube and blocking the end of the wire with a glass bead. Such electrodes have the disadvantage that convection is not prevented and that the cylindrical field of diffusion must be taken into account. Equations are available (*54*) for this purpose. The Sand equation for cylindrical diffusion is

$$\frac{i\tau^{\frac{1}{2}}}{C^0} = \frac{\pi^{\frac{1}{2}}nFAD^{\frac{1}{2}}}{2} \bigg/ \left[1 - \frac{\pi^{\frac{1}{2}}D^{\frac{1}{2}}\tau^{\frac{1}{2}}}{4r} + \frac{D\tau}{4r^2} - \frac{3\pi^{\frac{1}{2}}D^{\frac{3}{2}}\tau^{\frac{3}{2}}}{32r^3} + \cdots\right] \qquad (40)$$

where r is the radius of the electrode.

Graphite electrodes are also useful, especially since their anodic range is generally greater than that for either platinum or gold. Spectrographic graphite rods are often used, but even these must be impregnated with a wax (such as ceresin) to decrease their porosity (*55*). The end surface, used as the working area of the electrode, can be renewed by cutting. Pyrolytic graphite can also be used with fewer problems involving porosity (*56,57*).

Another type of carbon electrode is the carbon-paste electrode of Adams (*58*). A Teflon cup, with a wire contact at the bottom, is filled with a paste of powdered spectrographic carbon and Nujol, bromoform, or other viscous and inert organic liquid. The surface is smoothed with a spatula. The cup is then immersed in the solution to be electrolyzed. Although this electrode is generally characterized by a low background current over a wide range of potentials, certain disadvantages have been noticed during various studies. Many electrode reactions are more irreversible at carbon paste than at platinum (*59,60*) and also some substances—bromine, for instance (*60*)—can dissolve in and react with the electrode in an undesirable fashion.

Three types of mercury electrodes have found general use for chronopotentiometry: the mercury pool (Fig. 6), the hanging drop, and the dropping mercury electrode (DME). The mercury pool should be of uniform cross section so that reproducible electrode areas can be achieved without critical adjustment of the mercury level. The effect of curvature of the mercury surface must also be considered (*60a*). If a glass cup or cell is used, creeping of solution between the walls of the cup and the electrode can be avoided by coating the cell with silicone. Areas are usually reproducible to about 2 per cent. If a mercury pool of too small a diameter is used, the electrode area may change appreciably with potential as a result of variations in surface tension (*61*).

The hanging mercury-drop electrode has been used extensively because of its convenience, reproducible area (about 0.2 per cent), and the saving that results in the quantity of mercury consumed. The simplest electrode of this type was developed by Gerischer (*62*). He used a small plastic spoon to collect one or more drops of mercury from a dropping mercury electrode and then hung the accurately known amount of mercury on a small platinum tip. This electrode has been described in detail by Ross and co-workers (*63*). Its major analytical use is probably for stripping analysis rather than chronopotentiometry.

Hanging drops may also be formed by the expulsion of a small quantity of mercury from a microsyringe fitted with a micrometer. These devices are now commercially available. Although their area is not as reproducible as the Gerischer type and although they are subject to problems of temperature variation, difficulties resulting from the solubility of platinum in mercury are avoided.

The DME can also be used for chronopotentiometry, provided some means is used to synchronize the measurement with drop formation. Perhaps the best means of doing this is a small magnetic hammer connected to the timing circuit (*64*). If the drops are dislodged near the end of the drop life, variations of area with time are minimized.

C. Techniques

1. Measurement of Transition Times

The accuracy of chronopotentiometric methods depends on the exactness with which transition times can be measured. Assuming that the recorder or other timing device is accurate, the problem becomes how to measure τ's from the recorded curves. Because of the curvature caused by the charging of the double layer, this task is not always an easy one. Figures 8 and 9 can be used to illustrate some of the most used methods. Figure 8 shows a typical reversible chronopotentiogram without much distortion. The transition time can be measured in one of several ways. The simplest method is to draw the tangent lines *CF* and *AD* and locate point *Z*, where the potential-time curve departs from a straight line. The transition time is then taken from *Z* to *AD*. In a well-behaved system the potential at which *Z* is located will be very close to constant. In such systems one may obtain precise results simply by measuring τ at some

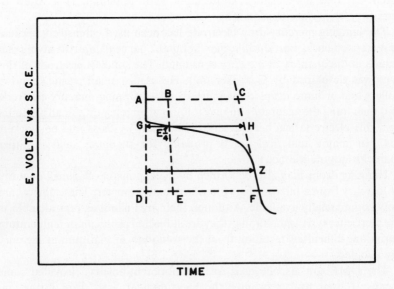

Fig. 8. Methods of transition-time measurement: "reversible" curve.

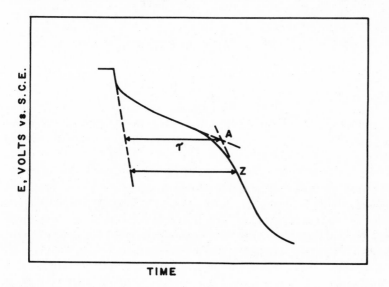

Fig. 9. Methods of transition-time measurement; irreversible or distorted curve.

constant potential near, but not necessarily at, Z. Precise results are sometimes sufficient for concentration measurements. However the method of finding Z discussed above is to be preferred if n values or, more important, if diffusion coefficients are of interest. In the author's experience this is superior to other methods in accuracy and simplicity.

Another method (*9*) for handling undistorted curves is illustrated in Fig. 8 and is inspired by polarographic practice. First, lines AD and CF are constructed (these lines are not necessarily parallel), and the horizontal lines DF and AC are drawn. The points B and E are marked such that $AB = \frac{1}{4}AC$ and $DE = \frac{1}{4}DF$. BE is then drawn and crosses the potential-time curve at the quarter-wave potential. GH is then taken as the transition time. Obviously these two methods give slightly different results, and no theoretical justification exists for choosing one over the other.

If the potential-time curve is quite distorted (Fig. 9), a practical method (*15*) is to construct the two tangent lines necessary to locate point A. The transition time is then taken from A to the tangent line for the beginning portion of the curve. The method discussed before of locating Z is preferred (*114*), τ being taken from Z to the extension of the starting portion of the potential-time curve. This method tends to yield too large τ's when the chronopotentiogram is very distorted.

2. Electrode Pretreatment

One of the main advantages of the hanging drop or dropping mercury electrodes is that a fresh surface is easily and often provided. Unfortunately, this is not practical with solid electrodes such as platinum. Cognizance should be taken of the fact that noble metal surfaces, such as platinum, can become oxidized (65,66) and that the presence of an oxide film or layer of oxygen can effect not only the value of the transition time (19) but also rates of electrochemical reactions (67) and thus the reversibility of the process.

Two problems thus must be considered. First, there is the correction necessary for the reduction (or oxidation) of the electrode, if this occurs concommittantly with the reaction under study. Methods for making this correction are discussed in Section II.G.1 [see also (19)]. The second problem is to establish methods of pretreatment which may be used to control the effects of electrode oxidation in such a way that meaningful results can be obtained.

Much discussion has appeared in the literature about whether an actual oxide of platinum is formed on the electrode surface or if oxygen is held by some other mechanism. Although this question will not be discussed in detail here, some of the points made by Anson (68) can be applied. He has found that platinum electrodes can be considered to have three different types of surface characteristics. Electrodes which have been subjected to oxidizing treatments such as anodization or soaking in a solution of a strong oxidant are classified as *oxidized* electrodes. This surface state is not desirable because almost all known electron-transfer processes are decreased in reversibility at such an electrode. In certain cases, it is necessary to use oxidized electrodes, as, for example, in the chronopotentiometric analysis of ceric solutions (69).

If an oxidized electrode is reduced to potentials not quite reducing enough to cause the evolution of hydrogen, a particularly active state of the electrode is achieved. Reduction of the oxidized electrode very likely produces a layer of freshly formed, finely divided platinum which provides the active surface. Electrodes of this type are very desirable, since most reactions achieve their highest state of reversibility on such surfaces. In addition, if the electrode is carefully brought into this state of prior oxidation followed by reduction, transition times are quite reproducible (70). Care must be taken to avoid oxidation of the active platinized electrodes in the course of a series of oxidations. Reversal of current after each oxidation usually accomplsihes the desired electrode surface in a reproducible manner. Hydrogen should not be generated at platinum electrodes to be used for other purposes, since the electrode can become deactivated, and at lower currents some noticeable quantity of electricity is consumed by the oxidation of hydrogen during subsequent experiments.

The final type of electrode surface may be achieved by chemically stripping the finely divided platinum from a "platinized" electrode. Indeed, the active electrode will revert to this form over a period of time either by rearrangement of the platinum atoms on the surface or the gradual adsorption of impurities. This type of surface is again undesirable, since the rates of many electrode reactions are slowed down.

The author recommends the following pretreatment for platinum electrodes to be used for chronopotentiometry:

1. Anodize the electrode at a potential oxidizing enough to cause oxygen evolution in the electrolyte to be used. This should be done at the start of each series of runs.

2. Reduce the electrode with several short-current pulses from the chronopotentiometric apparatus—not allowing the potential electrode to become reducing enough to reduce hydrogen ion or water.

3. Replace the electrolyte in the cell with the solution of sample and electrolyte. If, and only if, the sample is to be oxidized, reverse the current after each chronopotentiogram—again being careful not to evolve hydrogen.

Sometimes, especially in the case of organic substances, the electrode becomes fouled with oxidized reaction products. In such cases the pretreatment must be started over at step 1 after each chronopotentiogram, or else a comparable cleaning procedure developed.

3. Differential and AC Chronopotentiometry

Differentiation of potential-time curves has been suggested (*71*) as a method of determining transition times; such curves, if they are of the dE/dt type, show a maximum at the inflection of the potential-time curve or change sign if of the d^2E/dt^2 type. Several devices to accomplish differentiation of various curves have been reported (*71,72*), but the method is not much used at present, probably because of lack of sensitivity.

An alternative method is to superimpose a small sinusoidal current on the constant current through the cell (*73*). Differentiation is accomplished in reversible processes, since the magnitude of the sinusoidal component of the potential is proportional to dE/dt.

4. Thin-Layer Chronopotentiometry

When only small volumes of the solution to be analyzed are available, problems arise when cells of standard design are used. Chronopotentiograms can, however, be taken on single drops (*74*) with a volume of only

0.05 ml. The cell used consists of a platinum plate with a small depression or coating to hold the drop in place. This plate serves as the working electrode. Microreference and auxiliary electrodes are made to contact the drop from above. A mercury electrode of special design can also be used. The standard chronopotentiometric relationships apply to this apparatus.

Some interesting work has also been reported in which a thin layer of solution is confined next to the working electrode surface (75,76). The design of the special cells used for this work can be found in the literature (76). The interesting point is that the thickness of the solution is on the order of the diffusion-layer thickness, so that the assumption that semi-infinite diffusion exists is no longer valid. In such a situation the equation analagous to the Sand equation is:

$$\tau = \frac{nFAlC^0}{i} - \frac{l^2}{3D} \tag{41}$$

where l is the cell thickness, that is, the distance from the electrode surface to a parallel inert plate. Since conditions can be selected such that the second term on the right side of Eq. (41) is negligible, n values may be measured without the knowledge of diffusion coefficients. Thin-layer chronopotentiometry also has other advantages:

1. Chronopotentiograms are often less distorted than the corresponding semiinfinite-diffusion case.

2. Kinetic studies, particularly of slower reactions, can be carried out, since the products of the electrode reaction are held near the electrode and thus available for further study.

3. Studies of adsorbed reactants can be made, since in many cases an appreciable fraction of a species in the cell may be removed by adsorption.

The difficulty of this method is that the cells are hard to make and fill, but nevertheless the method shows great promise.

IV. APPLICATIONS

A. Concentration Measurements

Although the number of applications of chronopotentiometry is not large compared to polarography, a substantial amount of work has been accomplished. A number of substances that have been studied are summarized in Table II. Many of the references listed there are not necessarily only concentration studies. Since chronopotentiometry is just now emerging from the exploratory stage, much of this work has been involved in the

TABLE II

Chronopotentiometric Analysis of Various Substances

Substance	Electrode[a]	Remarks	Ref.
Acetic anhydride-acetic Acid	Pt: Ox, Red	Solvent study	77
p-Aminophenol	Pt: Ox and reversal	Hydrolysis kinetic study	30, 34
Anthracene	Pt: Ox	0.1 M NaClO$_4$	78
Antioxidants (various)	Graphite: Ox	Acetonitrile and ethanol solvents	79
Ascorbic acid	Graphite: Ox	pH 3.6	55
Benzaldehyde	DME (slow): Red	pH 4 Alternate 2-product reaction	80
Bromide	Pt: Ox		60
	Carbon paste: red		
	Ag: Ox	AgBr-produced	23
Bromine	Pt: Red		60
	Carbon paste: red	Br$_2$ reacts with carbon paste	
Cadmium	Hg: Red	1 M KNO$_3$ and others	10, 12, 35, 50
Cerium(IV)	Pt: Red	1 M H$_2$SO$_4$ with Fe(III)	17, 69
Chloride	Ag: Ox	AgCl-produced	23
Chromium(VI)	Pt: Red-Ox		97
	Hg: Red		12
	Pt: Red	1 M H$_2$SO$_4$	17
Cobalt(II)	Hg: Red	1 M KCl	12
	Pt: Ox	EDTA, adsorption effects	81
Cobalt(III)	Pt: Red	EDTA, adsorption effects	82
Copper(II)	Hg: Red-Ox	0.1 M KCl or 0.1 M NH$_3$–0.1 M NH$_4$Cl	10, 35
Cyanide	Ag: Ox	Ag(CN)$_2^-$-produced	23
Diphenylpicryl-hydrazyl	Pt: Ox-Red	Acetonitrile–0.1 M NaClO$_4$	83
Formaldehyde	Pt: Ox	Acid and base solutions	88
Formic acid	Pt: Ox	Oscillations noted	88
Ferric	Pt: Red	dil. HCl or H$_2$SO$_4$	17, 50, 68, 69
Ferricyanide	Pt: Red	1 M KCl	12, 50
Ferrocene	Hg: Ox	Acetonitrile–0.2 M LiNO$_3$	86, 87
	Pt: Ox	Substituent study	
Ferrocyanide	Pt: Ox		50
Ferrous	Pt: Ox		50, 84
Gallium(III)	Hg: Red-Ox	7.5 M KSCN	89
Hydrazine	Pt: Ox	0.1 M H$_2$SO$_4$ Oscillations noted	90
Hydrogen ion	Pt: Red	1 M KCl	85

TABLE II—continued*

Substance	Electrode	Remarks	Ref.
Hydrogen peroxide	Pt: Ox-Red	$0.5\ M\ HClO_4$ Oscillations noted	91
Hydroquinone	Pt: Ox	pH 7	44, 45
	Pt: Ox (also quin- one Red)	$1\ M\ H_2SO_4$	54
	Graphite: Ox	pH 5.5	55
	Pt: Ox	H_2SO_4 to 16 M	92
Hydroxylamine	Pt: Ox	$1\ M\ H_2SO_4$, pH5.9 $0.1\ M\ NaOH$	93
Iodate	Hg: Red	$1\ M\ NaOH$	13
	Pt: Red	Various solutions Effect of electrode	95
		Oxidation	109
Iodide	Pt: Ox	$1\ M\ H_2SO_4$	19, 44, 45, 94
Iodine	Pt: Red	$1\ M\ H_2SO_4$	94
Lead(II)	Hg: Red	Dilute acid	10, 35, 96
	Pt: Red	$0.2\ M\ NaNO_3$	44, 45
	Pt: Red-Ox	$0.1\ M\ HClO_4$	97
Manganese(VII)	Pt: Red		50
	Pt: Red	$1\ M\ H_2SO_4$–Mn in steel	17
Mercaptobenzo- thiazole	Pt: Ox	Borate buffer	84
Nickel	Hg: Red	$1\ M\ KCl$ or $0.5\ M\ KCNS$	12
Nitrogen oxides as NO, HNO$_2$	Pt: Ox-Red	Strong H_2SO_4	98
Oxalate	Pt and Au: Ox	$1\ M\ H_2SO_4$	99
	Pt: Ox	Adsorption occurs Electrode oxidation suppresses reaction	100
Oxygen	Hg: Red	$1\ M\ NaOH$ or acetate buffer	12
	Hg: Red (2 steps)	$1\ M\ LiCl$	16
	Pt: Red		101
	Au: Red		102
Phenylenediamines	Pt: Ox	H_2SO_4 Proton dissociation measured	103
Platinum(II) Platinum(IV)	Pt: Red-Ox	$1\ M\ HCl$ Pt(IV) reduced to Pt(III)	104
Silver(I)	Pt: Red	$0.2\ M\ NaNO_3$	44, 45, 50
	Pt: Red-Ox	$1\ M\ HClO_4$	97
Sulfanilamide	Graphite: Ox	pH 4.6	55
(S$_2$O$_8$)	Pt: Ox	$1\ M\ H_2SO_4$	78

TABLE II—*continued*

Substance	Electrode	Remarks	Ref.
Sulfate, peroxide	Pt: Red	Various solutions Oscillations	*105*
Thallium(I)	Hg: Red	1 M KNO$_3$	*12*
	Hg: Red	0.25 M NH$_4$OAc in glacial acetic acid	*106, 110, 111*
Titanium(IV) Hydroxylamine	Hg: Red	Catalytic	*34*
Toluene-2,4-diamine	Pt: Ox	pH 7.6	*84*
Triethylamine	Pt: Ox	0.1 M PbNO$_3$ in dimethyl-sulfoxide	*107*
Uranium(IV)	Pt: Ox	1 M HClO$_4$	*70*
Uranium(V)	Hg: Red (2 steps)	1 M KCl	*16*
	Hg: Red	NaClO$_4$–HClO$_4$ Dispro-portionation study	*108*
Vanadium(V)	Pt: Red	1 M H$_2$SO$_4$	*17, 109*
Zinc	Hg: Red	0.1 M KNO$_3$	*10*

ᵃ Ox = oxidation; Red = reduction.

testing of apparatus and electrodes, as well as studies on kinetics and adsorption discussed subsequently.

As a strictly analytical tool, chronopotentiometry is definitely not as useful as polarography. The concentration range that can be measured does not extend below about 10^{-4} M if accuracy better than a few per cent is desired. Under optimum conditions chronopotentiometry is capable of measuring concentrations with an accuracy of 0.5 to 1 per cent (*69*). Work with electrodes other than mercury can be carried out with results generally superior to polarography with rotating electrodes. Thus chronopotentiometry may be the method of choice for work at anodic potentials, although cyclic voltammetry and the rotatingdisk electrode are also becoming extensively used.

B. Electrode Kinetics

Chronopotentiometry has not been much used for standard studies of rates of electron transfer even though the theory has been well developed (see Sect. I.B). Exploratory work in the measurement of k^0's and α's, which was carried out by Delahay's group (*12,113*), included studies of iodate, oxygen, chromate, nickel, and cobalt. Comments have been reported (*14*) on the theory of irreversible electron transfer occuring at such a rate that the reverse reaction must be considered. Values for α have been measured for the bromine-bromide couple both on platinum and carbon-paste electrodes (*60*).

Various methods for measuring transition times of irreversible chronopotentiometric waves have been compared and useful comments on data handling reported (*112*). The best results were obtained with an algebraic method of calculating the transition time. A least-squares analysis of chronopotentiometric curves was accomplished with the aid of a digital computer.

The fact that chronopotentiometry has not been extensively used to measure the kinetic parameters k^0 and α for irreversible reactions is undoubtedly due to the development, shortly after chronopotentiometry was developed, of better and more rapid techniques for this purpose. Such techniques include voltammetry with linearly varying potential, pulse polarography, and faradaic rectification. These methods are all potentially capable of measuring k^0's larger than can be investigated with chronopotentiometry, even though chronopotentiometry is faster than polarography. In many cases, however, chronopotentiometry can be an important adjunct to other methods for the study of irreversible reactions (*89*).

C. Chemical Kinetics

One of the major applications of chronopotentiometry to date has been the study of chemical reactions either preceding or following electron-transfer reactions. The theory applicable to such cases has been discussed in Sec. II.D.

Certainly chronopotentiometry can be applied to chemical kinetic studies in a way analagous to any other concentration-measuring technique. This approach is illustrated by a study of the reaction between ceric ion and tartaric acid (*69*). The main advantages of chronopotentiometry, however, involve investigation of chemical reactions closely coupled with electron-transfer reactions.

Reactions often studied which precede the electron transfer include both the addition or loss of hydrogen or hydroxide ion and the dissociation of metal complexes. Mark and Anson (*103*) have investigated the oxidation of phenylenediamines in acid solution and measured the rate constants k_1 and k_{-1} for

$$H_2A^{2+} \underset{k_{-1}}{\overset{k_1}{\rightleftharpoons}} HA^+ + H^+ \tag{42}$$

where A signifies a phenylenediamine-type molecule. Similarly, the behavior of o-phthalic acid has been investigated in acidic solutions (*114*) and the rate of dissociation of HCN measured (*115*). The reaction of methanol with hydroxide ion has also been characterized (*88*).

Of great interest is the work which has been conducted with metal-ion complexes. It is possible to measure the rates of complex-ion dissociation, such as that for cadmium cyanide (*7*), but in other cases it can be concluded that some complexes (copper-ethylenediamine, for instance) are reduced directly without prior dissociation (*8,13*).

Reactions which follow the electron-transfer reaction can also be studied. Current-reversal chronopotentiometry is often helpful for such work. Testa and Reinmuth (*30*) have studied the hydrolysis reaction of p-benzoquinoneimine which follows the oxidation of p-aminophenol by current-reversal chronopotentiometry and reported a rate constant in agreement with previous work (*116*). This system has also been studied by cyclic chronopotentiometry (*34*) and by thin-layer chronopotentiometry (*75*).

Various catalytic reactions have also been investigated, but the common systems are the regeneration of ferric or titanic ions by hydroxylamine after their electrolytic reduction (*28,34,118*). Disproportionation reactions and monomerizations have also been the subject of chronopotentiometric investigations (*108,119*), as have substitution reactions involving metal complexes (*117*).

More complex reactions involving several steps, both electrochemical and chemical, have also been reported (*20,22*).

D. Adsorption

In Sec. II.G it was mentioned that adsorption could cause the chrono-potentiometric constant, $I\tau^{\frac{1}{2}}/C$ to vary with τ. Adsorption has been blamed as one of the more important factors in the inconsistency (*120*). Although subsequent work discussed below shows this effect to be smaller than originally thought, in many cases adsorption can be studied chrono-potentiometrically and seemingly anomalous phenomena explained in this way.

Lorenz and co-workers (*121–123*) have studied adsorption phenomena at platinized-platinum electrodes. They have demonstrated that triiodide, iodide, silver, and ferric ions are adsorbed, as well as iodine. Use was made of the fact that, when adsorption occurs, $I\tau^{\frac{1}{2}}/C$ is no longer independent of τ but increases as τ decreases, until, for very short τ's, $I\tau$ is a constant corresponding to the amount of material adsorbed. This principle has been considered by Laitinen (*124*) and the simplified model of Lorenz (*121*) suggested for use in some cases. This model, which assumes the adsorbed material undergoes complete electrolysis *before* the diffusing substance in solution, leads to

$$\frac{I^2\tau}{C^2} = \frac{n^2F^2\pi D}{4} + \frac{nF\Gamma I}{C^2} \tag{43}$$

where Γ is the amount of adsorbed material in moles per square centimeter. The assumption that the adsorbed species undergo electrode reactions more easily than the same material in solution is not necessarily true. Adsorbed molecules are in a lower free-energy state than those in solution and thus more difficult to reduce or oxidize (*125*).

The next reasonable possibility is then to assume that the adsorbed layer undergoes electrolysis only after the usual transition time for the diffusing species. This model is complicated by the fact that, while the adsorbed layer is reacting toward the end of the electrolysis, more soluble species diffuse up to the electrode thus reducing the current efficiency. Lorenz considered this model as well (*121*) and reported an approximate treatment. A more rigorous derivation has been given by Reinmuth (*21*) and Anson (*120*). Their treatments lead to the equation

$$\frac{nF\Gamma}{I} = \frac{\tau}{\pi}\arccos\frac{(\tau_1 - \tau_2)}{\tau} - \frac{2}{\pi}(\tau_1\tau_2)^{\frac{1}{2}} \tag{44}$$

where τ_1 is the transition time for the diffusing species and can be calculated from

$$\tau_1 = n^2F^2\pi DC^2/4I^2 \tag{45}$$

and τ_2 from

$$\tau_1 + \tau_2 = \tau \tag{46}$$

If two separate waves are observed, τ_1 and τ_2 could be measured directly.

It is also possible to treat the model which assumes that both adsorbed and diffusing species react simultaneously—each consuming a constant current—and get the following equation (*121*):

$$\frac{I\tau^{\frac{1}{2}}}{C} = \frac{nF(\pi D)^{\frac{1}{2}}}{2} + \frac{nF\Gamma}{C\tau^{\frac{1}{2}}} \tag{47}$$

Other more complicated situations can also be conceived involving various adsorption isotherms, but up to now the equations are hard to solve unless certain conditions pertain. A useful model of this type was developed for simultaneous electrolysis, assuming that the amount of adsorbed species present during the electrolysis was linearly related to the concentration of the diffusing species (*121*). If $C(D\tau)^{\frac{1}{2}} \gg \Gamma$, we have

$$\frac{I\tau^{\frac{1}{2}}}{C} = \frac{nF(\pi D)^{\frac{1}{2}}}{2} + \frac{\Gamma\pi^{\frac{1}{2}}I}{2D^{\frac{1}{2}}C^2} \tag{48}$$

Tatwawadi and Bard (*126*) have applied the various models represented by Eqs. (43), (44), (47), (48) to the adsorption of riboflavin at a mercury electrode and shown that Eq. (44) gives the best agreement with electrocapillary measurements. Laitinen and Chambers (*127*), however, have found it difficult to distinguish between the various models as to applicability, but recommend Eq. (47), since it usually gives results falling between the other models and thus is less likely to be too far off. They studied Alizarin Red S and cobalt-ethylenediamine-chloride adsorption. Munson (*128*) has applied the model of Eq. (48) to the adsorption of carbon monoxide on platinum and considered the various models (*129*) for the hydrogen electrode. Murray (*130*) has proposed chronopotentiometry with programmed current as a method for distinguishing between the various models.

Chronopotentiometric methods have also supplied information about adsorption in other ways. Current-reversal chronopotentiometry has been used to show that iodine is adsorbed at platinum electrodes (*131*). No adsorption of iodide, bromide, or bromine was detected in this study. A quantitative treatment for the effect of adsorption on current-reversal chronopotentiometry has been reported (*132*), but a computer program utilizing the iterative method is necessary for the calculation of Γ's.

The effect of adsorbed materials on the oxidation of oxalic acid at platinum electrodes has been investigated (*100*). The mechanism proposed for the oxidation is

$$H_2C_2O_4 \rightarrow (H_2C_2O_4)ads \tag{49}$$
$$(H_2C_2O_4)ads \rightarrow CO_2 + 2H^+ + 2e$$

If adsorption of another substance such as amyl alcohol or even oxygen (or platinum oxide) occurs, the adsorption of oxalic acid is prevented, and the oxidation is suppressed.

Similarly, the effect of adsorbed materials on the Co(II)–Co(III) couple in EDTA solution can be elucidated chronopotentiometrically (*81,82*). In this case both the Co(III)- and Co(II)-EDTA complexes are adsorbed on the platinum electrode and the electrolysis suppressed. Amyl alcohol also inhibits the reaction. Bromide ion, however, makes the electrode reaction more reversable by inhibiting electrode oxidation, displacing adsorbed chelate complexes, and acting as an electron-transfer catalyst.

Although the adsorption of iodine on platinum electrodes has been previously established (*121,131*) by current-reversal chronopotentiometry and variations of $i\tau^{\frac{1}{2}}$ with τ, disagreement appeared in the literature as to whether iodide was adsorbed. Recently, however, radiotracer methods and thin-layer chronopotentiometry have been used to show that iodide is indeed strongly adsorbed, even though the adsorbed iodide is not electrochemically active (*94*).

E. Other Applications

Chronopotentiometry has been used for various other applications. Perhaps one of the most interesting has been in fused-salt work, where several other standard methods such as polarography cannot be applied. Laitinen and co-workers (*133,134*) for instance, have investigated the reduction of cobalt, thallium, cadmium, and lead ions in the eutectic mixture of potassium and lithium chlorides. Other investigations include the oxidation of zinc and lithium from liquid bismuth in a chloride eutectic (*135*) and studies of the magnesium-magnesium chloride system (*136*).

It is interesting to note that interference by convection is more pronounced in molten salts than at room temperature, and so transition times should be limited to a few seconds (*133,134*).

Chronopotentiometric titrations, which are accomplished by measuring $i\tau^{\frac{1}{2}}$ as a function of titrant added, have been proposed (*137*). The method is similar to amperometric titrations but has been little used.

Membrane electrodes have been used for chronopotentiometry (*138*) and are advantageous since convection effects are minimized. The proper equations for this method may be found in the literature.

The thickness of thin films on various metals has been studied by essentially chronopotentiometric methods (*139,140*). The method is similar to that used to study electrode oxidation (*19*), although emphasis was placed on the calculations of thickness through the formula

$$T = 10^5 Mi\tau/AnFd \qquad (50)$$

where T is the thickness in angstroms, M is the gram-formula weight of the compound, d the density of the compound, and A the specimen area. The other symbols have standard meanings. Copper oxide films as thin as 10 A have been determined. Mixed oxide and sulfide films on copper and silver exhibit two inflections. Although this method suffers from some problems such as the unknown (in many cases) relation between microscopic and gross area and possible kinetic complications, it is of some practical significance.

REFERENCES

1. H. J. S. Sand, *Phil. Mag.*, **1**, 45 (1901).
2. J. A. V. Butler and G. Armstrong, *Proc. Roy. Soc. (London)*, **A139**, 406 (1933).
3. J. A. V. Butler and G. Armstrong, *Trans. Faraday Soc.*, **30**, 1173 (1934).
4. F. G. Cottrell, *Z. Physik. Chem. (Leipzig)*, **42**, 385 (1902).
5. A. Rius, J. Llopis, and S. Polo, *Anales Fis. y Quim. (Madrid)*, **45**, 1029 (1949).
6. N. Ibl and G. Trümpler, *Helv. Chim. Acta*, **33**, 2163 (1950); **35**, 363 (1952).
7. L. Gierst and A. Juliard, *Proc. Intern. Comm. Electrochem. Thermodyn. and Kinet., 2nd Meeting, Milan*, **1950**, pp. 117, 279; *J. Phys. Chem.* **57**, 701 (1953).
8. P. Delahay, *New Instrumental Methods in Electrochemistry*, Wiley (Interscience), New York, 1954, Chap. 8.
9. P. Delahay and G. Mamantov, *Anal. Chem.*, **27**, 478 (1955).
10. C. N. Reilley, G. W. Everett, and R. H. Johns, *Anal. Chem.*, **27**, 483 (1955).
11. Z. Karaoglanoff, *Z. Elektrochem.*, **12**, 5 (1906).
12. P. Delahay and C. C. Mattax, *J. Am. Chem. Soc.*, **76**, 874 (1954).
13. P. Delahay and T. Berzins, *J. Am. Chem. Soc.*, **75**, 2486 (1953).
14. W. H. Reinmuth, *Anal. Chem.*, **32**, 1514 (1960).
15. W. H. Reinmuth, *Anal. Chem.*, **33**, 485 (1961).
16. T. Berzins and P. Delahay, *J. Am. Chem. Soc.*, **75**, 4205 (1953).
17. D. G. Davis and J. Ganchoff, *J. Electroanal. Chem.*, **1**, 248 (1959/60).
18. J. J. Lingane, *Electroanalytical Chemistry*, 2nd ed., Wiley (Interscience), New York, 1958, p. 623.
19. F. C. Anson and J. J. Lingane, *J. Am. Chem. Soc.*, **79**, 1015 (1957).
20. A. C. Testa and W. H. Reinmuth, *Anal. Chem.*, **33**, 1320 (1961).
21. W. H. Reinmuth, *Anal. Chem.*, **33**, 322 (1961).
22. A. C. Testa and W. H. Reinmuth, *J. Am. Chem. Soc.*, **83**, 784 (1961).
23. P. Delahay, C. C. Mattax, and T. Berzins, *J. Am. Chem. Soc.*, **76**, 5319 (1954).
24. A. C. Testa and W. H. Reinmuth, *Anal. Chem.*, **32**, 1518 (1960).
25. C. C. Mattax, Doctoral Dissertation, Louisiana State Univ., 1954.
26. W. E. Palke, C. D. Russell, and F. C. Anson, *Anal. Chem.*, **34**, 1171 (1962).
27. O. Dračka, *Collection Czech. Chem. Commun.*, **25**, 338 (1960).
28. C. Furlani and G. Morpurgo, *J. Electroanal. Chem.*, **1**, 351 (1960).
29. R. M. King and C. N. Reilley, *J. Electroanal. Chem.*, **1**, 434 (1960).
30. A. C. Testa and W. H. Reinmuth, *Anal. Chem.*, **32**, 1512 (1960).
31. D. H. Geske, *J. Am. Chem. Soc.*, **81**, 4145 (1959).
32. A. C. Testa and W. H. Reinmuth, *Anal. Chem.*, **33**, 1324 (1961).
33. H. B. Herman and A. J. Bard, *Anal. Chem.*, **35**, 1121 (1963).
34. H. B. Herman and A. J. Bard, *Anal. Chem.*, **36**, 510 (1964).
35. H. B. Herman and A. J. Bard, *Anal. Chem.*, **36**, 971 (1964).
36. H. Hurwitz and L. Gierst, *J. Electroanal. Chem.*, **2**, 128 (1961).
37. H. Hurwitz, *J. Electroanal. Chem.*, **2**, 142 (1961).
38. H. Hurwitz, *J. Electroanal. Chem.*, **2**, 328 (1961).
39. R. W. Murray and C. N. Reilley, *J. Electroanal. Chem.*, **3**, 64 (1962).
40. R. W. Murray, *Anal. Chem.*, **35**, 1784 (1963).
41. R. W. Murray, *J. Electroanal. Chem.*, **7**, 242 (1964).
42. J. A. V. Butler, *Electrical Phenomena at Interfaces*, Methuen, London, 1951.
43. C. N. Reilley, "Fundamentals of Electrode Processes," in *Treatise on Analytical*

Chemistry (I. M. Kolthoff and P. Elving, eds.), Part I, Vol. 4, Wiley (Interscience), New York, 1963, pp. 2127–2129.
44. A. J. Bard, *Anal. Chem.*, **33**, 11 (1961).
45. A. J. Bard, *Anal. Chem.*, **35**, 340 (1963).
46, P. Delahay, "Chronoamperometry and Chronopotentiometry," in *Treatise on Analytical Chemistry* (I. M. Kolthoff and P. Elving, eds.), Part I, Vol. 4, Wiley (Interscience), New York, 1963, p. 2252.
47. F. C. Anson, *Anal. Chem.*, **33**, 1438 (1961).
48. W. H. Reinmuth, *Anal. Chem.*, **33**, 1438 (1961).
49. C. N. Reilley, *J. Chem. Educ.*, **39**, A853 (1962).
50. L. Gierst and P. Mechelynck, *Anal. Chim. Acta*, **12**, 79 (1955).
51. T. Kuwana and R. N. Adams, *J. Am. Chem. Soc.*, **79**, 3609 (1957).
52. T. Kuwana and R. N. Adams, *Anal. Chim. Acta*, **20**, 51 (1959).
53. H. B. Mark, *Anal. Chem.*, **36**, 958 (1964).
54. D. G. Peters and J. J. Lingane, *J. Electroanal. Chem.*, **2**, 1 (1961).
55. P. J. Elving and D. L. Smith, *Anal. Chem.*, **32**, 1849 (1960).
56. H. A. Laitinen and P. R. Rhodes, ACS Meeting, New York (1963).
57. A. L. Beilby, W. Brooks, and G. L. Lawrence, *Anal. Chem.*, **36**, 22 (1964).
58. R. N. Adams, *Anal. Chem.*, **30**, 1576 (1958).
59. Z. Galus and R. N. Adams, *J. Phys. Chem.*, **67**, 866 (1963).
60. D. G. Davis and M. E. Everhart, *Anal. Chem.*, **36**, 38 (1964).
60a. S. Bruckenstein and T. O. Rouse, *Anal. Chem.*, **36**, 2039 (1964).
61. J. G. Nikelley and W. D. Cooke, *Anal. Chem.*, **29**, 933 (1957).
62. H. Gerischer, *Z. physik. Chem.* (Leipzig), **202**, 302 (1953).
63. J. W. Ross, R. D. Demars, and I. Shain, *Anal. Chem.*, **28**, 1768 (1956).
64. L. Airey and A. A. Smales, *Analyst*, **75**, 287 (1950).
65. A. Hickling and W. Wilson, *J. Electrochem. Soc.*, **98**, 429 (1951).
66. H. A. Laitinen and C. G. Enke, *J. Electrochem. Soc.*, **107**, 773 (1960).
67. D. G. Davis, *Talanta*, **3**, 335 (1960).
68. F. C. Anson, *Anal. Chem.*, **33**, 934 (1961).
69. D. G. Davis, *Anal. Chem.*, **33**, 1839 (1961).
70. D. G. Davis, *Anal. Chim. Acta*, **27**, 26 (1962).
71. R. T. Iwamoto, *Anal. Chem.*, **31**, 1062 (1959).
72. P. Delahay, *Anal. Chem.*, **20**, 1212, 1215 (1948).
73. Y. Takemori, T. Kambara, M. Senda, and I. Tachi, *J. Phys. Chem.*, **61**, 968 (1957).
74. R. T. Iwamoto, R. N. Adams, and H. Lott, *Anal. Chim. Acta*, **20**, 84 (1959).
75. C. R. Christensen and F. C. Anson, *Anal. Chem.*, **35**, 205 (1963).
76. A. T. Hubbard and F. C. Anson, *Anal. Chem.*, **36**, 723 (1964).
77. W. B. Mather and F. C. Anson, *Anal. Chem.*, **33**, 1634 (1961).,
78. J. D. Voorhies and N. H. Furman, *Anal. Chem.*, **31**, 381 (1959).
79. G. A. Ward, *Talanta*, **10**, 261 (1963).
80. D. H. Evans, *J. Electroanal. Chem.*, **6**, 419 (1963).
81. F. C. Anson, *J. Electrochem. Soc.*, **110**, 436 (1963).
82. F. C. Anson, *Anal. Chem.*, **36**, 520 (1964).
83. E. Solon and A. J. Bard, *J. Am. Chem. Soc.*, **86**, 1926 (1964).
84. J. D. Voorhies and J. S. Parsons, *Anal. Chem.*, **31**, 516 (1959).
85. J. J. Lingane, *J. Electroanal. Chem.*, **2**, 46 (1961).
86. T. Kuwana, D. E. Bublitz, and G. Hoh, *J. Am. Chem. Soc.*, **82**, 5811 (1960).

87. G. L. K. Hoh, W. E. McEwen, and J. Kleinberg, *J. Am. Chem. Soc.*, **83**, 3949 (1961).
88. R. P. Buck and L. R. Griffith, *J. Electrochem. Soc.*, **109**, 1005 (1962).
89. E. D. Moorhead and N. H. Furman, *Anal. Chem.*, **32**, 1507 (1960).
90. A. J. Bard, *Anal. Chem.*, **35**, 1605 (1963).
91. J. J. Lingane and P. J. Lingane, *J. Electroanal. Chem.*, **5**, 411 (1963).
92. H. B. Mark and C. L. Aikin, *Anal. Chem.*, **36**, 514 (1964).
93. D. G. Davis, *Anal. Chem.*, **35**, 764 (1963).
94. R. A. Osteryoung and F. C. Anson, *Anal. Chem.*, **36**, 975 (1964).
95. F. C. Anson, *J. Am. Chem. Soc.*, **81**, 1554 (1959).
96. M. N. Nicholson and J. H. Karchmer, *Anal. Chem.*, **27**, 1045 (1955).
97. A. R. Nisbet and A. J. Bard, *J. Electroanal. Chem.*, **6**, 332 (1963).
98. G. Bianchi, T. Mussini, and C. Traini, *Chimica Ind.* (Milan), **45**, 1333 (1963).
99. J. J. Lingane, *J. Electroanal. Chem.*, **1**, 379 (1960).
100. F. C. Anson and F. A. Schultz, *Anal. Chem.*, **35**, 1114 (1963).
101. J. J. Lingane, *J. Electroanal. Chem.*, **2**, 296 (1961).
102. D. H. Evans and J. J. Lingane, *J. Electroanal. Chem.*, **6**, 283 (1963).
103. H. B. Mark and F. C. Anson, *Anal. Chem.*, **35**, 722 (1963).
104. J. J. Lingane, *J. Electroanal. Chem.*, **7**, 94 (1964).
105. H. B. Mark and F. C. Anson, *J. Electroanal. Chem.*, **6**, 251 (1963).
106. S. Bruckenstein, T. O. Rouse, and S. Prager, *Talanta.*, **11**, 337 (1964).
107. R. F. Dopo and C. K. Mann, *Anal. Chem.*, **35**, 677 (1963).
108. S. W. Feldberg and C. Auerbach, *Anal. Chem.*, **36**, 505 (1964).
109. F. C. Anson and D. M. King, *Anal. Chem.*, **34**, 362 (1962).
110. M. G. McKeon, *J. Electroanal. Chem.*, **4**, 93 (1962).
111. H. D. Hurwitz, *J. Electroanal. Chem.*, **7**, 368 (1964).
112. C. D. Russell and J. M. Peterson, *J. Electroanal. Chem.*, **5**, 467 (1963).
113. P. Delahay and C. C. Mattax, *J. Am. Chem. Soc.*, **76**, 5314 (1954).
114. R. P. Buck, *Anal. Chem.*, **35**, 1853 (1963).
115. N. Tanaka and T. Murayama, *Z. Physik. Chem.* (*Frankfurt*), **21**, 146 (1959).
116. W. K. Snead and A. E. Remick, *J. Am. Chem. Soc.*, **79**, 6121 (1957).
117. C. R. Christensen and F. C. Anson, *Anal. Chem.*, **36**, 495 (1964).
118. O. Fischer, O. Dračka, and E. Fischerová, *Collection Czech. Chem. Commun.*, **26**, 1505 (1961).
119. O. Fischer, O. Dračka, and E. Fischerová, *Collection Czech. Chem. Commun.*, **25**, 323 (1960).
120. F. C. Anson, *Anal. Chem.*, **33**, 1123 (1961).
121. W. Lorenz, *Z. Elektrochem.*, **59**, 730 (1955).
122. W. Lorenz and H. Mülhberg, *Z. Elektrochem.*, **59**, 736 (1955).
123. W. Lorenz and H. Mülhberg, *Z. Physik. Chem.* (*Frankfurt*), **17**, 129 (1958).
124. H. A. Laitinen, *Anal. Chem.*, **33**, 1459 (1961).
125. R. Brdička, *Collection Czech. Chem. Commun.*, **12**, 522 (1947).
126. S. V. Tatwawadi and A. J. Bard, *Anal. Chem.*, **36**, 2 (1964).
127. H. A. Laitinen and L. M. Chambers, *Anal. Chem.*, **36**, 5 (1964); **8**, 84 (1964).
128. R. A. Munson, *J. Electroanal. Chem.*, **5**, 292 (1963).
129. R. A. Munson, *J. Phys. Chem.*, **66**, 727 (1962).
130. R. W. Murray, *J. Electroanal. Chem.*, **7**, 242 (1964); **8**, 84 (1964).
131. R. A. Osteryoung, *Anal. Chem.*, **35**, 1100 (1963).

132. H. B. Herman, S. V. Tatwawadi, and A. J. Bard, *Anal. Chem.*, **35**, 2211 (1963).
133. H. A. Laitinen and W. S. Ferguson, *Anal. Chem.*, **29**, 4 (1957).
134. H. A. Laitinen and H. C. Gaur, *Anal. Chim. Acta*, **18**, 1 (1958).
135. J. D. VanNorman, *Anal. Chem.*, **33**, 946 (1961).
136. J. D. VanNorman and J. J. Egan, *J. Phys. Chem.*, **67**, 2460 (1963).
137. C. N. Reilley and W. G. Scribner, *Anal. Chem.*, **27**, 1210 (1955).
138. R. C. Bowers, G. Ward, C. M. Wilson, and D. D. DeFord, *J. Phys. Chem.*, **65**, 672 (1961).
139. U. R. Evans and L. C. Bannister, *Proc. Royal Soc.* (*London*), **A125**, 370 (1929).
140. W. E. Campbell and U. B. Thomas, *Bell Telephone System, Tech. Publ., Monograph* B-1170 (1939).

PHOTOELECTROCHEMISTRY AND ELECTROLUMINESCENCE

Theodore Kuwana

DEPARTMENT OF CHEMISTRY
CASE INSTITUTE OF TECHNOLOGY
CLEVELAND, OHIO

I. Introduction and Scope	197
II. Nomenclature	198
III. Some Photochemical Fundamentals	201
A. General Considerations	201
B. Characteristics of Excited States and Energy Transfer	204
IV. Photopotentials From Organic Systems	206
A. Mechanism	206
B. Photopotentials in the Presence of Oxygen	209
C. Photopotentials in the Presence of Acceptor Molecules	210
D. Photopolarography and Literature Summary of Photoeffects of Organic Systems	210
V. Irradiation of Metal Electrode Surfaces in "Nonabsorbing" Solutions	215
A. Mechanism	215
B. Hydrated Electrons	216
VI. Metal-Metal Oxide Surfaces and Semiconductors	219
VII. Thermal Effects	222
VIII. Electroluminescence	223
A. Mechanism of Chemiluminescence	223
B. Electroluminescence	226
Addendum	231
References	235

I. INTRODUCTION AND SCOPE

This chapter will review some recent studies of the effect of light irradiation on electrochemical cells and also others in which light emission is a consequence of an electrochemical process. The discussion will be limited to cells with electrodes immersed in solution and will not deal with solid-state systems. The literature survey, emphasizing organic systems, will not be completely comprehensive.

Historically, the discovery of the now well-known " Becquerel" effect occurred in 1839, when Becquerel (*1*) observed the flow of current between two unsymmetrically illuminated pieces of platinum metal in an acidic solution. The same phenomenon was also observed with gold, brass, and silver-silver halide electrodes. Similar effects were subsequently found for cells without liquid electrolyte. Because of their ease of fabrication and their commercial use as photosensitive devices, research and development centered around these " dry-type " cells for many years. Renewed interest in the potentialities of various photoelectrochemical cells has been prompted by the recent quest for a solar-power source for space vehicles. Although success to date has been mainly with solid-state devices, solution-phase work is of greater interest to the chemist. Literature to 1942 dealing with photovoltaic effects has been comprehensively reviewed by Copeland and co-workers (*2*). A summary of properties of some photovoltaic cells described in the literature from 1942 to 1955 has been compiled by Sancier (*3*).

Another area of photoelectrochemistry is the study of electrode processes accompanied by *emission of light*, that is, *electroluminescence*. As early as 1880, luminescence was observed from the electrolysis of water at high current densities using platinum electrodes (*4*). Bancroft and Weiser (*5*) reported emission of light from electrolysis of halide solutions at a mercury anode; similar emission has been reported with aluminum electrodes (*6*). In 1929, Harvey (*7*) reported the observation of light from electrolyses of luminol and luciferin. Investigation of similar phenomena has been dormant, with the exception of a few notable works. Several recent experiments have indicated that light emission can also arise from the electrolysis of polycyclic aromatic molecules; these fragmentary and preliminary studies will undoubtedly be the forerunners of greatly expanded activity in this area. Although electroluminescence in solid-state systems has been fairly well described and understood, solution-phase work still remains to be characterized and explained. Because of this, the part of this chapter dealing with electroluminescence will only indicate the scope of this subject by reporting what is available in the literature. Ivey (*8,9*) has reviewed electroluminescence of solid-state systems.

II. NOMENCLATURE

Rapid expansion of knowledge and methodology usually causes diffusion and alteration of old terms and introduction of new and often conflicting terminology. This constant change in the working language of a field leads to confusion between various groups, and hinders the rapid exchange of exact ideas and thoughts. The problem of terminology and nomenclature is further compounded today by the cross-fertilization of specialized fields. It is therefore periodically necessary to point out inaccuracies or inconsistencies in usage or to redefine certain terms. The "vocabulary" of the photochemist has been recently reviewed and clarified by Pitts and colleagues (*10*); the work of Reilley and Murray (*11*) should be consulted for the precise definition and usage of electrochemical terms.

In descriptions of work combining these two areas, a variety of terms has developed through the years. *Becquerel, photovoltaic,* and *photopotential effect* have all been commonly used to refer to the measurement of a potential difference between two electrodes, one in the "dark" and the other in the "light." (More commonly, today, a potential difference is measured between an electrode in either the dark or light and a reference half-cell electrode.) When a null-type (open circuit) or high-input-impedance instrument is used, the term *photopotential* is recommended. If the cell is operating under load, a photopotential can still be measured, but has little significance unless the conditions of the experiment and the current drawn from the cell are designated.

Rabinowich (*12*) and Eisenberg and Silverman (*13*) have made a distinction between photogalvanic and photovoltaic effects. To quote from the latter: "If the change in EMF is due to the action of light on a reversible galvanic process or a displacement of equilibrium it is photogalvanic; otherwise it is photovoltaic." Thus an attempt has been made to classify the areas according to whether measured potential change is due to change in the bulk concentration of the electroactive species (light-induced) or only to changes at the solution-electrode interface or in the surface of the electrode (for example, semiconductors). Since it is not always possible to determine the magnitude assignable to each effect, classification in such a manner seems meaningless. *Photopotential* seems adequate to describe all potential measurements in photoelectrochemical work.

If the only effect of light is to shift the equilibrium and/or change the concentration levels of one or both species of a redox reaction

$$Ox + ne^- = R \qquad (1)$$

where the oxidized and reduced species are designated as Ox and R, the potential in the dark E_d, and the potential in the light E_l are given by

$$E_d = E^{0'} + \frac{RT}{nF} \ln \frac{C_{Ox_1}}{C_{R_1}} \qquad (2)$$

$$E_l = E^{0'} + \frac{RT}{nF} \ln \frac{C_{Ox_2}}{C_{R_2}} \tag{3}$$

C_{Ox_1}, C_{R_1} and C_{Ox_2}, C_{R_2} are concentrations of the oxidized and reduced species in dark and light conditions, respectively, and $E^{0'}$ is the formal potential for reaction (1). The photopotential E_* is given by

$$E_* = E_l - E_d \tag{4}$$

Photopotential measurements of many photosensitive oxidation-reduction reactions of organic dye molecules and/or of inorganic ions adhere quite well to the preceding simple nernstian expressions.

Since the concentrations of species in reaction (1) are being altered by irradiation, the photopotential measured will depend on initial conditions, geometry, and type of cell; intensity, time, and wavelength of irradiation; and distribution of both the light and the photosensitive species from the point of light entry into the cell to the electrode surface. A complete description of these conditions is necessary for a meaningful value of a photopotential. When light impinges on an electrode whose surface is either a semiconductor or is coated with a photosensitive material, adequate (quantitative) description of the photopotential is still more complicated.

Photocurrent or *photoconduction* refers to the measurement of current under "light" conditions. Photoconduction has usually been used to refer to current or conduction measurements during irradiation of semiconductor materials or of liquids with high resistance. Since techniques of conduction measurements (for example, ac bridge) are primarily employed, the term *photoconduction* should be reserved for the designation of low-level (less than 10^{-6} amp) photoinduced currents, with the term *photocurrent* suggested for general usage. In electroluminescence, the term *galvanoluminescence* has been most commonly used with reference to light emission stimulated by an electrical current.

Systematic introduction and redefinition of terminology and nomenclature have, almost without exception, been a prologue to the development of a large body of facts in a given area. The possibility of systematizing terms with reference to methodologies will be suggested, however, in the hope of avoiding future confusion.

Under such a system, the generic names applicable to these two areas would be *photoelectrochemistry* and *electroluminescence*. In the former, light is an input signal; in the latter, an output signal. To designate experimental methods in photoelectrochemistry, the prefix *photo-* is used, followed by the name of the electrochemical method employed. Thus, if electrochemical current is measured as a function of time during irradiation and application of constant potential to the cell, the method would be designated as *photochronoamperometry*. The irradiation bandwidth, maximum wavelength, applied potential, etc., should be specified separately. The term *photopolarography* has been introduced and used by Berg and Schweiss (*14,15*), and, in keeping with the preferred electrochemical usage, it should be restricted to work at a dropping mercury electrode. In electroluminescence, the prefix is the electrochemical method followed by the suffix -luminescence. For example, chronopotentioluminescence would indicate light emission during a *chronopotentiometric* experiment.

III. SOME PHOTOCHEMICAL FUNDAMENTALS

A. General Considerations

Light incident on an electrode in a cell may induce a change in its electrochemical characteristics. Possible photochemical processes responsible for the change may be divided conveniently into two broad classifications: those which can be considered part of a *primary* photochemical process, and *secondary* ones that occur as a consequence of primary processes.

Primary processes

(a) light absorption by a molecule to produce an excited-state molecule;

(b) light absorption by a molecule to produce photodissociation or photoionization;

(c) light absorption by a metal, nonmetal, or semiconductor surface to produce photoionization, photoemission, or change in population of carriers between valence and conduction bands.

Secondary processes

(a) fluorescence and phosphorescence;

(b) photochemical reactions;

(c) energy-transfer reactions;

(d) other nonradiative deexcitation modes;

(e) thermal effects.

Some primary and secondary reaction processes will be illustrated and discussed for an organic absorber molecule A:

Primary process

$$A(S_0) + h\nu \rightarrow A^*(S_1) \tag{5}$$

Secondary processes

$$A^*(S_1) \rightarrow A(S_0) + h\nu \quad \text{(fluorescence)} \tag{6}$$

$$A^*(S_1) \rightarrow A^*(T) \tag{7}$$

$$A^*(T) \qquad \to A(S_0) + h\nu' \qquad \text{(phosphorescence)} \tag{8}$$

$$A^*(S_1) + M \to A(S_0) + M \tag{9}$$

$$A^*(T) + M \ \to A(S_0) + M \tag{10}$$

where the singlet ground state, the first excited singlet state, and the triplet state are denoted by $A(S_0)$, $A^*(S_1)$, and $A^*(T)$ respectively; M is any molecule capable of thermalizing the excitation energy of A^*. The energy level and state of the primary excitation are determined by the molecular structure. For example, the longer-wavelength-absorption band of aromatic ketones corresponds to a transition of an electron from the nonbonding orbital of the carbonyl group to the higher energy antibonding π orbital; such a transition is designated as $n \to \pi^*$. The shorter-wavelength-absorption band is a $\pi \to \pi^*$ transition involving the π electrons. The $n \to \pi^*$ bands are usually two or three orders of magnitude less intense than the $\pi \to \pi^*$ bands. The absorption shifts to shorter wavelengths with increased polarity of the solvent, owing to increased hydrogen bonding with the solvent which results in stabilization of the ground state with respect to the excited state.

Excitation from the ground-state singlet (S_0) to the first excited singlet (S_1) is an allowed transition; the spins of the electrons remain antiparallel. A Jablonski-type energy-level diagram is given in Fig. 1 for a molecule which possesses both $n \to \pi^*$ and $\pi \to \pi^*$ transitions. The vertical lines show transitions between electronic levels and superimposed vibrational-energy levels. If the molecule is excited to a higher vibrational level of S_1, rapid loss of energy (ca. 10^{-13} sec) to the lowest vibrational level takes place, and, with the loss of a photon, the molecule can return to the ground (S_0) state [reaction (6)]. This latter process corresponds to fluorescence; the lifetime of the (S_1) state is less than 10^{-8} sec. Transition to a state of different multiplicity or *intersystem crossing*, as illustrated by the horizontal wavy line in Fig. 1, goes from a vibrational level of (S_1) to a corresponding vibrationally excited level of the triplet state (T). The spins of the electrons are parallel in the triplet state (multiplicity of 3), and the transition [reaction (8)] from (T) to (S_0) is spin-forbidden. As a consequence, the lifetime of the triplet state is much greater than that of the singlet excited state, often several milliseconds. Emission from the triplet state is called *phosphorescence*.

Since the collision frequency between molecules is on the order of 10^{12} collisions/sec, nonradiative deactivation processes may predominate as the lifetime τ of the state becomes greater. These processes may proceed either through a photochemical reaction or through various modes of energy transfer. An example of the former is the classical case of

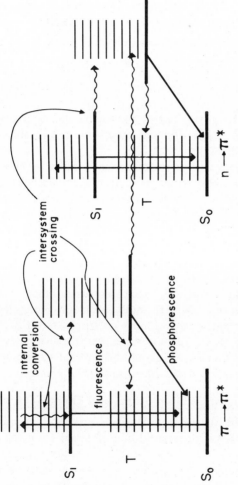

Fig. 1. Excited states of an organic molecule. Straight lines represent radiative and adsorptive processes; wavy lines show nonradiative processes.

203

hydrogen-atom abstraction by benzophenone from an alcohol (*16*). The photoreactive state is the triplet (*17,18*), and the end product is benzpinacol through the following sequence of reactions:

$$\phi_2CO + h\nu \rightarrow \phi_2CO^*(S_1) \rightarrow \phi_2CO^*(T) \tag{11}$$

$$\phi_2CO^*(T) + RCH_2OH \rightarrow \phi_2\dot{C}OH + R\dot{C}HOH \tag{12}$$

$$\phi_2\dot{C}OH + \phi_2\dot{C}OH \rightarrow \phi_2COHCOH\phi_2 \tag{13}$$

$$2R\dot{C}HOH \rightarrow RCH_2OH + RCHO \tag{14}$$

The quantum yield of the reaction is about unity, with rate of H-atom abstraction being controlled by rate of diffusion. Details of the mechanism and kinetics of the benzophenone photoreduction have been thoroughly investigated (*19–23*). For further discussion and examples of organic photoreactions, see (*24*) through (*26*).

The energy necessary for excitation to a higher level varies with the nature of the substituent group on an aromatic molecule and with the solvent. As a consequence, the reactivity in terms of quantum yield and the mechanism of the photoreaction may be subject to considerable variation. For example, certain *ortho*- and *para*-substituted benzophenones undergo little or no photoreaction in isopropyl alcohol (*27*). This lack of photoreactivity has been ascribed either to an intramolecular enol-keto isomerization reaction (*28*) or to an internal conversion of the $n \rightarrow \pi^*$ triplet to the $\pi \rightarrow \pi^*$ triplet (*27,29–31*). Such a conversion is supposedly found in *p*-aminobenzophenone (*27*). In most aromatic molecules the $n \rightarrow \pi^*$ triplet is lower in energy than the $\pi \rightarrow \pi^*$ triplet (see Fig. 1). Porter and Suppan (*32*) have recently demonstrated that the energy-level relationship between these two triplet levels could be reversed by appropriate solvents so that a photoreaction could occur, producing pinacol in good yields.

B. Characteristics of Excited States and Energy Transfer

Because of a shift in electron-density distribution, an excited state can be expected to exhibit properties quite different from the ground state. For example, the electron density of the carbonyl group is lowered in the excited state of an $n \rightarrow \pi^*$ transition (*32*). The acid-dissociation constants, expressed as pK_a's, of some aromatic molecules clearly reflect this electron-density shift. For example, the pK_a values of phenol, *p*-cresol, and *p*-chlorophenol in the ground state are 10.0, 10.3, and 9.4, whereas the excited-singlet-state pK_a values are 5.7, -0.6, and 3.1, respectively (*33*).

There has been much interest in the detection, characterization, and determination of the mechanism of energy-transfer processes involving triplet states of molecules. For example, triplet molecules (oriented) in a crystalline solid have been detected and studied (*34*) at the usual frequency and magnetic field used for free radicals in the electron-spin-resonance (ESR) spectrometer. Unoriented triplets in solutions or low-temperature glasses, on the other hand, produce ESR signals at half the usual magnetic field (*35,36*).

Excited-state triplets may transfer their energy to an acceptor molecule and return to the ground singlet state. A classical example of this process is the irradiation of benzophenone (sensitizer) in the presence of naphthalene (acceptor). At low temperatures, the observed phosphorescence spectrum is that of naphthalene, not benzophenone, although at the irradiation wavelength only benzophenone absorbs. The mechanism is

$$\text{benzophenone } (T) + \text{naphthalene } (S_0) \rightarrow$$
$$\text{naphthalene } (T) + \text{benzophenone } (S_0) \tag{15}$$

$$\text{naphthalene } (T) \rightarrow \text{naphthalene } (S_0) + h\nu \tag{16}$$

with the requirement that the naphthalene triplet be lower in energy than the benzophenone triplet. Terenin and Ermolaev (*37*) and Porter (*38*) have studied the energy relationships and detailed mechanism of several sensitizer-acceptor systems.

Mechanisms of other nonradiative collision-dependent quenching modes [reactions (9), (10)] of excited states are presently under intensive investigation in various laboratories.

The recent observation (*39,40*) of delayed fluorescence from irradiated polycyclic aromatic molecules has been of considerable interest. The proposed mechanism is

$$A(S_0) + h\nu \rightarrow A^*(S_1) \tag{17}$$

$$A^*(S_1) \rightarrow A(S_0) + h\nu \quad \text{(normal fluorescence)} \tag{18}$$

$$A^*(S_1) \rightarrow A^*(T) \tag{19}$$

$$A^*(T) \rightarrow A(S_0) + h\nu' \quad \text{(normal phosphorescence)} \tag{20}$$

$$A^*(T) + A^*(T) \rightarrow \text{``excited dimer''} \rightarrow A^*(S_1) + A(S_0)$$

$$A(S_0) + h\nu \tag{21}$$
$$\text{(delayed fluorescence)}$$

which was deduced from the facts that (a) the intensity of delayed fluorescence is proportional to the square of the rate of absorption of exciting light and (b) the lifetime of the component of delayed fluorescence is one-half the kinetic rate of decay of the triplet state. Cognizance of the many modes of deexcitation of excited states is important in interpretation of photochemical phenomena, particularly when they are coupled with another methodology, such as electrochemistry.

The preceding discussion briefly reviews some of the fundamentals and recent developments in photochemistry. Several excellent reviews (*24–26, 41–43*) and monographs (*44,45*) may be consulted for further details.

IV. PHOTOPOTENTIALS FROM ORGANIC SYSTEMS

A. Mechanism

The photoeffect of metal electrodes immersed in solutions containing organic fluorescent compounds or coated with organic dyes or fluorescent substances has been the subject of numerous investigations since early 1900. Only recently, because of advances in photochemistry, has the general effect been satisfactorily rationalized. The extent of the confusion that existed in the early literature may be recognized by consulting the review of photovoltaic effects to 1942 by Copeland et al. (*2*).

Early photopotential observations were characterized by a change in direction of polarity of the EMF measured during irradiation. For example, Levin and White (*46*) and Levin and co-workers (*47*) attempted to correlate photopotentials with fluorescent properties of several organic compounds. They observed a change in polarity of the potential from negative to positive or positive to negative during the irradiation. Only compounds with a carbonyl group adjacent to the aromatic ring consistently gave positive values. Surash and Hercules (*48,49*) in 1960 demonstrated convincingly that only *negative* values of photopotential should be obtained from the irradiation of these carbonyl compounds. These workers used a platinum electrode immersed in *well-degassed* solutions. That fluctuations of potential result from the presence of dissolved oxygen has also been confirmed in the author's laboratory. Photopotential values of the above two groups are compared in Table I for certain selected compounds. These values are dependent on irradiation time, intensity of light, wavelength, etc.; thus, only the magnitude and polarity should be noted.

The differences in the values of photopotentials may be understood by examining the species formed when these compounds are photolyzed. The molecule 9,10-anthraquinone (AQ) will be used as an example since its photochemistry is fairly well known (*27,50–53*) and its reactions apply to

TABLE I

PHOTOPOTENTIAL VALUES IN ETHANOL

Compound	E, volts $(46, 47)^a$	E, volts $(48, 49)^b$
Acetophenone	0.064	−0.290
Anthraquinone	0.163	−0.434
Benzaldehyde	0.045	−0.395
Benzoin	0.190	0.0
Benzophenone	0.053	−0.409

^{a} Photopotential measured between two platinum electrodes, one in the dark, the other in the light.

^{b} Photopotential measured between platinum (light) versus a reference silver-silver chloride electrode.

most aromatic ketones. The primary reaction is absorption of light to produce the excited singlet state:

$$AQ(S_0) + h\nu \rightarrow AQ^*(S_1) \tag{22}$$

which is followed by the secondary reactions

$$AQ^*(S_1) \rightarrow AQ + h\nu \tag{23}$$

$$AQ^*(S_1) \rightarrow AQ^*(T) \tag{24}$$

$$AQ^*(T) \rightarrow AQ + h\nu' \tag{25}$$

$$AQ^*(T) + RCH_2OH \rightarrow AQH \cdot + R\dot{C}HOH \tag{26}$$

$$AQ + R\dot{C}HOH \rightarrow AQH \cdot + RCHO \tag{27}$$

$$2AQH \cdot \rightarrow AQ + AQH_2 \tag{28}$$

$$2R\dot{C}HOH \rightarrow RCH_2OH + RCHO \tag{29}$$

Quantum yields (Φ) for the disappearance of AQ are close to unity in solutions of lower alcohols. The main mode of deactivation of the primary excited species apparently is through reactions (24) and (26). Surash and Hercules indicated that the electrode potential is determined by the half-reaction

$$AQ + e^- + H^+ = AQH \cdot \tag{30}$$

and follows the Nernst expression

$$E = K + \frac{RT}{nF} \ln \frac{C_{AQ}}{C_{AQH \cdot}} \tag{31}$$

where K is a constant which includes the formal potential E' of reaction (30) and the potential of the reference electrode. ESR work has shown that the equilibrium concentration of the free radical is less than 10^{-5} M at room temperature in solutions containing both AQ and AQH_2. The potential-determining half-reaction may, therefore, be better represented by a two-electron transfer involving AQ and AQH_2.

Results from the photolysis of AQ in well-degassed *alkaline* methanol solutions are more amenable to analysis. In addition to reactions (22) to (26), the following take place (*52*):

$$AQH \cdot + OH^- = AQ^{\overline{\cdot}} + H_2O \qquad (32)$$

$$AQ + \cdot CH_2OH \rightarrow AQH \cdot + CH_2O \qquad (33)$$

$$2 \cdot CH_2OH \rightarrow CH_3OH + CH_2O \qquad (34)$$

$$2AQ^{\overline{\cdot}} = AQ + AQ^{2-} \qquad (35)$$

The rate of formation of $AQ^{\overline{\cdot}}$ and AQ^{2-} is given by

$$[(AQ^{\overline{\cdot}}) + (AQ^{2-})] = I\Phi t / VA \qquad (36)$$

where I is intensity of light in quanta per second, Φ is quantum yield for disappearance of AQ, t is time in seconds, V is volume in liters, and A is Avogadro's number. Quantum-yield values for disappearance of AQ and for formation of CH_2O were found to be 1.14 ± 0.10 and 0.70 ± 0.12, respectively. The potential of the electrode followed the Nernst expression

$$E = E' + \frac{RT}{2F} \ln \frac{C_{AQ}}{C_{AQ^{2-}}} \qquad (37)$$

where $E' = \frac{1}{2}(E_1' + E_2')$ and has a value of -0.78 volt versus a reference saturated-calomel electrode (SCE). E_1' and E_2' are formal potentials versus SCE for the half-reactions

$$AQ + e^- = AQ^{\overline{\cdot}} \qquad (E_1') \qquad (38)$$

$$AQ^{\overline{\cdot}} + e^- = AQ^{2-} \qquad (E_2') \qquad (39)$$

The photopotential curve can be closely duplicated by monitoring the potential of a separate platinum indicator electrode during a coulometric reduction of AQ at a platinum electrode if the coulometric current level (amp sec^{-1}) is made equivalent to the light intensity level (quanta sec^{-1}). Evidence for species responsible for the *negative* value of the photopotential is quite convincing.

B. Photopotentials in the Presence of Oxygen

The inconsistency of earlier photopotential measurements was due to formation of hydrogen peroxide in the photochemical process. This can be illustrated by examining the photochemistry of AQ, now in the presence of oxygen. Following reactions (24) and (26), the additional reactions occur in neutral alcohol solutions:

$$AQH \cdot + O_2 \rightarrow AQ + HO_2 \cdot \tag{40}$$

$$2HO_2 \cdot \rightarrow H_2O_2 + O_2 \tag{41}$$

In *alkaline* methanol, the net reaction is

$$AQ^*(T) + O_2 + CH_3OH + OH^- \rightarrow AQ + HO_2^- + CH_2O + H_2O \tag{42}$$

Quantum-yield values (52) for disappearance of oxygen and for formation of formaldehyde were found to be 1.14 ± 0.08 and 1.1 ± 0.1, respectively. Because the potential is determined by hydrogen peroxide, positive potentials were always observed unless the solution was thoroughly purged of dissolved oxygen. Table II shows photopotential values for some carbonyl compounds in the presence and absence of oxygen.

The kinetic rates and mechanisms of H-atom transfer to anthraquinone-2-sulfonate ion in various solvents of alcohols, ethers, and ketones and in

TABLE II

EFFECT OF OXYGEN ON PHOTOPOTENTIALS[a]

Compound	Concentration,[b] M liter^{-1}	E_*,[c] mV	
		with oxygen	in N$_2$ (degassed)
Acetophenone	0.01	10	-434
4-Bromoaceto-phenone	0.01	60	-210
4-methoxyaceto-phenone	0.01	0	-350
9, 10-Anthra-quinone	0.0005	70	-395
Benzoin	0.01	47	-409
Benzophenone	0.01	60	-400

[a] Data taken from (49).
[b] In absolute ethanol.
[c] Platinum electrode; potential measured with respect to Ag/AgCl reference electrode.

the presence of oxygen have been determined by Wells (*54*). For photolysis
in isopropanol with oxygen, the yields of acetone and hydrogen peroxide
were nearly equal with a quantum yield of unity, substantiating the general
mechanism for AQ photolysis in the presence of oxygen.

C. Photopotentials in the Presence of Acceptor Molecules

The effect on the photopotential of an energy-transfer reaction competing with a H-atom-transfer reaction is demonstrated in the photolysis
of benzophenone with naphthalene. As previously stated, the energy level
of the triplet state of naphthalene (acceptor) is lower than that of benzophenone, so the energy of the excited triplet state of benzophenone is
transferred to naphthalene and the energy is lost to solution through
collisional deactivation processes (external conversion). The quantum yield
for H-atom abstraction by benzophenone is consequently lowered by naphthalene. Table III shows the effect of this on photopotential measurements.

TABLE III

Benzophenone Photopotentials in
Naphthalene

Naphthalene concentration	E_*, volt[a]
0.0	−0.47
4.0×10^{-3}	−0.21
1.0×10^{-2}	−0.15

[a] Irradiation wavelength, 3660 A;
thoroughly degassed 10^{-2} M benzophenone in ethanol; intensity and time
of irradiation same in every run.

D. Photopolarography and Literature Summary of Photoeffects of Organic Systems

1. Photopolarography

Berg has examined and considered in minute detail the effects of continuous and intermittent (flash) irradiation on current and potential
characteristics, using photopolarography. As a result of his studies, he has
suggested the following definitions and terms†:

† H. Berg, German Academy of Sciences, Institute of Microbiology and Experimental
Therapy, Jena; private communication to the author.

a. Photoresidual Current (i_v). This current, measurable during and after irradiation of a solution free of depolarizers, is generated by excitation, transfer (Durchtritts-reaction), solvation, and diffusion of electrons as follows:

$$(e^-)_{metal} \xrightarrow{h\nu} (e^-)^*_{metal} \tag{43}$$

$$(e^-)^*_{metal} \xrightarrow{\text{transfer-reaction}} e^- \tag{44}$$

$$e^- + n(\text{solvent}) \rightarrow e^-_{n\,(\text{solvent})} \tag{45}$$

b. Photodepolarizer (A_v). Photodepolarizers are formed by photo-chemical reactions and modify electrode processes.

(a) An excited-state depolarizer A* taking part in a transfer-reaction

$$A \xrightarrow{h\nu} A^* \tag{46}$$

$$A^* + ne^- \rightarrow A^{-n} \tag{47}$$

results in a more positive half-wave potential $(E^*_{\frac{1}{2}})$.

(b) A photoradical depolarizer A⁻ results from an excited molecule exchanging an electron with either the solution or the electrode:

$$A^* + e^- \xrightarrow{k_v} A^- \tag{48}$$

$$A^- \rightarrow A + e^- \tag{49}$$

(c) A product of the electron-transfer reaction undergoes a subsequent photochemical transformation

$$A + ne^- \rightarrow A^{-n} \tag{50}$$

$$A^{-n} \xrightarrow{k_v} B \tag{51}$$

with a resulting change in half-wave potential.

c. Photoreaction-Altered Diffusion Current (i_{pd}). This is a diffusion current which has been modified by the formation or consumption of a depolarizer A in a concurrent, fast photochemical reaction in the bulk of the solution or at the electrode surface:

$$A + ne^- \rightarrow A^{-n} \tag{52}$$
$$\downarrow k_v$$
$$B$$

d. Phototransfer Reaction. This is an electron-transfer reaction connected with excitation of one or more of the participants with consequent changes in the exchange rate and the half-wave potential:

$$\text{(a)}\quad n(e^-)^*_{\text{metal}} + A \underset{k_b}{\overset{k^*_f}{\rightleftharpoons}} A^{-n} \qquad (E_{\frac{1}{2}})^* \tag{53}$$

$$\text{(b)}\quad ne^- + A^* \underset{k_b}{\overset{k^*_f}{\rightleftharpoons}} A^{-n} \qquad (E^*_{\frac{1}{2}}) \tag{54}$$

$$\text{(c)}\quad n(e^-)^*_{\text{metal}} + A^* \underset{k_b}{\overset{k^{**}_f}{\rightleftharpoons}} A^{-n} \qquad (E^*_{\frac{1}{2}})^* \tag{55}$$

e. Photokinetic Current (i_{kv}). This may be an oxidation or reduction current limited by the rate of a photochemical reaction near or at the electrode surface.

(a) A photochemical generation of an excited state which undergoes a transfer reaction

$$A \underset{k}{\overset{hv}{\rightleftharpoons}} A^* \tag{56}$$

$$A^* + ne^- \to A^{-n} \tag{57}$$

or a photochemical reaction precedes a transfer reaction:

$$B \overset{k_v}{\to} A \tag{58}$$

$$A + ne^- \to A^{-n} \tag{59}$$

(b) A photochemical reaction regenerates the depolarizer, which produces the over-all effect of a catalytic-type wave:

$$A^{-n} \to A + ne^- \tag{60}$$

$$A \overset{k_v}{\to} A^{-n} \tag{61}$$

Both reaction schemes (a) and (b) can involve either cathodic or anodic photokinetic current, i_{kv}. (The limitation, scope or application of the above definitions must await further accumulation of data. Work to date has also been limited to the 2000- to 7000-A region, and the above definitions and terms may not describe fundamental properties of the X-ray- or infrared-energy regions.)

The photoresidual current i_v will be discussed in connection with irradiation of electrodes in nonabsorbing solutions (see Sec. V). Schemes **b** and **d** may perhaps be occurring, particularly with coated electrodes, but direct evidence is lacking. It is interesting to note that the reverse of reactions in scheme **b**(a) has been suggested as a possible route for electrochemical generation of light. There are plentiful examples of schemes **c** and **e**. Both the previously discussed cases of anthraquinone or benzophenone apply, and other examples of photokinetic currents using continuous (*14,15,48,49,52,55,56*) or flash (*57,58*) irradiation have been discussed. The latter technique provides possibilities of determining the kinetics of some photoreactions, and Berg has demonstrated its use for benzophenone in isopropyl alcohol.

2. Photoeffects of Organics Dissolved in Solutions and of Organic-Coated Electrodes

Photoeffects of organic compounds dissolved in solutions and of electrodes coated with organic substances are summarized in Table IV. The table also includes cases where the effect depends on the presence of an inorganic substance (for example, thionine-iron). In the case

TABLE IV

SUMMARY OF PHOTOEFFECTS

System	Measured parameter and reaction	Ref.
Thionine-iron	$E_* = -0.250$ volt; $Fe^{2+} +$ Thionine $\overset{h\nu}{\rightleftharpoons}$ $Fe^{3+} +$ leucothionine (1000-watt projection lamp or 20-amp carbon arc)	*12*
Thionine-iron	$E_* \doteq -0.185$	*59*
Thionine-iron	E_*: Pt -0.111 volt Ta -0.010 Ni -0.002 Monel -0.002 (1000-watt projection lamp)	*60*
Thionine-iron	$ThH^+ + 3H^+ + 2Fe^{2+} + h\nu \rightleftharpoons (ThH_4)^{2+} + 2Fe^{3+}$	*61*
Thionine-reducing agent	*Reducing agent* E_* Ferrous ion pH 2 -0.095 volt Ferrous-EDTA pH 6.8 -0.046 Chromic ion pH 4.0 0.015 EDTA pH 6.8 -0.132	*13, 62*
Nitrosyl chloride	E_* (theoretical) $= 0.21$ volt; $AlCl_3 + NOCl \rightleftharpoons NOAlCl_4 \rightleftharpoons NO^+ + AlCl_4^-$ $NOCl \rightleftharpoons NO + Cl$	*63*
Nitrosyl chloride	$E_* = 0.1$ volt at 1.5 amp ft^{-2}	*64*
Proflavin-ascorbic acid	Photocurrent: 30–200 μamp, $\Phi \cong 2.3\%$ of dye	*62*
Pyranthrene, pyrene, tetracene in benzene solutions	Photocurrent increased linearly with $I_{h\nu}$ below 3×10^{17} photons sec^{-1} cm^{-2}, 3700–5000 A	*65*
Anthracene-NaI-I	Photoconduction, 3650 A Anthracene crystal-NaI-I solution	*66*
Pyrimidines, purines, amino acids	Photopolarographic studies to follow photochemical reactions	*56*

TABLE IV—*continued*

System	Measured parameter and reaction	Ref.
Malachite green leuconitrile	Photocurrent and (CN^-) increase linearly with time, $\Phi_{CN^-} = 0.15 \pm 0.03$ mole Einstein^{-1} at 3130 A	67
Phloxine B (tetrabromotetrachloro-fluorescein)	E_* measurements of phloxine B and its silver salt in aqueous and nonaqueous solutions	68

System	*Reducing agent*	E_*	Ref.
Proflavin sulfate-reducing agent	Cuprous amino ion, pH 8.0	0.0 volt	62
	EDTA, pH 8.0	−0.476	
	Cobaltoushexamino ion, phH 8.0	0.0	
Phenosafranine-reducing agent	Ascorbic acid, pH 6.8	−0.292 volt	62
	EDTA, pH 6.8	−0.408	
	EDTA + ascorbic acid, pH 6.8	−0.365	
	Ethylenediamine, pH 6.8	−0.128	
	N,N,N',N'-tetramethyl-*p*-phenylenediamine, pH 6.8	−0.008	
	Citric acid, pH 6.8	−0.086	
	Oxalic acid, pH 6.8	−0.344	
	Hydroquinone	−0.008	
Euflavin-reducing agent	Ascorbic acid, pH 6.8	−0.355 volt	62
Riboflavin-reducing agent	Riboflavin, pH 6.8	−0.464 volt	62

System	*Coated Electrodes*		Ref.
	Electrolyte (pH 1.0)	E_*	
Victoria blue B-inorganic electrolytes	Cr^{3+}	0.06 volt	69
	Fe^{2+}	−0.03	
	Cu^+	−0.18	
	Sn^{2+}	−0.43	
Phthalocyanine and chlorophyll	E_* measurements as a function of λ, thin films of 0.005–1.0 μ in contact with electrolyte	70

of an electrode coated with a dye such as Victoria Blue B (*69*), the following interesting effects were noted: (a) The magnitude of E_* depends on the wavelength of absorption by Victoria Blue B. (b) The value and sign of E_* depends on the reducing agent (the metal ion) in solution (see Table IV); (c) E_* is an approximately linear function of light intensity at low levels. (d) The log of photocurrent is linear with the log of light intensity. These effects were interpreted to mean that the dye acts as a semiconductor surface with two phase boundaries: metal-dye and dye-solution. When the dye is illuminated, the electrons are raised to the "conduction" band of the dye. The direction of flow of the electron; consequently, the sign of E_* will depend on the oxidation potential of the electrolyte in solution. Strong reducing agents, such as ferrous ion, presumably repel electrons into the dye-metal boundary and create a negative photopotential. The converse will occur with strong oxidizing agents. This explanation coincides closely with that given by Williams (*71*) for the irradiation of CdS surfaces (see Sec. VI).

Coated electrodes present an interesting and challenging area for further study. Closer examination of the photoelectrochemical behavior and photochemical reactions in solutions is needed for understanding these systems.

V. IRRADIATION OF METAL ELECTRODE SURFACES IN "NONABSORBING" SOLUTIONS

Platinum (*72–75*), gold (*74,75,77*), silver (*74,76,77*), copper (*74,76,77*), cadmium (*76*), nickel (*74*), zinc (*76*), tungsten (*78*), and mercury (*72,72a, 74,79–81*) electrodes have been subjected to light irradiation in various aqueous solutions. Most of the earlier papers are now only of historical importance, since impurities, lack of proper equipment, and presence of oxygen in the solutions caused conflicting results. They did, however, present the essential ideas which have provided impetus for continued research. This has led to a partial understanding of the effects of light directed upon a metal-electrode surface.

A. Mechanism

In general, the change in measured photopotentials or photocurrents is dependent on (a) wavelength and intensity of the light beam, (b) nature of the electrolyte in solution, (c) metal used for the electrode, (d) presence or absence of oxygen, and (e) prepolarization or polarization conditions of the electrode. The possibilities generally advanced to explain these photoeffects are (i) photoemission of an electron from the metal, (ii) discharge of an electron from solvent or ions near the electrode, (iii) adsorption or desorption of molecules or ions from the metal surface, (iv) thermal effects, and (v) photoinduced chemical reactions.

Audubert (*75*) suggested that "photoelectrons" are emitted into solution when platinum, mercury, and copper electrodes were irradiated. Sihvonen (*82*) concluded that photoelectrons can be solvated or that cations or anions can be discharged when platinum, silver, copper, mercury, iron, chromium, zinc, aluminum, or nickel surfaces are illuminated. Duclaux (*78*) observed a decrease in the overpotential for oxygen evolution at a tungsten electrode upon illumination with ultraviolet light. Bowden (*72*) noted an increase in current at a cathodic mercury electrode or an anodic platinum electrode when light of 2000 to 4000 A was used for irradiation.

Hillson and Rideal (74), extending the work of Price (83), who investigated evolution of hydrogen and oxygen at a variety of electrodes under the influence of light, found that the log of quantum efficiency (the number of extra ions discharged per quantum of incident light) is directly proportional to the log of apparent current density. Quantum efficiency for evolution of both oxygen and hydrogen increases at shorter wavelengths. It was concluded that the primary mechanism for discharge of hydrogen involves the following sequence of reactions:

$$e^- + M\text{—}H + H_3O^+ \to M\text{—}H^-\text{—}H^+ + H_2O \to M + H_2 + H_2O \quad (62)$$

and also a rapid reaction

$$e^- + M + H_3O^+ \to M\text{—}H + H_2O \quad (63)$$

The potential required to evolve hydrogen is apparently determined by that necessary to displace adsorbed oxygen atoms or water molecules by hydrogen. The rate-determining step was concluded to be the electrochemical combination of an adsorbed atom with an ion in the double layer. Illumination supposedly activates some of the adsorbed hydrogen atoms; the possibility of activation by ejected photoelectrons was also suggested.

The suggested mechanism for enhancement by light of oxygen generation at a platinum electrode involves the release of an activated oxygen atom resulting from photolysis of a molecule of surface platinum oxide. (For further discussion of oxide photolysis, see Sec. VI.)

The idea of a photoelectron being ejected from the metal to discharge either an ion or an adsorbed atom or being solvated by the solvent was suggested but not experimentally verified by these early workers. The observed enhancement of anodic current during oxygen evolution was particularly disturbing, since photoemission of an electron should decrease rather than increase the current.

On the basis of theoretical consideration of photoeffects, it has been predicted that the photopotential should be a logarithmic function of the light intensity (84,85). Experimental verification of this relationship has been claimed for electrodes coated with oxides or sulfides of gold and copper (86). The problem should be reexamined with present-day understanding of photoeffects on semiconductor surfaces and consideration of possible photochemical decompositions at such surfaces.

B. Hydrated Electrons

The nature of photocurrents produced by light irradiation of mercury-electrode surfaces has been recently reexamined (79–81). Barker and

Gardner (*81*), using steady-state-irradiation methods, have concluded that the photocurrent is caused primarily by photoemission of electrons. These ejected electrons become thermalized and hydrated as they penetrate the solution, as indicated by the reaction

$$e^- + H_2O \rightarrow e^-_{H_2O} \tag{64}$$

and Barker has estimated a "cloud" of hydrated electrons to be at a distance of 30 to 300 A from the electrode surface. Under steady-state conditions, no net photocurrent is observed if the hydrated electrons return to the electrode, as is the case if no acceptors, such as hydrogen ions or other scavengers, are present. A small photocurrent may occur, nevertheless, because of the irreversible consumption of hydrated electrons by the reaction

$$e^-_{H_2O} \rightarrow H\cdot + OH^- \tag{65}$$

Barker used light at wavelengths below 3000 A from a 60-watt low-pressure mercury arc lamp. When there is a high concentration of electrolytes, a current i_p is observed at ca. -0.4 volt versus SCE, which in the presence of scavengers (reducible materials which capture electrons) is proportional to the square root of the scavenger concentration (C_a). A saturation (maximum) current i_p^s is obtained at high concentrations of the scavenger when the rate constant k_a is large for the reaction

$$M + h\nu \rightarrow e^- + A \xrightarrow{k_a} A^- \tag{66}$$

If A^- is immediately oxidized at the electrode, there is no net increase in the current. The current i_p is given by

$$i_p = i_{\text{photo emission}} - i_{\text{scavenger oxidized}} - i_{\text{electrons returned}} \tag{67}$$

If all hydrated electrons are assumed to be deposited at a hypothetical plane at distance δ from the electrode, we have

$$\frac{\partial C_e}{\partial t} = D_e \frac{\partial^2 C_e}{\partial x^2} - k_a C_a C_e \tag{68}$$

where C_e is the concentration of electrons, D_e the diffusion coefficient of the electron, and x the distance from the electrode. When $x = \delta$ and a pseudo-steady state is assumed,

$$i_p = F C_e^\delta (k_a C_a D_e)^{\frac{1}{2}} \tag{69}$$

and if k_a and D_e are independent of x and t,

$$i_p \propto (C_a)^{\frac{1}{2}} \tag{70}$$

This relationship apparently applies for low concentrations of efficient scavengers, such as H_3O^+, NO_2^-, and NO_3^-, but at higher concentrations (for example, hydrogen ion at 10^{-2} M or above) the photocurrent becomes relatively independent of the scavenger concentration. The saturation photocurrents i_p^s, on the other hand, were found to be proportional to the light intensity with efficient scavengers.

Berg, in a series of papers associated with photopolarographic studies, has diagnosed and defined three types of currents: (a) the photoresidual current i_v, (b) the photoreaction-limited diffusion current i_{pd}, and (c) the photokinetic current i_{kv}. The latter two have already been discussed in connection with organic photoelectrochemistry (see Sec. IV.D). The photoresidual current i_v occurs as a background current in nonabsorbing solutions. Berg and Schwiess (*80*) have investigated its source, using flash irradiation while simultaneously recording both light intensity and electrochemical current with a dual-trace oscilloscope. The photoresidual current,† i_v has been attributed to the flow of electrons from the electrode into the solution (i_a, the subscript *a* for *austritt* current) and to the capacitance current i_c. The conclusions of Berg are generally in agreement with Barker; however, the character of the i_v is more detailed in the work of Berg and Schwiess. The maximum i_v is found to coincide with maximum light intensity, and its magnitude is dependent on the potential applied to the electrode as well as on the type and concentration of electrolyte present in the solution. The current i_v is fairly independent of pH in the region of pH 7; however, it increases as the square root of hydrogen ion concentration in the pH region of 5 to 1, until a limiting value is attained at pH less than 1. The maximum i_v increases exponentially with more negative values of the electrode potential. Because the maximum of i_v is very dependent on the concentration of protons present in the solution, Berg and Schweiss suggest that, in addition to reactions

$$e^- + n H_2O = e^-_{(H_2O)_n} \tag{71}$$

$$e^-_{(H_2O)_n} = OH^-(aq) + H \cdot \tag{72}$$

and the possible Heyrovsky-type reaction

$$e^- + H \cdot + H_2O = H_2 + OH^- \tag{73}$$

a direct reaction may occur:

$$e^- + H_3O^+ = H \cdot + H_2O \tag{74}$$

---

† Berg's i_v corresponds to the i_p of Barker.

Such a direct reaction in the solution by the electron does not necessitate the postulation of a "stationary" layer of thermalized, hydrated electrons and would explain why the photocurrent did not always attain a high value during the flash irradiation. This conclusion is not in total agreement with Barker's.

The decay portions of i_v after the flash were interpreted as the drift of solvated electrons into the bulk solution and the restoration of the double layer which is perturbed by the electron passage. This latter current is related to the capacitance of the double layer and to its relaxation time, which is determined by the time constant of the circuit.

Advances in techniques for study of fast reactions coupled with development of high-intensity electron-pulse accelerators have allowed the spectroscopic identification of solvated electrons, the elucidation of the elementary reactions which they undergo, and the determination of absolute-rate constants of these elementary reactions. A hydrated electron has a considerably longer natural lifetime than a free electron, and decays in pure water by the process

$$H_2O + e^-_{H_2O} = OH^-(aq) + H\cdot \tag{75}$$

with a rate constant for the first-order decay process of 4.4×10^4 sec^{-1} at 23°C (*87*). Values between 0.6×10^4 and 3×10^4 sec^{-1} have been deduced by Barker from measurements at 22 to 25°C of the residual photocurrent observed at potentials in the neighborhood of -1.5 volt versus SCE in solutions containing 1 or 0.2 M inert salts such as KCl, NaCl, or Na$_2$SO$_4$. It is also interesting to note that cations or molecules which serve as efficient scavengers also have high reactivity (large bimolecular-rate constants) with the solvated electron.

The implications of the solvated electron, as pointed out by the previously mentioned investigators, are extremely interesting with respect to mechanism of electron transfer and of reduction of water by alkali metals. Clarification of the role of the solvated electron in electrochemical processes is needed, and future developments will be watched with great interest.

VI. METAL–METAL OXIDE SURFACES AND SEMICONDUCTORS

Platinum-platinum oxide (*88–92*), zinc-zinc oxide (*76,91,93,94*), copper-copper oxide (*76,77,95–101*), silver-silver oxide (*72,102,103*), gold-gold oxide (*104*), selenium-selenium oxide (*105–107*), lead-lead oxide (*88,108*),

tin-tin oxide (*101*), and semiconductor surfaces (*109–115*) have been
subjected to irradiation during electrochemical measurements. Usually,
photopotential or photocurrents were monitored, and changes were
ascribed to either a photochemical surface reaction involving the oxide or
to some effect of the current "carriers." For example, cupric oxide on
copper was thought to undergo the probable photodecomposition reaction (*95*)

$$2CuO \xrightarrow{h\nu} Cu_2O + \tfrac{1}{2}O_2 \tag{76}$$

while Blocher and Garrett (*76*) considered the primary process, based on a
mechanism proposed by Van Dijck (*96*), as

$$CuO \xrightarrow{h\nu} Cu + O \tag{77}$$

where the liberated oxygen determines the potential of the "oxygen
electrode."

Irradiation of platinum-platinum oxide surfaces has been widely studied
in an attempt to elucidate the nature of the surface and the mechanism of
oxygen evolution (*88*). Grube and Baumeiser (*89*) suggested that the change
in potential of preanodized platinum electrode under irradiation is due to
the photochemical decomposition of higher oxygenated platinum compounds. Veselovskii, in a series of papers (*88,90–92*), reported the following
observations: (a) the potential of a preanodized electrode changes in a
logarithmic fashion with intensity of the irradiation; (b) there are two
regions of potential arrests, at 0.9 and 1.45 volts (versus NHE); (c) the
change in potential from irradiation can also be simulated by a cathodic-
current polarization; and (d) the threshold wavelength is about 400 mμ,
with increasing effects at lower wavelengths. He suggested that oxygen is
photochemically evolved on the surface of the platinum electrode through
the decomposition of higher oxides (higher oxygenated complexes of
platinum) formed at a potential greater than 1.45 volts and that this
mechanism is similar to a sensitization process. A light quantum is absorbed by a particle of platinum oxide, and under the effect of a polarizing
field, the photoelectron and the electron defect are transferred to the
boundaries of the metal-semiconductor and the semiconductor-solution,
resulting in the electrochemical evolution of oxygen.

Potential shifts which arise from the photochemical decomposition of
oxides are slow, and quantum yields for such processes are usually low.
Presence of oxygen in the solution, as with electrodes in organic photo-
reacting systems, gives conflicting results. The interesting case of irradiation of small crystals of cadmium sulfide has been discussed by Williams

(*71*). The measured potential of the electrode changes in a negative direction under the influence of light. Williams proposed the reaction

$$CdS \xrightarrow{h\nu} Cd^{2+}(aq) + (S^{2-}) + 2e^- \qquad (78)$$

with the Cd^{2+} ion leaving the crystal and going into the solution. An excess charge is bound to the S^{2-} ions left in the crystal and, unless it can be transported out of the crystal into the external circuit, a potential difference between the crystal and solution will increase, preventing further Cd^{2+} ion dissolution. The extra electrons are not free to move out of the crystal, since their energy level is that of the fixed band, which is 2.5 eV below that of the conduction band. (If the semiconductor were of the *p* type, the excess negative charge could be neutralized by transport of holes through the valence band.) In the *n*-type CdS crystal, absorption of light raises the electrons into the conduction band so that the excess charge can flow from the crystal to the external circuit. Thus, if a photoinduced chemical reaction occurs with discharge of a positively charged cation into the solution, a photoeffect of a negative E_* will be observed only for an *n*-type semiconductor. On the other hand, if the photoreaction results in discharge of a negative ion into solution, the potential on the crystal would reside with positive ion excess and a photoeffect would be observed only for a *p*-type semiconductor.

Positive photopotentials of a millivolt or less have been observed (*100*) for single crystals of copper immersed in air-saturated water. The magnitude of the photopotentials increases with time of immersion (1 to 71 hr) and intensity of illumination. The behavior of the cuprous oxide which apparently formed on the surface (*98,99*) was explained in terms of the mechanism given by Williams for the case of cadmium sulfide (*71*). Single crystals of copper immersed in NaCl, NaF, NaBr, and NaI solutions give negative, negative, positive, and positive values of photopotential, respectively. Arbit and Nobe (*101*) have indicated that the sign of the photopotential corresponds to the correct *n*- or *p*-type semiconductor if the "semiconductor" interface is due to formation of CuCl, CuF, CuBr, and CuI. There is also a good correlation between the wavelength of response and the energy band-gap for Cu_2O, CuBr, and CuI.

Tin immersed in water gives a negative photopotential, which indicates an *n* type of semiconductor corresponding to SnO on the surface (*101*). Although the results seem consistent with Williams' mechanism, caution must be exercised until it is demonstrated that adsorbed materials from the solution are unimportant in instances where the photopotential is less than a millivolt.

The works of Wiesner (*109*), Brattain and co-workers (*110–112*), Gerischer (*113*), Dewald (*114*), and Gobrecht et al. (*115*) should be consulted for further discussion of photoeffects associated with semiconductor electrodes.

VII. THERMAL EFFECTS

There are examples in the literature where the photopotential shifts have been small in magnitude (few millivolts) and have occurred slowly (several minutes). Some of these cases may be attributable to thermal changes at the electrode surface. Hillson and Rideal (*74*), in their work on hydrogen and oxygen evolution at a platinum electrode during irradiation, have computed the magnitude of thermal changes that could take place as a result of the light. They concluded that thermal effects were small and could not account adequately for their observed results. Their conclusions have been criticized recently by Seiger (*116*), who considered thermal changes in a small volume immediately adjacent to the electrode and also convective currents due to thermal gradients. Seiger ascribed the entire observed photoeffect in the ferrous-ferric and iodide-iodine systems he studied at the platinum electrode to the absorbed input light being converted to heat. This heat has two effects: (a) a thermogalvanic effect created by nonuniform heating of the electrode while the reference compartment remains at the temperature of the bath, and (b) a convective motion of the electrolyte resulting from thermal gradients, which tend to warm the bulk of the electrolyte and to reduce the thickness of the convecto-diffusive layer at the electrode. Photochemical reactions, photoemission of electrons, and electrode surface effects were rejected as the source of these photoeffects on the following grounds: (i) the photopotential difference (between steady-state dark and light conditions) is proportional to the light at low intensities and correlates well with absorbancy of the solution as wavelength is varied; (ii) the slope of the current-voltage curve during illumination is greater than it is in the dark; (iii) the establishment of the photo-stationary state requires several minutes; (iv) the photogalvanic effect varies with the distance between window and electrode in the same way as with temperature, and it occurs even at large distances from the electrode; (v) Schelieren photographs of the solution near the electrode surface indicate convective movement of the solution during irradiation; and (vi) the sign of the photopotential difference is the same as that for a thermogalvanic effect.

In the above work, the role of photochemical reactions should have been considered in addition to thermal effects. It is known, for example, that photodissociation of iodine occurs throughout the entire visible region of the spectrum with fairly high quantum yields (*117,118*). Rideal and Williams (*119*) and Kistiakowsky (*120*) have shown that the reaction

$$2Fe^{2+} + I_3^- \overset{h\nu}{\rightleftharpoons} 2Fe^{3+} + 3I^- \tag{79}$$

occurs in the visible, and Dickinson and Ravitz (*121*) have studied the equilibrium

$$2Fe(CN)_6^{3-} + I_3^- \overset{h\nu}{\rightleftharpoons} 2Fe(CN)_6^{4-} + 3I^- \tag{80}$$

under illumination.

Furthermore, the photolysis in the visible region of an iodide-iodine solution to produce photodissociated iodine atoms or the photolysis of iodide at 2537 A to produce iodine atoms and solvated electrons (*122,123*) will result in a more-reducing solution than one in the dark. A photochemical reaction is also consistent with the data except for the convective movement of the solution, which could arise from thermal gradients. The case of the ferrous-ferric solution is less clear-cut. The quantum yield for a photoreaction at the wavelengths used (above 4000 A) would be extremely small if any reaction occurred. It is known, however, that at shorter wavelengths low yields of hydrogen may be produced from the photolysis of acidic ferrous solutions (*124,125*). Although thermal effects may be present, it is unlikely that the photopotential resulting from them would exceed a few millivolts in most systems.

VIII. ELECTROLUMINESCENCE

A. Mechanism of Chemiluminescence

Luminescence has been observed in biological systems (bioluminescence), upon heating of solid materials (thermoluminescence), upon applying a potential field across a material (electroluminescence), in chemical reactions (chemiluminescence), and in many other systems. The phenomenon is well understood in some systems, much less in others. The most widely examined and best understood systems are those in which light emission is the result of ion-ion, atom-atom, atom-electron, or ion-electron recombination reactions, for example, as in the following gas-phase reactions:

$$O + NO + Z \rightarrow NO_2 + Z + h\nu \tag{81}$$

$$H + H + Z \rightarrow H_2 + Z + h\nu \tag{82}$$

$$H + Cl + Z \rightarrow HCl + Z + h\nu \tag{83}$$

where Z is a third body in the thermolecular reactions. The greenish-yellow emission from reaction (81) has been used in the study of O-atom reactions (*126*). Reactions (82) and (83) may be found in hydrogen-oxygen flames with a band continuum in the ultraviolet [reaction (82)] or in the visible [reaction (83)]. Chemiluminescence of certain metal atoms (line spectrum) may be observed if the metal M is added to the flame from reactions (*127,128*)

$$H + H + M \rightarrow H_2 + M^* \tag{84}$$

$$H + OH + M \rightarrow H_2O + M^* \tag{85}$$

Recombination reactions of trapped radicals in low-temperature solid solutions and positive-hole annihilation by electrons in semiconductors are examples of luminescence in solids.

Less understood are mechanisms of luminescence in systems which involve molecular reactions, that is, bioluminescent or chemiluminescent ones. In these, several precursor reactions may precede the luminescent step, or, alternatively, the entities involved in the luminescent step may undergo further reactions which complicate their isolation and identification. A large amount of effort is presently underway to elucidate (a) reactions of biological significance, (b) hydrocarbon oxidations, and (c) reactions involving molecular oxygen or hydrogen peroxide. Many molecular chemiluminescent systems require oxygen or hydrogen peroxide; it is certainly the case in bioluminescence, as, for example, in the luciferin-luciferase system (*129,130*), in which the enzyme acts as an oxidative catalyst with oxygen as an electron acceptor. Additional requirements of the reaction are that the substrate can ionize and form a peroxide intermediate, which leads, presumably, to a fluorescent product.

Weak chemiluminescence often occurs in the oxidation of organic substances in nonaqueous solvents where short-lived free radicals may be present (*131–134*) and also in the reaction of oxygen with radicals such as triphenylmethyl or others (*135*). Many of these may involve a recombination of peroxy radicals of the type ROO·. Vasil'ev and Vichutinskii (*136*) have suggested its formation from hydrocarbon oxidation in the presence of oxygen through the following sequence of reactions:

$$R \cdot + O_2 \rightarrow ROO \cdot \tag{86}$$

$$ROO \cdot + RH \rightarrow ROOH + R \cdot \tag{87}$$

$$ROO \cdot + R \cdot \rightarrow \text{inactive products} \tag{88}$$

$$ROO \cdot + ROO \cdot \rightarrow \text{inactive products} + O_2 \tag{89}$$

where the hydrocarbon radical R· may be formed by a decomposition reaction. Reactions (86) and (87) are chain hydrocarbon reactions, and (88) and (89) are recombinations. The light is assumed to be from the last reaction. The catalyzed decomposition of ethylbenzene hydroperoxide to peroxide radicals (*137*) and of organic peroxides (*138*) also produces luminescence in the visible range. Stauff et al. (*134*) propose that excited $(O_2)_2^*$ is formed from recombination reactions such as

$$2ROO \cdot \rightarrow R_2 + (O_2)_2^* \qquad (90)$$

which would account for the weak chemiluminescence. A red chemiluminescence has been reported from the reaction between hydrogen peroxide and sodium hypochlorite or chlorine by Seliger (*139*) and has been further studied by Seliger (*140*), Khan and Kasha (*141*), and Browne and Ogryzol (*142*). The over-all reaction is

$$Cl_2 + HO_2^- + OH^- \rightarrow 2Cl^- + H_2O + O_2^* \qquad (91)$$

with the following emission peaks now resolved: 5750, 6350, 7032, 7619, 7700, 8645, 10,700, and 12,000 A. These have been assigned to transitions of $O_2(^1\Sigma_g^+)$, $O_2(^1\Delta_u)$ and ground state $O_2(^3\Sigma_g^-)$. McKeown and Waters (*143*) have recently suggested chemiluminescence as a diagnostic feature of heterolytic reactions which produce oxygen. Vasil'ev and Vichutinskii (*144*) have used chemiluminescence to measure kinetic rates of hydrocarbon oxidations.

Chemiluminescence which appears as fluorescence of a product molecule or ion must fulfill two general requirements: (a) The reaction must liberate an amount of chemical energy at least equivalent to the energy difference between the product and its excited state, and (b) the product must either be capable of fluorescing itself or transfer its energy to a fluorescent molecule or ion present in the system. Few systems possess these requirements and of those that do, fewer are able to convert efficiently the chemical energy through a suitable mechanistic path. In the best known chemiluminescent systems (*145*), luciginen [10,10'-dimethyl-9,9'-biacridinium nitrate(III)], luminol (3-aminophthalhydrazide), and lophine (1,3,5-triphenylimidizole), the light is characteristic of fluorescence from a product.

Activation by energy transfer may be taking place in cases where the chemiluminescence is characteristic of fluorescence from one of the reactants or added substances. For example, fluorescence has been reported for dye molecules reacting with hydrogen peroxide (*146*) and for fluorescent materials added to hydrocarbon oxidations (*133*) or to oxalyl peroxide with hydrogen peroxide (*147,148*). If energy transfer is actually occurring, many new possibilities arise for the study of energy-transfer mechanisms and of energetics and kinetics of reactions.

B. Electroluminescence

When one considers the various manners in which chemiluminescence occurs, it is not surprising to find low levels of light emission as a consequence of electrode reactions. Earliest reports of luminescence in the vicinity of an electrode are probably those of Sluginov (4), who observed light from electrolysis of water at a platinum electrode, and of Braun (149), who found light at an aluminum anode of a carbon-aluminum rectifier. Arc-like discharges associated with high-current-density electrolytes were also described about the same time. These latter phenomena were summarized by Harvey (150) and more recently by Hickling and Ingram (151). Harvey has discussed the emission of light from luminol (152) and luciferin (153), when they are reacted with electrolytically generated hydrogen peroxide or oxygen. Most of these early works on luminescence at electrodes are now primarily of historical significance because their explanations are not consistent with presently accepted electrochemical concepts.

The following classification of types of systems is suggested for light emission initiated by electrochemical pulse (electroluminescence):

Type I. Two solid conducting surfaces separated by appropriate materials, such as

    A. Gases: neon, argon, mercury vapor, etc.

    B. Solids: phosphors, oxides, semiconductors, etc.

Type II. Two solid conducting surfaces immersed in an electrolyte, resulting in

    A. Film-semiconductor interface processes.

    B. Charge-transfer and solution-reaction processes.

Classification of type I depends on material between the conducting surfaces, whereas type II depends on the process of electroluminescence. Although not self-consistent, the classification is justified on the basis of convenience. The above classification does not include emission from incandescent bodies, which depends only on temperature. Both the type I and II systems are "cold" emissions dependent on the chemical and physical properties.

Electroluminescence from systems of type IB has been recently reviewed by Ivey (8,9). Type-IA systems, which encompass conventional gas-filled bulbs and lamps, will not be discussed here.

Descriptions of studies of type IIA, in which luminescence is observed on anodic polarization, are quite numerous. The luminescence has been attributed to electroexcitation of a semiconducting layer formed on the anode surface. It has been observed at aluminum, magnesium, tantalum, tungsten, and zinc electrodes and appears to be restricted to those metals which form semiconducting oxide films in solutions where the oxide is not soluble. Luminescence can also be induced in semiconducting materials such as germanium and silicon. The paper by Anderson (*153a*) should be consulted for a detailed description and analysis of the emission process as well as a review of earlier work. The more recent publications have been tabulated by Ivey under the heading of "galvanoluminescence" (*8,9*).

Type IIB systems fall into three possible categories which are distinguishable by the mechanism of the activation: (a) direct electron-transfer reaction, (b) recombination reaction of atoms or radicals, and (c) further chemical reaction of a product of an electron-transfer reaction to produce an excited state, usually through interaction with oxygen or hydrogen peroxide.

The suggested, theoretically possible, formation (*154*) of an excited state through direct heterogeneous electron-transfer process has been reported (*148*) but not fully proved. Possible mechanisms for fluorescence or phosphorescence of an aromatic molecule A may be through the reduction of a radical cation A^+ or the oxidation of a radical anion A^-, as follows:

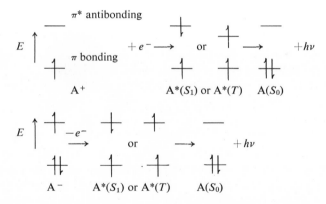

where the net over-all effect is an electron being abstracted from the more stable bonding orbital in the latter case. Chandross and co-workers (*155, 156*) have suggested a similar mechanism but involving a homogeneous electron-transfer between the chemically or electrolytically generated radical anions and either the radical cations or some other oxidants. The removal of an electron from the bonding π orbital is energetically less favored than from the antibonding π^* orbital, but it might possibly occur if the radical anions were in a high potential field, as may be the situation for molecular encounters between charged radicals. Shlyapintokh et al. (*157*) have observed chemiluminescence in the electrolysis of 0.1 N NaOH in the presence of fluorescein or eosin at a platinum electrode. Hercules (*158*) has reported electroluminescence from the alternating polarization of a platinum electrode in acetonitrile or dimethylformamide solutions which contain polycyclic aromatics. He suggested that the luminescence, which is characteristic of the fluorescent color of the particular aromatics (as shown in Table V), is a consequence of the reaction between the positive (perhaps associated or dimeric) and negative radical ions.

The hydrocarbon-free-radical cation is known to be a short-lived species in solution at room temperature so it is doubtful whether it is one of the reactants. Santhanam and Bard (*159*) have observed in the case of 9,10-diphenylanthracene (DPA) that the DPA$^+$ ion is not necessarily the oxidant that reacts with DPA$^-$ to produce chemiluminescence. This is also the case with anthracene in acetonitrile (*160*).

Bader and Kuwana (*160*) have investigated electroluminescence from polarizations of a platinum electrode in acetonitrile and dimethylsulfoxide (DMSO) in the presence and absence of oxygen and/or anthracene (AN). Their results indicate that, in addition to luminescence from hydrocarbon radicals as discussed, luminescence also occurs from: (a) reduction of oxygen, with enhancement of light intensity in the presence of AN, (b) reduction of background electrolyte or solvent, (c) oxidation of CH_3CN, and (d) oxidation at about 0.0 volt versus SCE of a product from reduction of oxygen in DMSO (supporting electrolyte, $LiClO_4$). The following reactions were suggested:

$$RH \rightarrow R \cdot + H^+ + e^- \tag{92}$$

$$R \cdot + O_2 \rightarrow ROO \cdot \tag{93}$$

$$ROO \cdot + ROO \cdot \rightarrow ROOR + O_2 + h\nu \tag{94}$$

for luminescence from hydrocarbon, RH, oxidation in the presence of oxygen. In degassed solutions at anodic potentials where background solution is being oxidized, oxygen is undoubtedly being formed from the concurrent oxidation of traces of water present in the solvent. Upon reduction of oxygen, chemiluminescence may appear through

$$O_2 + e^- \rightarrow O_2^- \tag{95}$$

$$O_2^- + RH \rightarrow R \cdot + HO_2^- \tag{96}$$

followed by reactions (93) and (94). Since annihilation of the radical R· may proceed through several alternate routes (such as $R \cdot + R \cdot \rightarrow R - R$), reaction (93) is undoubtedly of secondary importance.

TABLE V[a]

AROMATIC HYDROCARBONS SHOWING CHEMILUMINESCENCE PRODUCED BY ELECTROCHEMICALLY GENERATED SPECIES

Aromatic hydrocarbon	Solvent[c]	Electrolyte	Color	
			Chemiluminescence	Solution fluorescence
Anthracene	CH_3CN	TEAB	Blue-white (w)	Blue-violet
Chrysene	CH_3CN	TEAB	Blue-white (m)	Blue-violet
Chrysene	DMF	TEAP	Blue-white (m)	Blue-violet
Pyrene	DMF	TEAB	Blue-white (s)	Blue
Naphthalene	DMF	TEAB	Green (s)	Green
Perylene	DMF	TEAP	Blue (s)	Blue
Perylene	CH_3CN	TEAB	Blue (s)	Blue
Perylene	DMF	TEAB	Blue (s)	Blue
Coronene	DMF	TEAB	Blue (m)	Blue
Rubrene	DMF	TEAB	Orange-red (vs)	Orange-red
Decacyclene	DMF	TEAB	Green (w)	Green
1,2,5,6-Dibenzanthracene	DMF	TEAB	Blue-violet (m)	Violet

[a] From D. M. Hercules, *Science*, **145**, 808 (1964).

[b] The hydrocarbon concentrations are 10^{-3} *M*. The supporting electrolyte concentrations are 0.1 or 0.01 *M*.

[c] DMF, dimethyl formamide; TEAB, tetraethyl ammonium bromide; TEAP, tetraethyl ammonium perchlorate; w, weak emission; m, moderate intensity; s, strong emission; vs, very strong emission.

Chemiluminescence from reduction of lithium ions in degassed acetonitrile solutions probably results from radical recombination reactions similar to processes described by Hickling and co-workers (*151,161,162*). Lithium metal can reduce both acetonitrile and traces of water as follows:

Primary process

$$Li^+ + e^- \rightarrow Li^0 \tag{97}$$

Secondary process

$$Li^0 + H_2O \rightarrow H\cdot + OH^- + Li^+ \tag{98}$$

$$Li^0 + H_2O \rightarrow e^-_{H_2O} + Li^+ \tag{99}$$

$$e^-_{H_2O} + H_2O \rightarrow H\cdot + OH^- + H_2O \tag{100}$$

$$H\cdot + H\cdot \rightarrow H_2 \tag{101}$$

or

$$Li^0 + CH_3CN \rightarrow Li^+ + H:^- + \cdot CH_2CN \tag{102}$$

$$H:^- + H_2O \rightarrow H_2 + OH^- \tag{103}$$

$$2\cdot CH_2CN \rightarrow CN-CH_2-CH_2-CN \tag{104}$$

Reactions (99) and (100) suggest the possibility of hydrated electrons participating in the reduction of water. Reactions (101), (103), and (104) are sufficiently energetic to produce light in the visible region. Radical recombinations within a reaction "cage" formed in the solvent and within micro gas pockets formed by hydrogen were suggested as possible means for energy loss by emission of a photon rather than by collision processes. Electroluminescence from the oxidation of a product from O_2 reduction in DMSO is believed to involve peroxides, but an adequate explanation has not been given. The identity of the emitting species in the above reactions is presently unclear, and it should be emphasized that, at such low levels of light emission, the possibility of presence of impurities which may participate in reactions to produce excited states should not be discounted.

Luminol (LH_2) oxidation in the presence of dissolved oxygen in alkaline solutions is an example of emission with the luminescent color characteristic of a product. Vojir (*163*) reported electroluminescence from the cathodic polarization of an alkaline luminol solution and attributed this to the reaction of electrogenerated hydrogen peroxide with luminol. Anodic polarizations at platinum (*152*) and carbon (*164*) electrodes were reported earlier, but detailed examination was not made until recently (*154, 165*). The essential features of luminol electroluminescence are as follows (*166*): (a) the anion LH^- undergoes an irreversible electron-transfer reaction at about 0.4 volt versus SCE in alkaline aqueous solutions; (b) the product of the oxidation reacts with dissolved oxygen and produces an excited molecule through a sequence of reaction steps; (c) the spectrum of the emitted light is identical to fluorescence of aminophthalate; (d) the rate of emission follows a first-order decay law with a rate constant of 1.6×10^4 sec^{-1}; (e) the rate-determining step is believed to be an intra-molecular process with an activation energy of 2.9 kcal mole^{-1}; (f) the light intensity is proportional to the electrochemical-current level in oxygen-saturated solutions; and (g) the quantum yield (based on molecules of luminol oxidized in air-saturated alkaline solutions) is on the order of 3×10^{-4} Einstein mole^{-1}. The rate constant is several orders of magnitude larger than the rate for the light-intensity decay of purely chemically generated chemiluminescence. The quantum yield, however, is of the same order of magnitude (*167*).

Electroluminescence in solutions constitutes an " exciting " and challenging area of investigation. Further progress must await determination of spectral distribution, quantum and fluorescent yields (in solvent used), and identification of species generated in the electron-transfer processes.

ADDENDUM

In the published materials discussing photopotential measurements, it seems clear that *photochemical reactions* occurring in the vicinity of an electrode surface consistently produce the largest potential changes. Smaller changes are observed for photopotentials as a result of *thermal* effects or ejection of *photoelectrons* from the surface of an electrode. Photopotentials caused by the presence of *excited states* have not been proved conclusively, although there is good evidence that such states do affect measured potentials.

Delahay and Srinivasan (*168*) recently examined the question of photo-electrons from mercury electrodes using pulsed irradiation and, in-geniously, a differential coulostatic technique. The variation of the electrode charge Q (irradiated electrode with respect to a similar electrode in the dark) during pulsed irradiation was determined at different potentials for a solution containing hydrogen ion as scavenger of solvated electrons. Barker and Gardner (see Sec. V.B) calculated the steady-state photo-current by assuming thermalization of photoelectrons at a fixed distance δ from the electrode surface. Transposing in terms of Q and recognizing the approximateness of the treatment, Delahay and Srinivasan obtained

$$Q/(Q^s - Q) = \delta(kC_a/D)^{1/2} \tag{1'}$$

where Q^s is the saturation value of Q which would be observed for $C_a \to \infty$; k is the rate constant for the forward reaction of the solvated electron, $e_{H_2O}^-$, C_a is the bulk concentration of H^+, and D is the diffusion coefficient $e_{H_2O}^-$. The results were in excellent agreement with the above relationship and also verified the conclusions of Barker and Gardner that the photocurrent is a function of $C_a^{1/2}$ at low concentration of C_a and becomes independent of C_a at high concentrations (Barker and Gardner: ~ 0.01 M; Delahay and Srinivasan: ~ 0.5 M). The difference in the concentration value for saturation may result not only from differences in the experimental conditions but also because Q may be more precisely measured than small values of photocurrent.

The field dependence of photocurrent was first examined by Berg and Schweiss (80), who found that the maximum photocurrent (peak value) varied exponentially as a function of the potential applied. The theoretical justification of the experimental observation was based on a potential dependence of an "activated" cathodic charge-transfer reaction which is, in effect, a lowering of the cathodic hydrogen overvoltage. Such a mechanism suggests a direct reaction:

$$e^- + H_3O^+ \to H \cdot + H_2O \quad \cdot \tag{2'}$$

rather than the electron being thermalized first. Delahay and Srinivasan found $Q^{1/2}$ to vary linearly with applied potential and suggested *tentatively* that Q varied with the square of the field intensity. A theoretical justification was not given. The lines of $Q^{1/2}$ versus E were observed to intersect the abscissa at potentials between -0.2 and -0.3 volt versus SCE for given concentrations of C_a. A possible explanation was that the difference between the intercept (~ -0.2 volt) and the point of zero charge corresponded to the surface potential generated by the adsorption of water at highly negative potentials.

Some double-layer effects (different electrolytes were used) were noted to influence Q versus E but no definite conclusions were reached. As the authors recognized, further theoretical and experimental work is necessary to understand the detailed features of photocurrents and photopotentials.

In electroluminescence Chandross and co-workers (169) have reported that the spectral distribution of the emission from the annihilation of electrochemically generated aromatic hydrocarbon radical cation (A^+) and anion (A^-) resulted for many polycyclic hydrocarbons, most interestingly, in a broad structureless band shifted toward the red from the normal fluorescence of $A^*(S_1)$. For several planar hydrocarbons such as perylene, 3,4-benzyprene, and anthracene, which are unencumbered by bulky substituents, the spectral emission supported excimer formation through reaction:

$$A^+ + A^- \rightarrow (A_2)^* \qquad (3')$$

whereas 9,10-diphenylanthracene and rubrene, which are sterically hindered, produced only normal monomer fluorescence under all conditions. The results imply that the mechanism for emission requires a close approach of A^+ and A^- before electron transfer takes place. Furthermore, the results indicate that, in addition to excimer formation, in some cases such as anthracene, there may be another type of excimer capable of direct fluorescent emission. These authors have suggested that this excimer may be a charge-transfer state.

With the aid of computor analysis, Feldberg (*170*) has derived a relationship which quantitatively relates the light intensity produced in aromatic electroluminescence to the current, time, and kinetic parameters based on the generalized reaction mechanism

$$A \longrightarrow A^+ + e^- \qquad \text{oxidation at electrode} \qquad (4')$$

$$A + e^- \longrightarrow A^- \qquad \text{reduction at electrode} \qquad (5')$$

$$A^+ + 2e^- \longrightarrow A^- \qquad (6')$$

$$A^+ + A^- \xrightarrow{k'_7} A^* + A \qquad \text{in bulk of solution} \qquad (7')$$

$$A^* \xrightarrow{k'_8} A + h\nu \qquad \text{radiative decay} \qquad (8')$$

$$A^* \xrightarrow{k'_9} A \qquad \text{nonradiative decay} \qquad (9')$$

$$A + A^* \xrightarrow{k'_{10}} 2A \qquad \text{self-quenching} \qquad (10')$$

$$\log \frac{IF}{\pi^{1/2}\Phi_e i_f} = -1.45(t_r/t_f)^{1/2} + 0.71 \qquad (11')$$

where either species A^+ or A^- is produced at controlled potential for a time t_f, which is followed immediately by a potential step to a value where the other species is produced. The total luminescent intensity I is then measured with time t_r at the latter potential. The quantum efficiency Φ_e is the ratio of the number of the excited species undergoing radiative decay to the number which could be formed from the reaction of A^+ and A^-. The symbol F is faraday and i_f is the current at time t_f. The relationship predicts the logarithmic term to be linear function of $(t_r/t_f)^{1/2}$ under the conditions that (*a*) radiative and nonradiative decays are very fast, (*b*) mass transport is diffusion-controlled ($D_A = D_{A^+} = D_{A^-}$), (*c*) $k'_{10} = 0$, and (d) the potential pulses drive the electrode reactions so that the concentration of the species undergoing electron transfer at the electrode surface is instantaneously reduced to zero at the moment of potential application or switching (double potential step mode of generation). The slope of the plot of the logarithmic term versus $(t_r/t_f)^{1/2}$ approaches a maximal negative value of -1.45 as indicated in equation (11′) when the parameter $k_7' t_f C_A \geqslant 10^3$.

Lansbury et al. (171) have found, from experiments with rubrene in solvents of acetonitrile and dimethylformamide using tetrabutylammonium perchlorate (TBAP) as supporting electrolyte, that the plots of log I versus $(t_r/t_f)^{1/2}$ were linear over a time of 1.5 to 10 sec with the slopes of the lines averaging -1.5 and corresponding closely to the predicted values. These results have clearly indicated that $k_7' \geqslant 10^5$ liters mole^{-1} sec^{-1} and that, within the limitations of the experiments, the formation of the excited species was diffusion controlled. Table I′ summarizes some quantum-yield data obtained by these workers.

TABLE I′

QUANTUM YIELD OF FLUORESCENCE (Φ_f) AND ELECTROLUMINESCENCE (Φ_e) FOR RUBRENE[a] (171)

Rubrene conc., M	Solvent	Φ_e	Φ_f
10^{-3}	Dimethylformamide		0.48
10^{-4}	Acetonitrile		0.40
10^{-4}	Acetonitrile	0.11[b]	
10^{-4}	Acetonitrile	0.01[c]	
10^{-4}	Acetonitrile	0.01[d]	

[a] Standard $\Phi_f = 1.00$, rubrene in hexane.
[b] $t_f = 0.7$ sec, O_2 conc. 10^{-6} M; A$^-$ produced first.
[c] $t_f = 10$ sec, O_2 conc. 5×10^{-5} M; A$^-$ produced first.
[d] $t_f = 0.7$ sec, O_2 conc. 10^{-6} M; A$^+$ produced first.

Visco (172), for rubrene in benzonitrile, and Maricle (173), for rubrene in dimethylformamide, have also obtained linear plots for the Feldberg relationship, except that their slopes were -1.87 and -3.8, respectively. It should be pointed out that these results represent preliminary experiments and that the differences may be due to instrumental reasons and solution purity, particularly the level of dissolved oxygen present.

In addition to electroluminescence from the annihilation reaction of radical cation and anion, Maricle (*173*) has observed light emission from solutions of 1,3,6,8-tetraphenylpyrene and 9,10-diphenylanthracene in dimethylformamide and from 0.1 M TBAP during oxidation of the anion or reduction of the cation at voltages too low to generate the oppositely charged ion, and has suggested a possible mechanism involving electrochemical production of triplet states which subsequently undergo triplet-triplet annihilation to produce an excited singlet state:

$$A + e^- \rightarrow A^- \tag{12'}$$

$$A^- \rightarrow A^*(T) + e^- \tag{13'}$$

$$A^*(T) + A^*(T) \rightarrow A^*(S_1) + A \tag{14'}$$

$$A^*(S_1) \rightarrow A(S_0) + h\nu \tag{15'}$$

Electron transfer to produce directly the excited singlet was ruled out on the basis of energy arguments.

The author wishes to express sincere appreciation to Drs. Srinivasan, Delahay, Roe, Feldberg, Visco, and Maricle for providing results of their work prior to publication.

ACKNOWLEDGMENT

The author gratefully acknowledges financial support during the period of this work by Grant No. 11670 from the Public Health Service, National Institute of Health. Helpful comments on the manuscript by J. Bader and D. Bell are greatly appreciated.

REFERENCES

1. E. Becquerel, *Compt. Rend.*, **9**, 58, 561, 711 (1839).

2. A. W. Copeland, O. D. Black, and A. B. Garrett, *Chem. Rev.*, **31**, 177 (1942).

3. K. W. Sancier, *Conference on the Use of Solar Energy, The Scientific Bases*, Vol. 5, University of Arizona, 1955, Chap. 6.

4. N. P. Sluginov, *J. Phys. Theor. and Appl. (Ser. 2)*, **3**, 465 (1884).

5. W. D. Bancroft and H. B. Weiser, *J. Phys. Chem.*, **18**, 762 (1914).

6. V. V. Mikho and L. V. Levkovtseva, *Opt. and Spectr. (USSR)*, **12**, 365 (1962).

7. N. Harvey, *J. Phys. Chem.*, **33**, 1456 (1929).

8. H. F. Ivey, Part I in *IRE Trans. Electron Devices*, ED6, 203 (1959); Part II in *J. Electrochem. Soc.*, **108**, 590 (1961); Part III in *Electrochem. Technol.*, **1**, 42 (1963).

9. H. F. Ivey, *Electroluminescence and Related Effects; Advances in Electronics and Electron Physics*, Academic Press, New York, 1963.

10. J. N. Pitts, Jr., F. Wilkinson, and G. S. Hammond, The "Vocabulary" of Photochemistry, in *Advances in Photochemistry*, W. A. Noyes, G. S. Hammond, and J. N. Pitts, Jr., eds.), Wiley (Interscience), New York, 1963.

11. C. N. Reilley and R. W. Murray, Introduction to Electrochemical Techniques, in *Treatise on Analytical Chemistry*, Part 1, Vol. 4 (I. M. Kolthoff and P. J. Elving, eds.), Wiley (Interscience), 1963, Chap. 43.

12. E. Rabinowitch, *J. Chem. Phys.*, **8**, 560 (1940).

13. M. Eisenberg and H. P. Silverman, *Electrochim. Acta*, **5**, 1 (1961).

14. H. Berg, *Naturwiss.*, **47**, 320 (1960).

15. H. Berg and H. Schweiss, *Naturwiss.*, **47**, 513 (1960).

16. G. Ciamician and P. Silber, *Ber.*, **33**, 2911 (1900); **34**, 1541 (1901).

17. G. Porter and F. Wilkinson, *Trans. Faraday Soc.*, **57**, 1686 (1961).

18. H. L. J. Bäckström and K. Sandros, *Acta Chem. Scand.*, **14**, 48 (1960); **12**, 823 (1958).

19. J. N. Pitts, Jr., R. L. Letsinger, R. P. Taylor, J. M. Patterson, G. Recktenwald, and R. B. Martin, *J. Am. Chem. Soc.*, **76**, 1068 (1954).

20. G. S. Hammond and W. M. Moore, *J. Am. Chem. Soc.*, **81**, 6334 (1959).

21. W. M. Moore, G. S. Hammond, and R. P. Foss, *J. Am. Chem. Soc.*, **83**, 2789 (1961).

22. G. S. Hammond, W. P. Baker, and W. M. Moore, *J. Am. Chem. Soc.*, **83**, 2795 (1961).

23. A. Beckett and G. Porter, *Trans. Faraday Soc.*, **59**, 2038 (1963) and pertinent references therein.

24. R. M. Hochstrasser and G. B. Porter, *Quart. Rev. (London)*, **14**, 146 (1960).

25. O. L. Chapman, "Photochemical Rearrangements of Organic Molecules" in *Advances in Photochemistry*, Vol. 1 (W. A. Noyes, Jr., G. S. Hammond, and J. N. Pitts, Jr., eds.), Wiley (Interscience), New York, 1963, pp. 323–420.

26. G. S. Hammond and N. J. Turro, *Science.* **142**, 1541 (1963).

27. J. N. Pitts, Jr., H. W. Johnson, Jr., and T. Kuwana, *J. Phys. Chem.*, **66**, 2456 (1962).

28. N. C. Yang and C. Rivas, *J. Am. Chem. Soc.*, **83**, 2213 (1961).

29. G. S. Hammond and P. A. Leermakers, *J. Am. Chem. Soc.*, **84**, 207 (1962).

30. F. Wilkinson, *J. Phys. Chem.*, **66**, 2569 (1962).

31. A. Beckett and G. Porter, *Trans. Faraday Soc.*, **59**, 2051 (1963).

32. G. Porter and P. Suppan, *Proc. Chem. Soc.* (1964), 191.

33. W. Bartok, P. J. Lucchesi, and N. S. Snider, *J. Am. Chem. Soc.*, **84**, 1842 (1962).

34. C. A. Hutchinson and B. W. Mangrum, *J. Chem. Phys.*, **29**, 952 (1958); **34**, 908 (1961).

35. J. H. van der Waals and M. S. deGroot, *Mol. Phys.*, **2**, 233 (1959); **3**, 190 (1960).

36. L. H. Piette, J. H. Sharp, T. Kuwana, and J. N. Pitts, Jr., *J. Chem. Phys.*, **36**, 3094 (1962).

37. A. Terenin and V. Ermolaev, *Trans. Faraday Soc.*, **52**, 1042 (1956).

38. G. Porter, *Proc. Chem. Soc.* (1959), 291.

39. C. A. Parker and C. G. Hatchard, *Proc. Chem. Soc.*, (1962) 147, 386; *Proc. Roy. Soc. (London)*, (1962) A269, 574; *Trans. Faraday Soc.*, **59**, 284 (1963).

40. B. Stevens, M. S. Walker, and E. Hutton, *Proc. Chem. Soc.*, (1963) 62, 181.

41. J. P. Simons, *Quart. Rev. (London)*, **13**, 1 (1959).

42. P. De Mayo and S. T. Reid, *Quart. Rev. (London)*, **15**, 393 (1961).

43. C. Reid, *Quart. Rev. (London)*, **12**, 205 (1958).

44. L. J. Heidt, R. S. Livingston, and E. Rabinowitch, and F. Daniels, eds., *Photochemistry in the Liquid and Solid States*, Wiley, New York, 1960.

45. W. A. Noyes, Jr., G. S. Hammond, and J. N. Pitts, Jr., eds., *Advances in Photo chemistry*, Vol. 1, Wiley (Interscience), New York, 1963; Vol. II, 1964,

46. I. Levin and C. E. White, *J. Chem. Phys.*, **18**, 417 (1950); **19**, 1079 (1951).

47. I. Levin, J. R. Wiebush, M. B. Bush, and C. E. White, *J. Chem. Phys.*, **21**, 1654 (1953).

48. J. J. Surash, "Studies on Photo-Induced Luminescence and Electrode Potentials," Ph.D. Thesis, Lehigh University, Bethlehem, Pa., 1960.

49. J. J. Surash and D. M. Hercules, *J. Phys. Chem.*, **66**, 1602 (1962).

50. J. L. Bolland and H. R. Cooper, *Proc. Roy. Soc. (London)*, **A255**, 405 (1954).

51. E. W. Abrahamson, I. Panik, and K. V. Sarkaneis, *Proc. Cellulose Conf. 2nd*, *Syracuse*, May 1959.

52. T. Kuwana, *Anal. Chem.*, **35**, 1398 (1963).

53. N. K. Bridge and G. Porter, *Proc. Roy. Soc. (London)*, **A244**, 259 (1958).

54. C. F. Wells, *Trans. Faraday Soc.*, **57**, 1719 (1961).

55. H. Berg, *Naturwiss.*, **49**, 11 (1962).

56. D. Kalab, *Experientia*, **19**, 392 (1963); *CA*, **59**, 8294h (1963).

57. H. Berg, Fifth International Symposium on Free Radicals, Preprint 8, Uppsala, Sweden, 1961.

58. H. Berg and H. Schweiss, *Nature*, **191**, 1270 (1961).

59. A. F. Potter, Jr. and L. H. Thaller, *Solar Energy*, **3**, 1 (1959).

60. L. J. Miller and W. E. McKee, "A Feasibility Study of a Thionine Photogalvanic Power Generation System," Proj. #3145 Tech. Progress Report #3, contract AF33(616)-7911, Sunstrand Aviation, Denver, 1961.

61. F. Juston-Coumat, *Rev. Gen. Electron.*, **16**, 39 (1962).

62. H. Silverman, W. R. Momyer, and M. Eisenberg, *Ann. Proc. Power Sources Conf.*, **15**, 53 (1961).

63. J. N. Pitts, Jr., J. D. Margerum, and W. E. McKee, *Am. Rocket Soc.*, (1961) 890.

64. W. E. McKee, J. D. Margerum, E. Findl, and W. B. Lee, *Soc. Auto. Eng.*, *Preprint*, **179c**, 8 (1960).

65. M. Sano and H. Akamatsu, *Bull. Chem. Soc. Japan*, **36**, 480 (1963).

66. H. P. Kallman and M. Pope, *J. Chem. Phys.*, **32**, 300 (1960).

67. J. Szychlinski and Z. Pawlak, *Roczniki Chem.*, **37**, 1547 (1963); *CA*, **60** 15281 (1964).

68. S. Paszyc, *Zeszyty Nauk. Uniw. Poznaniu, Mat., Fiz. Chem.*, No. 6, 68 (1962); *CA*, **60**, 11481e (1964).

69. S. A. Greenberg and H. P. Silverman, private communication, 1965.

70. E. K. Putseiko, *Dokl. Akad. Nauk SSSR*, **150**, 343 (1963); *CA*, **59** 8238e (1963).

71. R. Williams, *J. Chem. Phys.*, **32**, 1505 (1960).

72. F. P. Bowden, *Trans. Faraday Soc.*, **27**, 505 (1931).

72a. M. Heyvrosky and R. G. W. Norrish, *Nature*, **200**, 880 (1963).

73. V. I. Ginzburg and V. I. Veselovskii, *Zh. Fiz. Khim.*, **24**, 366 (1950).

74. P. J. Hillson and E. K. Rideal, *Proc. Roy. Soc. (London)*, **A199**, 295 (1949).

75. R. Audubert, *Compt. Rend.*, **189**, 800 (1929).

76. J. M. Blocker, Jr. and A. B. Garrett, *J. Am. Chem. Soc.*, **69**, 1594 (1947).

77. P. E. Clark and A. B. Garrett, *J. Am. Chem. Soc.*, **61**, 1805 (1939).

78. J. P. E. Duclaux, *Compt. Rend.*, **200**, 1838 (1935).

79. H. Berg, *Z. Chem.*, **12**, 479 (1963).

80. H. Berg and H. Schweiss, *Electrochim. Acta*, **9**, 425 (1964).

81. G. C. Barker and A. W. Gardner, preprint of paper presented at CITCE Meeting Moscow 1964.

82. V. Sihvonen, *Ann Acad. Sci. Fennincae*, **24A**, 3 (1925); *CA*, **20**, 3270 (1926).

83. G. Price, Ph.D. dissertation, Cambridge University (England) 1938 (see Ref. 74).

84. E. Adler, *J. Chem. Phys.*, **8**, 500 (1940).

85. J. O'M. Bockris, ed., *Modern Aspects of Electrochemistry*, Butterworths, London, Vol. 1, 1954, pp. 243–246.

86. S. Glasstone, K. J. Laidler, and H. Eyring, *The Theory of Rate Processes*, McGraw-Hill, New York, 1941, p. 599.

87. L. M. Dorfman and I. A. Taub, *J. Am. Chem. Soc.*, **85**, 2370 (1963).

88. V. I. Veselovskii, *Tr. Soveshch. po Electrokhim.*, *Akad. Nauk SSSR, Otd. Khim. Nauk, Moscow 1950*, **1953**, 47.

89. G. Grube and L. Baumeiser, *Z. Elektrochem.*, **30**, 332 (1924).

90. V. I. Veselovskii, *Zh. Fiz. Khim.*, **22**, 1427 (1948).

91. V. I. Veselovskii, *Zh. Fiz. Khim.*, **23**, 1095 (1949).

92. V. I. Ginzburg and V. I. Veselovskii, *Zh. Fiz. Khim.*, **24**, 366 (1950).

93. V. I. Veselovskii, *Zh. Fiz. Khim.*, **26**, 509 (1952).

94. V. I. Veselovskii, *Zh. Fiz. Khim.*, **21**, 983 (1947).

95. N. Hayami, *Rev. Phys. Chem. Japan*, **2**, No. 3, 166 (1937).

96. W. J. D. Van Dijck, *Trans. Faraday Soc.*, **21**, 630 (1925).

97. V. E. Kasgkarev, K. M. Kosonogoya, *Zh. Eksperim. i Theor. Fiz.*, **18**, 927 (1948); transl., see Nucl. Sci. Abstr., AD-45886.

98. J. Kruger, *J. Appl. Phys.*, **28**, 1212 (1957).

99. D. J. G. Ives and A. E. Rawson, *J. Electrochem. Soc.*, **109**, 447 (1962).

100. D. A. Chance and K. Nobe, *J. Appl. Phys.*, **34**, 1824 (1963).

101. H. A. Arbit and K. Nobe, "Effect of Monochromatic Illumination on the Electrode Potential of Cu and Sn in Aqueous Solutions," Rept. No. 62-40, Water Resources Center Contribution No. 57, Univ. of California, Los Angeles, Calif., 1962.

102. V. I. Veselovskii, *Zh. Fiz. Khim.*, **15**, 145 (1941).

103. V. I. Veselovskii, *Zh. Fiz. Khim.*, **22**, 1302 (1948).

104. V. I. Veselovskii, *Zh. Fiz. Khim.*, **20**, 269 (1946).

105. R. W. Pittman, *J. Chem. Soc.*, **1949**, 1811 (Part II).

106. R. W. Pittman, *J. Chem. Soc.*, **1953**, 855.

107. R. W. Pittman, *J. Chem. Soc.*, **1953**, 3888.

108. V. I. Ginzburg and V. I. Veselovskii, *Zh. Fiz. Khim.*, **26**, 60 (1952).

109. R. Wiesner, "Der p, n Photoeffect," in *Halbleiter Probleme* (W. Schottky, ed.), Friedrl Vieweg, Braunschweig, 1956, p. 59–74.

110. W. L. Brown, W. H. Brattain, C. G. B. Garrett, and H. C. Montgomery, "Field Effect and Photoeffect Experiments on Germanium Surfaces," *Semiconductor Surface Physics* (R. H. Kingston, ed.), Univ. Penn. Press, Philadelphia, 1957, p. 111.

111. W. H. Brattain and J. Bardeen, *Bell System Tech. J.*, **32**, 1 (1953).

112. W. H. Brattain and C. G. B. Garrett, *Bell System Tech. J.*, **34**, 129 (1955).

113. H. Gerischer, "Semiconductor Electrode Reactions" in *Advances in Electrochemistry and Electrochemical Engineering*, Vol. 1 (P. Delahay, ed.), Wiley (Interscience), 1961, p. 139–232.

114. J. F. Dewald, "Semiconductor Electrodes," in ACS Monograph No. 140, *Semiconductors* (N. B. Hannay, ed.), Reinhold Publ. Corp., New York, 1959, Chapter 17.

115. H. R. Gobrecht, R. Kuhnkies, and A. Tausend, *Z. Electrochem.*, **63**, 541 (1959).
116. H. N. Seiger, The Photogalvanic Effect in Inorganic Oxidation-Reduction Systems Ph.D. Thesis, Polytechnic Institute of Brooklyn, N.Y., 1962. (University Microfilms, Inc., Ann Arbor, Mich. #62-5635.)
117. D. Booth and R. M. Noyes, *J. Am. Chem. Soc.*, **82**, 1868 (1960).
118. L. F. Meadows and R. M. Noyes, *J. Am. Chem. Soc.*, **82**, 1872 (1960).
119. E. K. Rideal and E. G. Williams, *J. Chem. Soc.*, **1925**, 258.
120. G. B. Kistiakowsky, *J. Am. Chem. Soc.*, **52**, 4770 (1930).
121. R. G. Dickinson and S. F. Ravitz, *J. Am. Chem. Soc.*, **49**, 976 (1927).
122. F. S. Dainton and S. A. Sills, *Bull. Soc. Chim. Belges*, **71**, 801 (1962).
123. J. Jortner, M. Ottolenghi, and G. Stein, *J. Am. Chem. Soc.*, **67**, 1271 (1963).
124. G. Stein, *Bull. Res. Council Israel*, **A10**, 127 (1961).
125. J. Jortner and G. Stein, *J. Phys. Chem.*, **66**, 1258 (1962).
126. F. Kaufman, *Progr. Reaction Kinetics*, **1**, 1 (1961).
127. R. W. Reid and T. M. Sugden, *Discussions Faraday Soc.*, **33**, 213 (1962).
128. L. F. Phillips and T. M. Sugden, *Trans. Faraday Soc.*, **57**, 2188 (1961).
129. F. H. Johnson, E. H. C. Sie and Y. Haneda, "The Luciferin-Luciferase Reaction," in *Light and Life* (W. D. McElroy and B. Glass, eds.), Johns Hopkins Press, Baltimore, 1961, pp. 206–218.
130. W. D. McElroy and H. H. Seliger, "Mechanisms of Bioluminescent Reactions," in *Light and Life* (W. D. McElroy and B. Glass, eds.), Johns Hopkins Press, Baltimore, 1961, pp. 219–257.
131. R. Audubert, *Trans. Faraday Soc.*, **35**, 197 (1939).
132. R. F. Vasil'ev, *Dokl. Akad. Nauk SSSR*, **124**, 1258 (1959).
133. R. F. Vasil'ev, *Nature*, **194**, 1276 (1962); **196**, 668 (1962).
134. J. Stauff, H. Schmidkunz, and G. Hartmann, *Nature*, **198**, 281 (1963).
135. J. Stauff and H. Schmidkunz, *Z. Physik. Chem. Neue Folge* (*Frankfurt*), **33**, 273 (1962).
136. R. F. Vasil'ev and A. A. Vichutinskii, *Dokl. Akad. Nauk SSSR*, **142**, 615 (1962).
137. I. V. Zakharov and V. Ya. Shlyapintokh, *Dokl. Akad. Nauk SSSR*, **150**, 1069 (1963).
138. E. J. Bowen and R. A. Lloyd, *Proc. Roy. Soc.* (*London*), **A275**, 465 (1963).
139. H. H. Seliger, *Anal. Biochem.*, **1**, 60 (1960).
140. H. H. Seliger, *J. Chem. Phys.*, **40**, 3133 (1964).
141. A. U. Khan and M. Kasha, *J. Chem. Phys.*, **39**, 2105 (1963).
142. R. J. Browne and E. A. Ogryzlo, *Proc. Chem. Soc.*, (1964) 117.
143. E. McKeown and W. A. Waters, *Nature*, **203**, 1063 (1964).
144. R. F. Vasil'ev and A. A. Vichutinskii, *Nature*, **194**, 1276 (1962).
145. For a list of pertinent references see E. H. White in *Light and Life* (W. D. McElroy and B. Glass, eds.), Johns Hopkins Press, Baltimore, 1961; and R. M. Acheson in "Acridines," in *The Chemistry of Heterocyclic Compounds* (A. Weissberger, ed.). Wiley (Interscience), New York, 1956, p. 281.
146. R. Iwaki and I. Kamiya, *Nippon Kagaku Zasshi*, **78**, 1613 (1957).
147. E. A. Chandross, *Tetrahedron Letters*, **1963** (12), 761.
148. American Cyanamid Co., "Chemiluminescent Materials," Technical Report No. 5, Contract NONR 4200(00); Ad 606989, U.S. Department of Commerce, Office of Technical Services; Washington D.C.
149. F. Braun, *Ann. Physik. Chem.*, **65**, 361 (1898).

150. E. N. Harvey, *A History of Luminescence*, The American Philosophical Society, Philadelphia, 1957, pp. 251–309.
151. A. Hickling and M. D. Ingram, *J. Electroanal. Chem.*, **8**, 65 (1964).
152. E. N. Harvey, *J. Phys. Chem.*, **33**, 1456 (1929).
153. E. N. Harvey, *J. Gen. Physiol.*, **5**, 275 (1923).
153a. S. Anderson, *J. Appl., Phys.* **14**, 601 (1943).
154. T. Kuwana, *J. Electroanal. Chem.*, **6**, 164 (1963).
155. E. A. Chandross and F. I. Sonntag, *J. Am. Chem. Soc.*, **86**, 3179 (1964); *Chem. Eng. News*, **42**, 32 (1964).
156. E. A. Chandross and R. Visco, *J. Am. Chem. Soc.*, **86**, 5350 (1964).
157. V. Ya. Shlyapintokh, I. M. Postnikov, O. N. Karpukhin, and A. Ya. Veretil'nyi, *Zh. Fiz. Khim.*, **37**, 2374 (1963).
158. D. M. Hercules, *Science*, **145**, 808 (1964).
159. K. S. V. Santhanam and A. J. Bard, *J. Am. Chem. Soc.* **87**, 121 (1965).
160. J. Bader and T. Kuwana, *J. Electroanal. Chem.*, **10**, 104 (1965).
161. A. Hickling and G. R. Newns, *J. Chem. Soc. (London)*, (1961) 5186.
162. A. Hickling and M. D. Ingram, *Trans. Faraday Soc.*, **60**, 783 (1964).
163. V. Vojir, *Collection Czech. Chem. Commun.*, **19**, 862 (1954).
164. A. Bernanose, T. Bremer, and P. Goldfinger, *Bull. Soc. Chim. Belges*, **56**, 269 (1947).
165. T. Kuwana, B. Epstein, and E. T. Seo, *J. Phys. Chem.*, **67**, 2243 (1963).
166. B. Epstein and T. Kuwana, *Photochem. Photobiol.*, **4**, 1157 (1965).
167. For comparison, see table in J. Lee and H. H. Seliger, preprint of papers, "Symposium on Chemiluminescence," Duke University, Durham, N.C., April 1965, pp. 129–196.
168. P. Delahay and V. S. Strinivasan, *J. Phys. Chem.*, in press.
169. E. A. Chandross, J. W. Longworth, and R. E. Visco, *J. Am. Chem. Soc.*, **87**, 3259 (1965).
170. S. W. Feldberg, *J. Am. Chem. Soc.*, in press (presented at Winter Meeting of American Chemical Society, January 1966, Phoenix, Arizona).
171. R. C. Lansbury, D. M. Hercules, and D. K. Roe, "Experimental Verification of Diffusion Theory of Controlled Potential Electrogeneration of Chemiluminescence" (presented at Winter Meeting of American Chemical Society, January 1966, Phoenix, Arizona).
172. R. E. Visco, private communication, 1966.
173. D. L. Maricle, private communication, 1966.

THE ELECTRICAL DOUBLE LAYER

PART I.

ELEMENTS OF DOUBLE-LAYER THEORY

<hr>

David M. Mohilner†

DEPARTMENT OF CHEMISTRY
UNIVERSITY OF TEXAS
AUSTIN, TEXAS

I. Introduction 242
  A. Definition and Scope 242
  B. Qualitative Description of the Double Layer 243
  C. Usefulness of Double-Layer Studies in Electroanalytical
    Chemistry 248
II. Double-Layer Thermodynamics 249
  A. Classification of Electrodes and Electrode Processes 249
  B. Electrocapillary Curves: Qualitative Discussion 251
  C. Derivation of the Gibbs Adsorption Equation 254
  D. Derivation of the Electrocapillary Equation for Ideal Polarized
    Electrodes 263
  E. Physical Implications of the Electrocapillary Equation 275
  F. The Electrocapillary Equation for Reversible Charge-Transfer
    Electrodes 293
III. Adsorption and Double-Layer Models 298
  A. Introduction 298
  B. Classification of Measures and Types of Adsorption 301
  C. The Diffuse Layer 306
  D. Specific Ionic Adsorption 331
  E. Adsorption of Neutral Compounds 352
Appendixes
  A. Equivalence of the Gibbsian and the Interphase Approaches .. 391
  B. Examples of the Elimination of the Electrochemical Potential of
    the Indicator Ion 393
  C. Permittivity and Electrical Units 396

† Welch Postdoctoral Fellow, University of Texas, 1964–1965. Present Address: Department of Chemistry, Colorado State University, Fort Collins, Colorado.

Appendixes—*continued*
   D. Effect of Varying Ionic Size on the Predictions of Diffuse-Layer
      Theory   397
   Glossary of Symbols   401
   References   405

I. INTRODUCTION

A. Definition and Scope

Electrical double layer is a term used to denote the arrays of charged particles and/or oriented dipoles believed to exist at every material interface. In this chapter, the term will be restricted to interfaces formed by metals in contact with electrolyte solutions. For information on electrical double layers at semiconductor electrodes or in fused salts, the reader is referred to recent reviews (*1–5*).

Strictly speaking, electrical double layer is a misnomer, in view of modern ideas about the structure of the interfacial region. A more descriptive term would be electrochemical multilayer, since (a) the forces which lead to its formation include, in addition to long-range electrostatic forces, shorter-range forces of types usually considered chemical, and (b) the interfacial region consists of not two, but at least three, and sometimes more, distinct subregions or layers. However, the old name has persisted for so long and has been used so widely in the electrochemical literature that an attempt to change it now would lead to confusion. Hence electrical double layer will refer in this chapter to the interfacial region formed by a metal electrode dipping into an electrolyte solution, regardless of the actual structural complexity of the phase boundary.

The aim of this chapter is to provide an elementary, but reasonably rigorous, treatment of those parts of the theory of the electrical double layer which may be of interest to electroanalytical chemists. The thermodynamic theory of electrocapillarity will be treated in detail, because an understanding of this theory is prerequisite to an appreciation of the physical significance of double-layer parameters, such as *relative surface excess* and *differential capacitance*, which are useful in the discussion of electrode kinetics. An elementary discussion of the structural theory of the double layer will be given as background for a study of the role of adsorption in the kinetics of faradaic processes. However, neither an exhaustive nor a critical review of the more recent refinements of the structural theory will be attempted. Emphasis will be placed on those aspects of double-layer structure which seem useful in accounting for double-layer effects in electrode kinetics. For discussions of aspects of the electrical double layer either not covered in this chapter or covered in more detail elsewhere, as well as for extensive bibliographies, the reader is referred to Delahay's recent book, *Double Layer and Electrode Kinetics* (*5*) and to the reviews of Grahame (*6,7*), Parsons (*8,9*), Delahay (*10,130*), Breiter (*11*), Frumkin (*12,13*), Frumkin and Damaskin (*14*) Macdonald and Barlow (*15*), Reilley and Stumm (*131*), and Devanathan (*132*). Part II of this discussion, to be published in a later volume of this series, will cover double-layer effects in electrode kinetics and experimental methods for studying the double layer.

B. Qualitative Description of the Double Layer

The modern qualitative picture of the electrical double layer at a metal-electrolyte solution interface is based on a model proposed by Grahame (*6,16,17*), which is a modification of an earlier model due to Stern (*18*). A schematic diagram is shown in Fig. 1. Briefly, the double layer consists of three main parts: (a) the *metallic phase*, and, on the solution side of the interface, (b) an *inner layer* only a few molecular diameters thick located next to the metal surface, and (c) an outer, or *diffuse layer*, which is really a three-dimensional region extending all the way into the bulk of the solution.

In general, the metallic phase bears a net electrical charge on its surface because of an excess or deficit of electrons. This electronic excess or

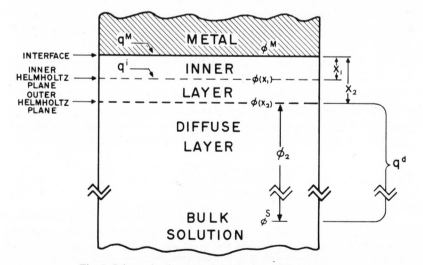

Fig. 1. Schematic diagram of the electrical double layer.

deficit may either be imposed on the metal by means of an external source of electric current, or it may be produced on the metal by the action of a faradaic (charge-transfer) electrode process. The excess charge resides in so thin a layer in the metallic surface that it may be considered to be effectively two-dimensional.[†] The density of excess electronic charge on the metal surface is denoted by the symbol q^M and is usually expressed in microcoulombs per square centimeter. $q^M < 0$ implies an excess of electrons; $q^M > 0$ implies a deficit. It will be shown (Sec. II.E.1.a and c) that q^M has a well-defined thermodynamic significance and can be measured absolutely. In contrast, the *inner electric potential* ϕ^M of the metallic phase cannot be measured, although differences in it can be.[‡]

The inner layer (also called the *compact, rigid, Helmholtz*, or *Stern layer*) on the solution side of the interface contains solvent molecules and sometimes other neutral molecules adsorbed on the metal surface. In addition, in most electrolyte solutions the inner layer contains a fractional monolayer of ions (usually anions) which are said to be *specifically adsorbed*. The locus of the electrical centers of these adsorbed ions is called the *inner Helmholtz plane* (IHP). Occasionally, the IHP is referred to simply as the Helmholtz plane (*20,21*). The electric charge density resulting from this layer is denoted by q^i (usually μcoulombs cm^{-2}). The electric potential at the IHP is denoted by $\phi(x_1)$, where the distance of the IHP from the metal surface is given by x_1, which is approximately equal to the radius of the (unsolvated) ion which is specifically adsorbed.

The detailed nature of the forces of interaction between an ion and the metal surface which bring about specific adsorption has not yet been wholly resolved. [For a critical review, see MacDonald and Barlow (*15*).] What is known from experiment is that some ionic species can (under given conditions) be adsorbed in the inner region, whereas others cannot. The ability of an ion to penetrate into the inner layer and become adsorbed at the IHP thus, obviously, depends on some peculiar properties of the ion; this, simply, is the origin of the term *specific adsorption*. Grahame et al. (*6,22*) suggested that specific adsorption involves the formation of some sort

[†] In contrast, the excess charge in a semiconductor phase is a three-dimensional-space charge which extends an appreciable distance into the bulk of the semiconductor. Because of this difference between metals and semiconductors, double-layer phenomena at semiconductor electrodes are conveniently treated as a separate subject.

[‡] For a lucid discussion of the definitions and physical meanings of *inner electric potential, outer electric potential, surface potential, real potential, chemical potential*, and *electrochemical potential* see Parsons' review (*8*). An introduction to the application of electrochemical and inner potentials to the thermodynamics of galvanic cells and to electrode kinetics has been published by Van Rysselberghe (*19*).

of covalent bond between the ion and the metal surface, at least in the case of the specific adsorption of anions at mercury. Recently, however, objections have been raised to such covalent bonding as the "cause" of specific adsorption, and it has been suggested instead that electrical-image energy (*23*) or the degree and type of ionic solvation (*21,24*) are the chief factors which must be considered. Whatever the actual nature of the forces of interaction which bring about specific adsorption† may be, there does appear to be general agreement that, before an ion can penetrate into the inner layer and become adsorbed at the IHP, it must first lose its solvation sheath, at least in the direction toward the metal. This idea is illustrated pictorially in Fig. 2. There the small circles containing arrows represent solvent dipoles. The large circles represent specifically adsorbed anions, and it will be noticed that they have lost their solvation sheaths in the direction toward the metal.

Whenever the interaction between an ion and the metal is not strong enough to bring about desolvation, the ion will not be able to come as close to the metal as specifically adsorbed ions. The closest the electrical centers of solvated ions can get is a distance $x_2 > x_1$. The imaginary plane passing through the electrical centers of the closest approaching solvated ions is known as the *outer Helmholtz plane* (OHP). [The OHP is sometimes called the *Gouy plane* (*20,21*).] In Fig. 2 the shaded clusters represent solvated cations at the OHP. Further, it is now generally believed that in the case of aqueous electrolyte solutions the closest approaching solvated ions will be separated from the metal surface by a layer of oriented solvent molecules on the metal surface. This is also illustrated in Fig. 2.

In contrast to the complicated case of specific adsorption, in which various types of short-range forces of interaction must be taken into account, the interaction of the solvated ions with the metal may be accounted for by ordinary long-range coulombic forces. The interaction of these ions with the metal is essentially independent of their chemical properties, except for their ionic charges. Such ions are therefore said to be *nonspecifically adsorbed*. Positive nonspecific adsorption implies electrostatic attraction of the ion toward the metal; negative nonspecific adsorption implies electrostatic repulsion. Unlike specifically adsorbed ions at the IHP which form a two-dimensional monolayer, the nonspecifically adsorbed ions are not all located at the OHP but are contained in a three-dimensional region, the *diffuse layer*, which extends all the way from the OHP into the bulk of the solution (cf. Fig. 1). The origin of the diffuse

---

† Some workers (*21,24,25*) now prefer to use the term *superequivalent adsorption* in place of specific adsorption.

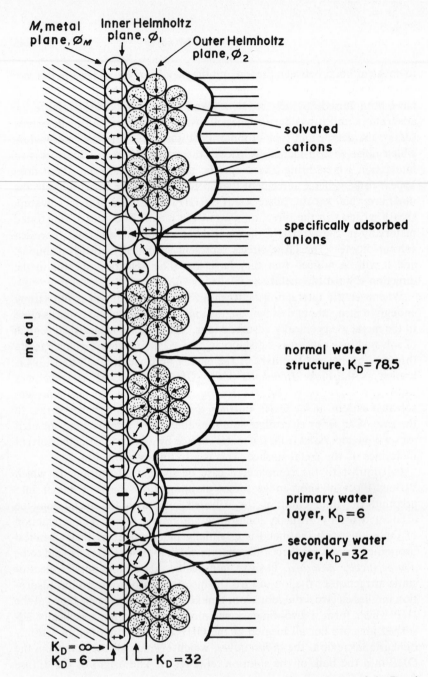

Fig. 2. Model of the electrical double layer [from (*21*)]. (By permission of the Royal Society.)

246

character of this region is thermal agitation, which provides a disordering force opposed to the ordering coulombic forces of attraction and repulsion. The net result is that the diffuse layer behaves as an "ionic atmosphere" of the electrode, and indeed, the equations which describe its behavior greatly resemble the Debye-Hückel equations for an electrolyte (cf. Sec. III.C).

The excess charge density in the diffuse layer is denoted q^d. Although this quantity is expressed in units of charge per unit area of interface (μcoulombs cm^{-2}), q^d is really a three-dimensional-space charge. q^d is the total excess charge in a column of solution of unit cross section extending from the OHP all the way into the bulk of the solution. As a practical matter, however, it may be noted that the effective perturbation of the structure of the electrolyte which occurs inside the diffuse layer actually dies out quite rapidly as one recedes from the electrode surface. For example, the thickness of the region next to the electrode inside which 99.99 per cent of the diffuse-layer perturbation occurs ($\tau_{99.99}$) is shown in Table I

TABLE I

EFFECTIVE THICKNESS OF THE DIFFUSE LAYER
FOR $z - z$ ELECTROLYTES AT 25°C

Concentration, gram-formula weight liter^{-1}	Ionic charge z	$\tau_{99.99}$ cm
10^{-6}	1	2.8×10^{-4}
	2	1.4×10^{-4}
	3	9.4×10^{-5}
10^{-5}	1	8.9×10^{-5}
	2	4.4×10^{-5}
	3	3.0×10^{-5}
10^{-4}	1	2.8×10^{-5}
	2	1.4×10^{-5}
	3	9.4×10^{-6}
10^{-3}	1	8.9×10^{-6}
	2	4.4×10^{-6}
	3	3.0×10^{-6}
10^{-2}	1	2.8×10^{-6}
	2	1.4×10^{-6}
	3	9.4×10^{-7}
10^{-1}	1	8.8×10^{-7}
	2	4.4×10^{-7}
	3	3.0×10^{-7}

for $1:1, 2:2$, and $3:3$ electrolytes at $25°C$. $\tau_{99.99}$ has the following meaning: If we denote the electric potential at the OHP by $\phi(x_2)$ and the inner potential in the bulk of the solution by ϕ^S, the total difference of potential across the diffuse layer is $\phi_2 = \phi(x_2) - \phi^S$. Now, at a distance $\tau_{99.99}$ from the OHP, 99.99 per cent of this potential difference will have occurred. Since the thickness of the inner layer is extremely small, the thickness of the entire double layer is essentially given by $\tau_{99.99}$. Note that this effective thickness, which is only about 3×10^{-4} cm for a 10^{-6} F solution of a $1:1$ electrolyte, decreases rapidly with increasing concentration and with increasing charge type (cf. Sec. III.C.4). Thus in a 10^{-2} F solution of a $1:1$ electrolyte, the effective thickness of the double layer has diminished to about 3×10^{-6} cm. Compared to the thickness of the diffusion layer produced at an electrode by a faradaic process, the thickness of the double layer is negligibly small in most cases. This has the important consequence that it will be permissible to *neglect the existence of the double layer* in the solution of boundary-value problems for mass transport in most cases of electroanalytical interest. After mass transport has been accounted for in the usual way, a double-layer correction can be made on the electrode kinetics as a separate step.

The entire interfacial region at an electrode must, of course, be electrically neutral. This means that the total excess charge density on the solution side of the double layer must be equal and opposite in sign to the excess charge density q^M on the metal. The total excess charge density on the solution side q^S is equal to the sum of the charge density in the inner layer and in the diffuse layer. Thus

$$q^M = -(q^i + q^d) = -q^S \tag{1}$$

C. Usefulness of Double-Layer Studies in Electroanalytical Chemistry

The electrical double layer is of interest in electroanalytical chemistry primarily for two reasons. First, measured values of the fundamental parameters of electrode kinetics such as rate constant and transfer coefficient can be strongly dependent on the structure of the double layer. Consequently, elucidation of the mechanisms of electrode processes, which is a major preoccupation of modern electroanalytical research, requires knowledge of double-layer structure. Neglect of double-layer effects in electrode kinetics can lead to erroneous mechanistic interpretations.

Second, the development in recent years of several highly sensitive techniques such as ac polarography (Chapter 1), chronopotentiometry (Chapter 2), pulse polarography, and coulostatic analysis has brought about an increasing emphasis on the detection and determination of trace constituents by electroanalysis. In trace analysis, adsorption of the constituent sought at the electrode may become a predominant consideration, for even a fractional monolayer can markedly change the electrode response. For example, in chronopotentiometry adsorption of an electroactive constituent can give rise to an increased transition time. If the constituent sought is a major one, such effects are usually negligible. However, for trace analysis, this increase may represent an appreciable portion of the total measured transition time, and failure to take account of adsorption can lead to serious errors. On the other hand, if adsorption is properly interpreted, it may enable one to extend the sensitivity of an electroanalytical method. Therefore, whether one's interest in electroanalytical chemistry is primarily in electrode kinetics and mechanisms or in electroanalysis per se, knowledge of the properties of the electrical double layer should be useful. It is hoped that this chapter will provide a convenient introduction.

II. DOUBLE-LAYER THERMODYNAMICS

A. Classification of Electrodes and Electrode Processes

1. Ideal Polarized Electrodes

Most of what is known today about the structure and behavior of the electrical double layer has been learned from studies of so-called ideal polarized electrodes (6,26,27). These are defined as electrodes at which no charge transfer across the metal-solution interface can occur, regardless of the potential imposed on the electrode from an outside source of voltage. Strictly speaking, no real electrode system can meet this stringent requirement, but certain systems can approach *ideal polarizability* very closely, provided the range of imposed electrode potentials is sufficiently restricted. We shall refer to such real electrodes also as ideal polarized, bearing in mind the practical limitation on the potential range. Preeminent among the attainable ideal polarized electrodes is mercury in contact with any of a large variety of carefully deaerated aqueous electrolyte solutions. A familiar example is the dropping mercury electrode (DME) in contact with a typical polarographic supporting electrolyte solution in the absence of any oxidizable or reducible species. Thus, although the terminology has not often been used in electroanalytical literature, ideal polarized electrodes themselves are actually quite familiar in electroanalytical practice.

In a given solution at any fixed potential within the permissible range, the double layer at an ideal polarized electrode attains a true state of equilibrium which can be described precisely in terms of classical equilibrium thermodynamics. However, this equilibrium is not of the familiar nernstian type. Rather, it is a state of *electrostatic equilibrium* in the electrical double layer. Therefore, to define the state of an ideal polarized electrode at equilibrium, it is necessary to specify not only the temperature, pressure, and composition (chemical potentials) of each phase, but also the value of an additional electrical variable. This electrical variable expresses the degree of charge separation across the interface. Depending on convenience, one may choose for the electrical variable either the excess charge density on the metal q^M or the potential E of the ideal polarized electrode with respect to a reference electrode. Thus, an ideal polarized electrode at equilibrium is a system having one more degree of freedom than it would have were it in a state of nernstian equilibrium. This means that an ideal polarized electrode has the unique capability of being in thermodynamic equilibrium at any potential whatever (within a certain range), although the temperature, pressure, and composition of its phases remain fixed.

2. Charge-Transfer Electrodes

Electrodes which are not ideal polarized may be called *charge-transfer electrodes*. At these electrodes the familiar electrochemical processes of oxidation and reduction take place. In terms of the electrical double layer, a charge-transfer electrode is one at which electrically charged particles, ions or electrons, can be transferred across the metal-solution interface. In electrical terminology, a *conduction current* can flow across the interface of a charge-transfer electrode, but only a *displacement current* can flow at the interface of an ideal polarized electrode.

For fixed temperature, pressure, and composition of each phase, there is one, and only one, value of the electrode potential for which a charge-transfer electrode may be at equilibrium. This is the potential specified by Nernst's equation. In contrast, for an ideal polarized electrode to be at equilibrium under the same conditions, any of a continuously infinite set of potentials will suffice.

3. Faradaic and Nonfaradaic Processes

The familiar electrode processes of oxidation and reduction which take place at charge-transfer electrodes obey Faraday's laws; hence they are called *faradaic*. At an ideal polarized electrode, faradaic processes are prohibited. Whenever a real electrode behaves as an ideal polarized electrode, it is because, within a certain range of potentials, all the faradaic processes which might conceivably take place there fall into either of two categories (*27*); (a) The activation energy is so high that the faradaic process occurs at a negligible rate. For example, the cathodic hydrogen evolution reaction at mercury electrodes is so slow in many solutions that, over a considerable potential range, it essentially does not occur [cf. Koenig (*28*)]. (b) Even though the activation energy is low, the equilibrium constant for the faradaic process is such that the concentration of either reactants or products is so low as to be meaningless (except in a statistical sense). Therefore, any charge transfer accompanying a change of electrode potential is entirely negligible. For example, at the electrocapillary maximum (see Sec. II.B), the value of the mercurous ion concentration necessary to satisfy Nernst's equation for a mercury electrode in $1F$ KCl solution is of the order of 10^{-36} F [cf. Grahame (*6*)].

The processes of adsorption and desorption which take place whenever the structure of the electrical double layer changes are not described by Faraday's laws; hence they are called *nonfaradaic*. At ideal polarized electrodes, only nonfaradaic processes can take place, but at charge-transfer electrodes, both faradaic and nonfaradaic processes occur simultaneously.

B. Electrocapillary Curves: Qualitative Discussion

A common observation in polarography is the change of drop life of the DME during the recording of a polarogram. The explanation of this phenomenon lies in the fact that the interfacial tension γ of the mercury-solution interface varies with the electrode potential E [for details, cf. Meites (*29*)]. The larger the interfacial tension, the "stronger" will be the neck of the drop; hence, the greater a drop weight it can support. As E is scanned, γ changes, and thus the drop life varies. Typical results are illustrated in Fig. 3. From the standpoint of polarographic analysis, this variation of interfacial tension with electrode potential is merely a nuisance, for which appropriate correction must be made to obtain reliable analytical

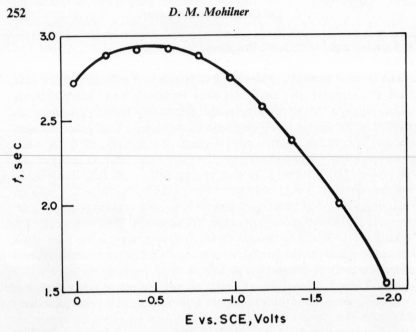

Fig. 3. Variation of drop time of DME with potential in 0.1 M KCl [from (30)]. (By permission of Wiley.)

results (*30*). However, this *electrocapillary effect*† is of paramount importance from the standpoint of developing a quantitative theory of the electrical double layer. The remainder of Sec. II will be devoted to the thermodynamic theory of electrocapillarity. Before beginning the thermodynamic derivation, it will be useful to discuss qualitatively some of the salient features of electrocapillary curves.

A plot of the interfacial tension γ of the metal-solution interface of an electrode versus the electrode potential E is known as an *electrocapillary curve*. A typical set of electrocapillary curves is shown in Fig. 4. These curves, which were obtained by Devanathan and Peries (*33*) for an ideal

† *Electrocapillary effect* pertains quite generally to variations of interfacial tension of a metal-electrolyte interface with the electrode potential without regard to the presence or absence of a capillary. The name *electrocapillary* is derived from the fact that the first measurements of the variation of interfacial tension with electrode potential, which were reported in 1873 by the French physicist Gabriel Lippmann, were obtained using his invention—the capillary electrometer (*31*). Lippmann was, in fact, the first to realize that a relationship existed between surface phenomena and electrical phenomena (*32*). He derived the equation which relates the interfacial tension and the excess charge density q^M (Lippmann equation, cf. Section II.E.1.a), and he also first derived the relationships between the differential capacity of the double layer and q^M and γ. Lippmann was awarded the Nobel Prize in physics in 1908.

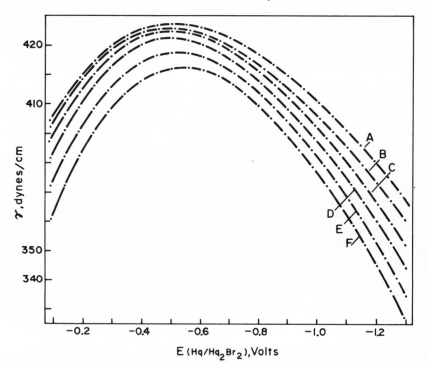

Fig. 4. Electrocapillary curves for the mercury-aqueous KBr interface. Concentration of KBr solution: (a) 0.01 F, (b) 0.03 F, (c) 0.1 F, (d) 0.3 F, (e) 1.0 F, (f) 3.0 F. E versus Hg-Hg$_2$Br$_2$ electrode [from (33)]. (By permission of the Faraday Society.)

polarized mercury electrode in aqueous KBr solutions, illustrate several typical features of electrocapillary curves.

The interfacial tension of the ideal polarized electrode in a given solution is a continuous single-valued function of the electrode potential. Provided the range of potential is sufficiently great, as E changes from positive to more negative values, γ first increases (*ascending branch*), then passes through an *electrocapillary maximum* (ECM), and finally decreases (*descending branch*). The ECM is the most important feature of a single electrocapillary curve; it corresponds to the unique situation in the double layer for which the electronic charge density on the metal is zero $[q^M = -(q^i + q^d) = -q^S = 0]$. The value of the electrode potential at the ECM is called the *point of zero charge* and is denoted by E_z. In general, E_z is different for the same metal in different solutions, or for different metals in the same solution. On the ascending branch $(E > E_z)$, q^M is positive, and on the descending branch $(E < E_z)$, it is negative. (Most workers in electrocapillarity employ the usual polarographic convention of plotting increasingly negative values of E to the right.) An electrocapillary curve is therefore a kind of map of the way the double layer changes with electrode potential. We shall prove later that the slope of the curve at any point is numerically equal, and opposite in sign, to the value of q^M. Another quite general feature of electrocapillary curves, which is shown in Fig. 4, is the lowering of the curves with increasing concentration. We shall show (Secs. II.E.1.b and d) that the slope of a plot of γ versus the logarithm of concentration at fixed E is a measure of the amount of adsorption in the double layer. More precise statements about the significance of electrocapillary curves must await the thermodynamic derivations.

All electrocapillary curves may be described by the *electrocapillary equation*, which is a differential equation relating the interfacial tension to the temperature, pressure, electrode potential, and chemical potentials of the components. The starting point of the derivation is the Gibbs adsorption equation.

C. Derivation of the Gibbs Adsorption Equation

Thermodynamic derivations of the electrocapillary equations for both ideal polarized and charge-transfer electrodes are most conveniently carried out by means of the Gibbs adsorption equation (*34–36*). Because the Gibbs equation in its original form was restricted to systems of neutral components, it is necessary to make certain modifications in the language to allow consideration of the charged components of the electrical double layer. The way to do this was indicated by Grahame and Whitney (*26*), who published the first rigorous thermodynamic derivation of the electrocapillary equation in 1942.

1. The Real System and the Gibbs Model

Consider the schematic diagram in Fig. 5(*a*). This diagram represents a real electrode system. We select for special consideration a macroscopic subsystem whose trace on the plane of the paper is the area $WXYZ$. Let the area of metal-solution interface enclosed in the system be A. The surfaces whose traces on the paper are the lines WX and ZY are placed sufficiently far from the interface that they are located entirely within homogeneous metal or solution, respectively. The surface whose traces on the paper are the lines WZ and XY is everywhere normal to the real interface, but its shape is otherwise arbitrary. For example, when the interface is planar, the subsystem may be shaped as a right cylinder or a rectangular parallelopiped of cross-section A. The region of nonhomo-

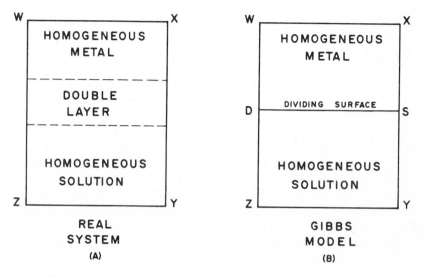

Fig. 5. Schematic diagram of the real system (*a*) and Gibbs model (*b*).

geneity (double layer) is contained essentially between the two dotted lines. We assume that the electrode is in thermal, mechanical, electrostatic, and physicochemical equilibrium.

The subsystem just defined is itself a macrocsopic system at equilibrium. Therefore, it has a well-defined thermodynamic state. We shall imagine that this subsystem is abstracted from the larger electrode system, and henceforth, when we refer to the "real system," we shall imply only this abstracted system. The advantage of such a procedure, which is due to Gibbs, is that it will allow us to concentrate our attention on the essential properties of the metal-solution interface without the necessity of considering the possible effects of other interfaces which exist in the entire electrode system. The real system we shall consider is therefore one possessing a single interface and having no "edge effects." It is electrically neutral when considered as a whole, but there is a charge separation across the interface.†

† The real system just defined is identical, by definition, with the "interphase" of Guggenheim (*37*). It is possible to derive the Gibbs adsorption equation and the electrocapillary equation directly from the interphase concept without resort to the Gibbs model (*27*), as is done below, provided that the curvature of the interface is not too large (*38*). However, the use of the Gibbs model helps to bring out certain features which are involved in the structural theory of the double layer (cf. Sec. III-B), which might not otherwise be sufficiently clear. A proof of the equivalence of the two approaches is given in Appendix A.

Consider next the schematic drawing in Fig. 5(*b*). This diagram depicts a wholly imaginary, model system which we shall call the *Gibbs model*. The Gibbs model has precisely the same volume and geometric shape as the real system, but it has no region of inhomogeneity. Rather, the two phases of the system are assumed to be completely homogeneous up to a purely mathematical *dividing surface* (DS). There is no charge separation at the DS. The metal and solution portions of the Gibbs model are individually macroscopic, electrically neutral systems at equilibrium having well-defined thermodynamic states. We assume their temperature, pressure, and composition are identical to those of the corresponding homogeneous portions of the real system. The DS is chosen parallel to the real interface, but otherwise its position is arbitrary within the model. [This statement is not true for the case of highly curved interfaces (*38*).] Of course, the individual volumes (V_{MG} and V_{SG}) of the metallic and solution phases of the model depend on the location of the dividing surface, but their sum is independent of it, being by definition equal to the total volume V of the real system.

2. The Differential Equation of Excess Internal Energy

The essence of the Gibbs method is to describe the properties of the real interface in terms of the amounts by which the extensive properties of the real system exceed those of the model system. Let U and U_G stand for the internal energies of the real system and of the Gibbs model, respectively. Then define the *excess internal energy*

$$U_x = U - U_G$$

The first step is to derive an expression for the total differential of excess internal energy

$$dU_x = dU - dU_G \qquad (2)$$

To do this we must specify the variables upon which U and U_G depend. Actually there is considerable freedom of choice, since thermodynamics requires only that there be the proper number of independent variables. Convenient choices are entropy, volume, interfacial area, and mole numbers of the components. (The *mole number* n_i^α of a component i, in a phase α, of finite dimensions is simply the number of moles of component i in α.) These variables and their contributions to the internal energies of the real system and the Gibbs model will be examined in the next three subsections.

a. Entropy Terms. Since the real system is specified to be in thermal equilibrium, it follows (*34*) that the absolute temperature T is uniform throughout. The heat absorbed by the real system during an infinitesimal reversible change will be given, therefore, by the term $T\,dS$, where S is the total entropy of the real system. Similarly, the contributions for the Gibbs model will be $(T\,dS_{MG} + T\,dS_{SG})$, where S_{MG} and S_{SG} are the entropies of its metallic and solution phases, respectively.

b. Mechanical-Work Terms. The increase in the internal energy of the real system during an infinitesimal reversible expansion depends on both the increase of the system's volume V and on the accompanying increase of the area A of the interface. When the interface is planar, the condition of mechanical equilibrium requires that the pressure p inside the real system be uniform throughout. Thus when the volume changes by an amount dV, the internal energy of the real system changes by an amount $-p\,dV$ from this cause. However, it is found experimentally that the total work which must be done on a real interfacial system exceeds that which would be calculated from pressure-volume considerations alone. The excess of experimental work over calculated work is proportional to the increase of the interfacial area. In fact, a real interfacial system behaves mechanically as if the two phases in contact were separated by an infinitely thin, uniformly stretched membrane which resists further stretching. Thus the change of internal energy of the real system resulting from an infinitesimal reversible expansion may be expressed by

$$dW = -p\,dV + \gamma\,dA$$

where γ is a positive quantity called the *interfacial tension*. The interfacial tension is an intensive property of the real system.

The corresponding expansion contribution to the internal energy of the Gibbs model is given by

$$dW_G = -p_M\,dV_{MG} - p_S\,dV_{SG} \tag{3}$$

where p_M and p_S are the pressures inside the homogeneous metallic and solution phases of the real system. No term involving interfacial area is required, since the Gibbs model system has no interfacial properties, by definition. Equation (3) may be simplified by again noting for a planar interface that the pressure is uniform in the real system:

$$p_M = p_S = p$$

Moreover, although the individual volumes V_{MG} and V_{SG} depend on the position of the dividing surface DS in the Gibbs model, their sum is equal to the volume of the real system and

$$dV_{MG} + dV_{SG} = dV$$

Thus it follows that the work of expansion of the corresponding Gibbs model is simply

$$dW_G = -p\,dV$$

c. Composition Variables. Having completely specified the entropy and mechanical work terms, all that remains is to specify the terms in the equations for dU and dU_G which depend on chemical composition.

Let the metallic phase of the electrode be a single-phase alloy containing m different metals, each species of which is indicated by a member of a set of subscripts $\{i\}$. We can consider that in phase M there is a formal electrochemical (dissociation) equilibrium between atoms of metal M_i and the corresponding cations $M_i^{z_i}$, of ionic charge z_i and the electrons. Thus we have a set of m chemical equations of the form (see second footnote, p. 244)

$$M_i(M) \rightleftharpoons M_i^{z_i}(M) + z_i e(M)$$

To each of these chemical equations corresponds an equation of the form

$$\mu_i = \bar{\mu}_i + z_i \bar{\mu}_e \tag{4}$$

where μ_i is the chemical potential (per mole) of neutral metal atoms M_i, and $\bar{\mu}_i$ and $\bar{\mu}_e$ are the electrochemical potentials† of ions $M_i^{z_i}$, and of the electrons in the metallic phase.

Let phase S be a solution of salts and neutral compounds. One of the latter will be the solvent. Let there be a total of c cationic species, a anionic species, and b species of neutral molecular substances. We shall designate each cationic species by a member of a set of subscripts $\{j\}$. The chemical symbol for each cation in solution will be $C_j^{z_j}$, where z_j is the ionic charge including sign. Similarly, we shall define each species of anion in solution by a symbol $A_k^{z_k}$, where k is a member of another set of subscripts $\{k\}$. Finally, each neutral molecular species in solution will be designated by a member of a set of subscripts $\{h\}$. For convenience we specify that the sets of integer subscripts $\{i\}$, $\{j\}$, $\{k\}$, and $\{h\}$ are nonintersecting, that is, they have no members in common. Thus, designation of any variable by one of these subscripts will be unambiguous. The electrochemical potentials of the cations and anions in solution will be designated by the sets of symbols $\{\bar{\mu}_j\}$ and $\{\bar{\mu}_k\}$, respectively. Similarly, the chemical potentials of the

---

† The *electrochemical potential* $\bar{\mu}_i^{\alpha}$ (8) per mole of any component i, in a phase α is given by $\bar{\mu}_i^{\alpha} = \mu_i^{\alpha} + z_i F \phi^{\alpha}$, where μ_i^{α} is the *chemical potential* of that component in phase α, z_i is its ionic charge (including sign), F is the faraday, and ϕ^{α} is the inner potential of the phase. For neutral components $\bar{\mu}_i^{\alpha} = \mu_i^{\alpha}$.

neutral molecular species in solution will be denoted by the set of symbols $\{\mu_h\}$.

In the real system, we shall designate the mole number of the electrons in the metallic phase by the symbol n_e. We shall denote the mole numbers of the metal ions in the metallic phase, and of the cations, anions, and neutral substances in the solution phase, by the sets of symbols $\{n_i\}$, $\{n_j\}$, $\{n_k\}$, and $\{n_h\}$, respectively. Similarly, for the Gibbs model, the corresponding mole numbers will be n_{eG}, $\{n_{iG}\}$, $\{n_{jG}\}$, $\{n_{kG}\}$, and $\{n_{hG}\}$.

The contribution to the internal energy of the real system resulting from an infinitesimal reversible change of the mole number of each neutral component is given by a term $\mu\,dn$. For each charged component, one would be tempted to write down a corresponding term $\bar{\mu}\,dn$. This would imply that the mole numbers of each charged component were independent. However, we know that the whole system is electrically neutral, and, therefore, the mole numbers of all the charged components cannot be independent. We could, of course, apply the condition of electroneutrality at this stage of the derivation by substituting for the mole number of one of the charged components (for example, electrons) in terms of the mole numbers of the remaining charged components. This was the method used by Grahame and Whitney (26). Alternatively, we could temporarily relax the condition of electroneutrality, carry out certain mathematical operations which are required in the derivation of the Gibbs adsorption equation, and later impose the condition of electroneutrality. Since the mathematical operations involved commute, the two methods are, in fact, equivalent and of equal thermodynamic rigor. For convenience, we choose the latter approach, which is essentially the method of Parsons and Devanathan (27). Thus temporarily relaxing the condition of electroneutrality, we can write the contribution to the internal energy of the real system because of infinitesimal reversible changes of the mole numbers of the components as

$$\bar{\mu}_e\,dn_e + \sum_i \bar{\mu}_i\,dn_i + \sum_j \bar{\mu}_j\,dn_j + \sum_k \bar{\mu}_k\,dn_k + \sum_h \mu_h\,dn_h$$

Similarly, the corresponding contributions to the internal energies of the metallic and solution phases of the Gibbs model are expressed by

$$\bar{\mu}_e\,dn_{eG} + \sum_i \bar{\mu}_i\,dn_{iG}$$

and

$$\sum_j \bar{\mu}_j\,dn_{jG} + \sum_k \bar{\mu}_k\,dn_{kG} + \sum_h \mu_h\,dn_{hG}$$

d. The Equation for dU_x. We can now write equations for the differentials of internal energy, dU, dU_{MG}, and dU_{SG}, of the real system and of the metallic and solution phases of the Gibbs model, respectively:

$$dU = T\,dS - p\,dV + \gamma\,dA + \bar{\mu}_e\,dn_e + \sum_i \bar{\mu}_i\,dn_i + \sum_j \bar{\mu}_j\,dn_j$$
$$+ \sum_k \bar{\mu}_k\,dn_k + \sum_h \mu_h\,dn_h \tag{5}$$

$$dU_{MG} = T\,dS_{MG} - p_M\,dV_{MG} + \bar{\mu}_e\,dn_{eG} + \sum_i \bar{\mu}_i\,dn_{iG} \tag{6}$$

$$dU_{SG} = T\,dS_{SG} - p_S\,dV_{SG} + \sum_j \bar{\mu}_j\,dn_{jG} + \sum_k \bar{\mu}_k\,dn_{kG} + \sum_h \mu_h\,dn_{hG} \tag{7}$$

Let the *excess mole numbers* be defined by the sets of equations:

$$n_{ex} = n_e - n_{eG}$$
$$n_{ix} = n_i - n_{iG}$$
$$n_{jx} = n_j - n_{jG} \tag{8}$$
$$n_{kx} = n_k - n_{kG}$$
$$n_{hx} = n_h - n_{hG}$$

Subtract Eqs. (6) and (7) from Eq. (5), and substitute Eqs. (8) into the result to obtain the following differential equation for the excess internal energy:

$$dU_x = T\,dS_x + \gamma\,dA + \bar{\mu}_e\,dn_{ex} +$$
$$\sum_i \bar{\mu}_i\,dn_{ix} + \sum_j \bar{\mu}_j\,dn_{jx} + \sum_k \bar{\mu}_k\,dn_{kx} + \sum_h \mu_h\,dn_{hx} \tag{9}$$

Note that in the combination of Eqs. (5), (6), and (7) to obtain Eq. (9) the pressure-volume terms have cancelled.

3. The Adsorption Equation

a. The Equation Itself. The Gibbs adsorption equation is a differential equation in which the interfacial tension is expressed as a function of the intensive variables, temperature, and the chemical potentials. Equation (9), just derived, expresses the interfacial tension implicitly in terms of the excess extensive properties of the interphase. By means of Euler's theorem (*39*), we can convert from the expression in terms of extensive variables to the desired expression in terms of intensive properties. Since the condition of electroneutrality has been temporarily relaxed, all the extensive variables appearing in Eq. (9) may be treated as independent. Therefore, U_x is a homogeneous function of degree one in these variables. Thus, applying Euler's theorem to Eq. (9), we obtain

$$U_x = TS_x + \gamma A + \bar{\mu}_e n_{ex} + \sum_i \bar{\mu}_i n_{ix} + \sum_j \bar{\mu}_j n_{jx} + \sum_k \bar{\mu}_k n_{kx} + \sum_h \mu_h n_{hx} \quad (10)$$

Differentiation of Eq. (10) yields a new expression for dU_x:

$$dU_x = T\,dS_x + S_x\,dT + \gamma\,dA + A\,d\gamma + \bar{\mu}_e\,dn_{ex} + n_{ex}\,d\bar{\mu}_e$$
$$+ \sum_i \bar{\mu}_i\,dn_{ix} + \sum_i n_{ix}\,d\bar{\mu}_i + \sum_j \bar{\mu}_j\,dn_{jx} + \sum_j n_{jx}\,d\bar{\mu}_j \quad (11)$$
$$+ \sum_k \bar{\mu}_k\,dn_{kx} + \sum_k n_{kx}\,d\bar{\mu}_k + \sum_h \mu_h\,dn_{hx} + \sum_h n_{hx}\,d\mu_h$$

Equating the two expressions for dU_x [Eqs. (9) and (11)] and rearranging, we obtain the interfacial analog of the familiar Gibbs-Duhem equation:

$$A\,d\gamma = -S_x\,dT - n_{ex}\,d\bar{\mu}_e - \sum_i n_{ix}\,d\bar{\mu}_i - \sum_j n_{jx}\,d\bar{\mu}_j$$
$$- \sum_k n_{kx}\,d\bar{\mu}_k - \sum_h n_{hx}\,d\mu_h \quad (12)$$

It is convenient to normalize Eq. (12) to unit area of the interface by dividing both sides of the equation by A. Let

$$\sigma = S/A$$
$$\Gamma_e = n_{ex}/A$$
$$\Gamma_i = n_{ix}/A$$
$$\Gamma_j = n_{jx}/A \quad (13)$$
$$\Gamma_k = n_{kx}/A$$
$$\Gamma_h = n_{hx}/A$$

The quantity σ is the surface excess of entropy; it is expressed in eu cm^{-2}. Each Γ is called the *surface excess* or *adsorption* of the respective components. Combination of Eqs. (12) and (13) yields the *Gibbs adsorption equation*

$$d\gamma = -\sigma\,dT - \Gamma_e\,d\bar{\mu}_e - \sum_i \Gamma_i\,d\bar{\mu}_i - \sum_j \Gamma_j\,d\bar{\mu}_j - \sum_k \Gamma_k\,d\bar{\mu}_k - \sum_h \Gamma_h\,d\mu_h \quad (14)$$

The values of each Γ in Eq. (14) depend on the arbitrary choice of location of the dividing surface since the excess quantities, the n_x's [cf.

Eq. (8)], were obtained by subtracting the mole numbers of the particular component in the Gibbs model from the corresponding mole numbers of the real system. The mole numbers of the Gibbs model, in turn, depend on the location of the dividing surface. A similar statement can be made about σ.†

b. Electroneutrality. The condition that the entire interfacial region is electrically neutral was given in Sec. I by Eq. (1):

$$q^M = -q^S$$

It will be convenient to express this condition in terms of surface excesses.

The total excess (positive) charge on the metallic phase of the real system is given by

$$\left(\sum_i z_i n_i - n_e \right) F$$

where F is the faraday. The total excess charge of the solution phase of the real system is

$$\left(\sum_j z_j n_j - \sum_k |z_k| n_k \right) F$$

Hence the condition of electroneutrality for the real system is expressed by the equation

$$\left(\sum_i z_i n_i - n_e \right) = -\left(\sum_j z_j n_j - \sum_k |z_k| n_k \right) \tag{15}$$

The condition that phases M and S of the Gibbs model are each independently electrically neutral is

$$\left(\sum_i z_i n_{iG} - n_{eG} \right) = \left(\sum_j z_j n_{jG} - \sum_k |z_k| n_{kG} \right) = 0 \tag{16}$$

---

† Since the Γ's and σ are not absolute quantities any use of these terms always carries the implicit statement " reckoned at a dividing surface located at ..." For purely thermodynamic purposes the location of the dividing surface is unimportant because, as is shown below, only relative surface excesses (cf. Sec. II.E.1.b and d) are thermodynamically accessible. However, in structural discussions (cf. Sec. III) it is often necessary to speak not of relative surface excesses but of *nonrelativised surface excesses*, that is, surface excesses reckoned at a particular dividing surface. For example, from the theory of the diffuse layer (Sec. III.C), one calculates surface excesses at a dividing surface placed at the same location in the Gibbs model as is the OHP in the real system. Such a surface excess is called the *surface excess reckoned at the OHP*.

Combining Eqs. (15), (16), and (8), we obtain a single equation expressing the simultaneous electroneutrality of the real system and of each phase of the Gibbs model:

$$\left(\sum_i z_i n_{ix} - n_{ex}\right) = -\left(\sum_j z_j n_{jx} - \sum_k |z_k| n_{kx}\right) \tag{17}$$

Now, normalizing Eq. (17) to unit area and using the definitions of the surface excesses [Eqs. (13)], we have

$$(q^M/F) = \left(\sum_i \Gamma_i z_i - \Gamma_e\right) = -\left(\sum_j \Gamma_j z_j - \sum_k \Gamma_k |z_k|\right) = -(q^S/F) \tag{18}$$

Equation (18) gives a precise definition of the excess charge densities on the metal and solution sides of the double layer. *Note that the definition of q^M (and q^S) just given is totally independent of the arbitrary location of the dividing surface in the Gibbs model.*

D. Derivation of the Electrocapillary Equation for Ideal Polarized Electrodes

Actually, the Gibbs adsorption equation [Eq. (14)] is a form of the electrocapillary equation. However, it is not yet in a usable form, because its variables are not directly applicable to experimental measurements, and, moreover, three of them are not independent. One of these three dependent variables can be eliminated through the electroneutrality condition, Eq. (18); the other two can be eliminated through application of Gibbs-Duhem equations for each phase. The variables which can be measured in the laboratory are temperature, pressure, electrode potential, and the concentrations (activities) of neutral components, that is, metals in phase M and salts and neutral molecular species in phase S. (In the sense of this discussion, the terms *salt* includes acids and bases.) To recast the equation in terms of such variables, we must consider in detail the type of cell used in experimental measurements of electrocapillary curves.

1. The Cell

Electrocapillary measurements, like any other electrochemical measurements, require the use of a complete cell, that is, one with two electrodes. One electrode is the ideal polarized electrode; the other electrode is a reversible charge-transfer electrode (*6,27*). However, it is important to realize from the outset that this second electrode is not an ordinary "constant-potential" reference electrode like the saturated calomel electrode (SCE). Rather, it is simply some electrode dipping into the solution *S*, which is reversible (in the nernstian sense) to one of the ions of that solution. When the concentration of the solution changes, the potential of this electrode also changes. In electroanalytical discussions (for example, in potentiometric titrations) such an electrode is usually referred to as an indicator electrode. That terminology will be adopted here. The second electrode of the electrocapillary cell will be called the *indicator electrode* and denoted by the symbol IN. The particular ion of the solution to which electrode IN is reversible will be called the *indicator ion*. For simplicity, we shall exclude from consideration as indicator electrodes those which are reversible simultaneously to two ions of the solution. (For example, a platinum electrode dipping into a solution containing the ferrocyanide-ferricyanide redox couple would not qualify as an indicator electrode for our purposes.)

The complete cell is represented by the diagram

$$Cu|M|S|IN|Cu'$$

Cu is a metal terminal connected to phase *M* of the ideal polarized electrode. *S* is the solution into which the indicator electrode IN also dips. Cu' is a metal terminal attached to the metallic phase of the indicator electrode. The terminals Cu and Cu' are considered part of the cell (*19*) and must be of identical chemical composition (copper or any other metal). A potential-measuring device is assumed to be connected across the two terminals. The measured potential difference (electric tension) is defined by the equation [cf. (*8,19,27*)]

$$E^{\pm} = \phi^{Cu} - \phi^{Cu'} = (\bar{\mu}_e^{Cu'} - \bar{\mu}_e^{Cu})/F \qquad (19)$$

where ϕ^{Cu} and $\phi^{Cu'}$ are the *inner potentials* (see second footnote, p. 244) of the indicated terminals, $\bar{\mu}_e^{Cu}$ and $\bar{\mu}_e^{Cu'}$ are the electrochemical potentials of the electrons in the two terminals, and $E^{\pm}$ is the *potential of the ideal polarized electrode with respect to the indicator electrode* (*27,40*). E^{+} or E^{-} is used, depending on whether the indicator electrode is reversible to a cation or an anion of the solution, respectively.

It is important to emphasize that the indicator electrode IN must be an electrode which is reversible to one of the ions of the solution *S*.

It is, of course, true that one is often unable to employ an indicator electrode of the type described above for practical reasons. For example, electrodes reversible to perchlorate ion are not readily available. In such cases measurements are made using a conventional *reference electrode*, for example, SCE. The potential of the ideal polarized electrode with respect to such a reference electrode would be designated E_{ref}. Electrocapillary or differential capacitance measurements on the E_{ref} scale must be converted to the $E^{\pm}$ scale before a complete analysis can be made.

2. Replacement of Electrochemical Potentials

a. Phase M. When two different metals are in contact, the condition of electrochemical equilibrium between them is expressed by the equality of the electrochemical potentials of their electrons (*19*). Thus, for the contact between terminals Cu and the metallic phase *M* of the ideal polarized electrode, we have

$$\bar{\mu}_e^{Cu} = \bar{\mu}_e \tag{20}$$

By Eq. (20) we can substitute the electrochemical potential of the electrons in the terminal Cu for the electrochemical potential of the electrons in phase *M*. Hence the second term of the right side of Eq. (14) becomes

$$\Gamma_e \, d\bar{\mu}_e^{Cu} \tag{21}$$

For each of the *m* metals comprising phase *M*, we have written a formal "dissociation" equilibrium expression relating the chemical potential of the metal atoms and the electrochemical potentials of the corresponding metal ions and of the electrons in phase *M* [Eq. (4)]. Thus

$$d\bar{\mu}_i = d\mu_i - z_i \, d\bar{\mu}_e$$

Using Eq. (20) again, we obtain

$$d\bar{\mu}_i = d\mu_i - z_i \, d\bar{\mu}_e^{Cu} \tag{22}$$

Substitution of Eq. (22) into the third term on the right side of Eq. (14) gives

$$\sum_i \Gamma_i \, d\mu_i - \sum_i \Gamma_i z_i \, d\bar{\mu}_e^{Cu} \tag{23}$$

Combining expressions (21) and (23), we obtain

$$\sum_i \Gamma_i \, d\mu_i - \left(\sum_i \Gamma_i z_i - \Gamma_e \right) d\bar{\mu}_e^{Cu}$$

Finally, recalling the definition of the excess charge density on the metal q^M [cf. Eq. (18)], the second and third terms of the Gibbs adsorption equation become

$$\sum_i \Gamma_i \, d\mu_i - (q^M/F) \, d\bar{\mu}_e^{Cu} \tag{24}$$

which contains chemical potentials of the metals in phase *M* instead of electrochemical potentials of the ions in phase *M*.

b. Phase S.   To replace the electrochemical potentials of the ions of the solution by chemical potentials of neutral species, we must first decide what the neutral species, that is, salts, are to be.

(1) *Salt Components of Phase S.*   It was specified (Sec. II.C.2.c) that the solution S contains c cationic species and a anionic species. Such a solution could be prepared in many different ways. However, for the purpose of the thermodynamic treatment, the most general electrocapillary equation can be derived if we assume that the ions of the solution are furnished by neutral *binary* salts (*27*). Of the ca different binary salts that could be chosen, we shall select $c + a - 1$ binary salts in the following way.

If the indicator electrode IN is reversible to cation j', we arbitrarily select an anion, say k'. If the indicator electrode is reversible to anion k', we arbitrarily select a cation, say j'. In either case we have selected a binary salt containing ions j' and k'. We shall call this salt the *indicator salt*. The electrolyte solution is then considered to have been made up by dissolving $c + a - 2$ additional binary salts of which $c - 1$ have anion k' in common with the indicator salt; the remaining $a - 1$ salts have cation j' in common with the indicator salt.

EXAMPLE

Let the solution S contain the following ions: Cs^+, Zn^{2+}, Al^{3+}, ClO_4^-, SO_4^{2-}, and I^-, and let IN be reversible to Zn^{2+}. We can choose any anion to make the indicator salt. Let us choose ClO_4^-. The solution is then considered made up of $Zn(ClO_4)_2$, the indicator salt, and $ZnSO_4$, ZnI_2, $CsClO_4$, and $Al(ClO_4)_3$. On the other hand, if IN were reversible to SO_4^{2-} we could arbitrarily choose a cation to make the indicator salt. Suppose we choose Al^{3+}. Then the same solution is considered made up of $Al_2(SO_4)_3$, the indicator salt, and $Al(ClO_4)_3$, AlI_3, $ZnSO_4$, and Cs_2SO_4.

(2) *Introduction of the Chemical Potentials of the Salts.*   For each salt in S we may write a formal (dissociation) equilibrium

$$(C_j)_{v_{jk}^+}(A_k)_{v_{jk}^-} \rightleftharpoons v_{jk}^+ C_j^{z_j} + v_{jk}^- A_k^{z_k}$$

where the v's indicate the number of moles of cations (or anions) per formula weight of the salt. Since each salt is electrically neutral,

$$v_{jk}^+ z_j = v_{jk}^- |z_k| \tag{25}$$

Corresponding to each of the above equilibria, there is an equation relating the chemical potentials of the neutral salt to the electrochemical potentials of its constituent ions. Therefore, we have the following $c + a - 1$ equations:

$$d\mu_{j'k'} = v_{j'k'}^{+} \, d\bar{\mu}_{j'} + v_{j'k'}^{-} \, d\bar{\mu}_{k'} \tag{26}$$

$$d\mu_{jk'} = v_{jk'}^{+} \, d\bar{\mu}_{j} + v_{jk'}^{-} \, d\bar{\mu}_{k'} \qquad (j \neq j') \tag{27}$$

$$d\mu_{j'k} = v_{j'k}^{+} \, d\bar{\mu}_{j'} + v_{j'k}^{-} \, d\bar{\mu}_{k} \qquad (k \neq k') \tag{28}$$

Equations (26) to (28) allow the electrochemical potentials of the individual ionic species (that is, single-ion activities) which appear in Eq. (14) to be replaced by the measurable chemical potentials of the neutral salts.

In the j summation of Eq. (14), substitute for the electrochemical potential of each cation, using Eqs. (26) and (27), to obtain

$$\sum_{j} (\Gamma_{j}/v_{jk'}^{+}) \, d\mu_{jk'} - \left[\sum_{j} \Gamma_{j}(v_{jk'}^{-}/v_{jk'}^{+}) \right] d\bar{\mu}_{k'}$$

Introduction of Eq. (25) for each salt in the preceding expression yields

$$\sum_{j} (\Gamma_{j}/v_{jk'}^{+}) \, d\mu_{jk'} - \left[\sum_{j} \Gamma_{j}(z_{j}/|z_{k'}|) \right] d\bar{\mu}_{k'} \tag{29}$$

Similarly, substituting Eqs. (26) and (28) into the k summation of Eq. (14) and introducing Eq. (25) for each salt, we obtain

$$\sum_{k} (\Gamma_{k}/v_{j'k}^{-}) \, d\mu_{j'k} - \left[\sum_{k} \Gamma_{k}(|z_{k}|/z_{j'}) \right] d\bar{\mu}_{j'} \tag{30}$$

The electrochemical potentials of all the ions in the solution, except the cation j' and anion k' of the indicator salt, have now been eliminated.

(3) *Ions of the Indicator Salt.* Since the indicator electrode is reversible to an ion of the solution, the electrochemical potential of that ion may be expressed in terms of the electrochemical potential of the electrons in the terminal attached to the indicator electrode and the chemical potentials of the pure substances comprising the indicator electrode. The following equations apply for indicator electrodes reversible to cation j' [Eq. (31a)] or to anion k' [Eq. (31b)]:

$$d\bar{\mu}_{j'} = -z_{j'} \, d\bar{\mu}_{e}^{Cu'} + d\mu_{+}^{*}$$
$$= -z_{j'} \, d\bar{\mu}_{e}^{Cu'} - s_{+}^{*} \, dT + v_{+}^{*} \, dp \tag{31a}$$

$$d\bar{\mu}_{k'} = |z_{k'}| \, d\bar{\mu}_{e}^{Cu'} + d\mu_{-}^{*}$$
$$= |z_{k'}| \, d\bar{\mu}_{e}^{Cu'} - s_{-}^{*} \, dT + v_{-}^{*} \, dp \tag{31b}$$

The symbols $d\mu_{+}^{*}$ and $d\mu_{-}^{*}$ in Eqs. (31a) and (31b), respectively, are shorthand notations for the differentials (or appropriate linear combinations of differentials) of the chemical potentials of the pure components of the indicator electrode. Similarly, s_+°, s_-°, v_+°, and v_-° represent the molar entropies and molar volumes of the pure indicator electrode components (or appropriate linear combinations of these). (For details see Appendix B.)

The actual elimination of the electrochemical potentials of the ions of the indicator salt is divided into two cases depending on whether IN is reversible to the cation j' or the anion k'. (From this point on, the derivation will follow two parallel tracks, depending on whether the indicator ion is a cation or an anion, and we shall eventually arrive at two equivalent but different forms of the electrocapillary equation.)

Case a. The indicator electrode is reversible to cation j'. Substitution of Eq. (31a) into Eq. (26) and introduction of Eq. (25) yields

$$d\bar\mu_{k'} = \left(\frac{1}{v_{j'k'}^-}\right) d\mu_{j'k'} + |z_{k'}|\, d\bar\mu_e^{Cu'} - \left(\frac{v_{j'k'}^+}{v_{j'k'}^-}\right) d\mu_+^\circ \tag{32}$$

Substitution of Eq. (32) into expression (29) eliminates the electrochemical potential of anion k' and gives

$$\sum_{j \neq j'} (\Gamma_j/v_{jk'}^+)\, d\mu_{jk'} - \left[\left(\frac{1}{|z_{k'}|v_{j'k'}^-}\right) \sum_{j \neq j'} \Gamma_j z_j\right] d\mu_{j'k'}$$
$$- \left(\sum_j \Gamma_j z_j\right) d\bar\mu_e^{Cu'} + \frac{1}{z_{j'}} \left(\sum_j \Gamma_j z_j\right) d\mu_+^\circ \tag{33}$$

Substitution of Eq. (31a) into expression (30) eliminates the electrochemical potential of cation j' and gives

$$\sum_{k \neq k'} (\Gamma_k/v_{j'k}^-)\, d\mu_{j'k} + (\Gamma_{k'}/v_{j'k'}^-)\, d\mu_{j'k'}$$
$$+ \left(\sum_k \Gamma_k |z_k|\right) d\bar\mu_e^{Cu'} - \frac{1}{z_{j'}} \left(\sum_k \Gamma_k |z_k|\right) d\mu_+^\circ \tag{34}$$

Adding expression (33) and (34), we obtain

$$\sum_{j \neq j'} (\Gamma_j/v_{jk'}^+)\, d\mu_{jk'} + \sum_{k \neq k'} (\Gamma_k/v_{j'k}^-)\, d\mu_{j'k}$$
$$+ \left[(\Gamma_{k'}/v_{j'k'}^-) - \left(\frac{1}{|z_{k'}|v_{j'k'}^-}\right) \sum_{j \neq j'} \Gamma_j z_j\right] d\mu_{j'k'}$$
$$- \left[\sum_j \Gamma_j z_j - \sum_k \Gamma_k |z_k|\right] d\bar\mu_e^{Cu'}$$
$$+ \left[\sum_j \Gamma_j z_j - \sum_k \Gamma_k |z_k|\right] \left(\frac{1}{z_{j'}}\right) d\mu_+^\circ \tag{35}$$

Recalling the definition of q^S, the charge density on the solution side of the double layer [cf. Eq. (18)], one simplifies expression (35) to yield

$$\sum_{j \neq j'} (\Gamma_j / v_{jk'}^+) \, d\mu_{jk'} + \sum_{k \neq k'} (\Gamma_k / v_{j'k}^-) \, d\mu_{j'k}$$

$$+ \left[(\Gamma_{k'} / v_{j'k'}^-) - \left(\frac{1}{|z_{k'}| v_{j'k'}^-} \right) \sum_{j \neq j'} \Gamma_j z_j \right] d\mu_{j'k'} \qquad (36a)$$

$$- (q^S / F) \, d\bar{\mu}_e^{Cu'} + (q^S / z_{j'} F) \, d\mu_+^*$$

The electrochemical potentials of all the ions of the solution have now been eliminated from the Gibbs adsorption equation. They have been replaced by the chemical potentials of the $c + a - 1$ neutral salts and two terms involving the excess charge density q^S on the solution side of the double layer.

Case b. The indicator electrode is reversible to anion k'. Proceeding in the same way as in case a, we obtain the following replacement for the original j and k summations in Eq. (14):

$$\sum_{j \neq j'} (\Gamma_j / v_{jk'}^+) \, d\mu_{jk'} + \sum_{k \neq k'} (\Gamma_k / v_{j'k}^-) \, d\mu_{j'k}$$

$$+ \left[(\Gamma_{j'} / v_{j'k'}^+) - \left(\frac{1}{z_{j'} v_{j'k'}^+} \right) \sum_{k \neq k'} \Gamma_k |z_k| \right] d\mu_{j'k'}$$

$$- (q^S / F) \, d\bar{\mu}_e^{Cu'} - (q^S / |z_{k'}| F) \, d\mu_-^* \qquad (36b)$$

Note that expressions (36a) and (36b) differ only in the coefficients of the differential of the chemical potential $d\mu_{j'k'}$ of the indicator salt and in the last term, which depends on the particular nature of the indicator electrode (cf. Appendix B).

3. Elimination of the Dependent Variables

All that remains to complete the conversion of the Gibbs adsorption equation into a usable form of electrocapillary equation is to eliminate the three dependent variables by means of the electroneutrality condition and the two Gibbs-Duhem equations.

a. The Electroneutrality Condition. Substitution of expressions (24) and either (36a) or (36b) into Eq. (14) yields an equation containing no electrochemical potentials except for those of the electrons in the cell terminals Cu and Cu'. These electrochemical potentials may be eliminated by applying the electroneutrality condition for the whole double layer [Eq. (18)], that is,

$$q^M = -q^S \qquad (18)$$

Using this relationship, and recalling the definitions of E^+ [Eq. (19)] and of μ_+^*, s_+^* and v_+^* (Appendix B), we obtain (for the cation-reversible indicator electrode):

$$
\begin{aligned}
d\gamma = &- \left[\sigma + \left(\frac{q^M s_+^*}{z_{j'} F} \right) \right] dT + \left(\frac{q^M v_+^*}{z_{j'} F} \right) dp - q^M \, dE^+ \\
&- \sum_i \Gamma_i \, d\mu_i - \sum_{j \neq j'} (\Gamma_j / v_{jk'}^+) \, d\mu_{jk'} \\
&- \sum_{k \neq k'} (\Gamma_k / v_{j'k}^-) \, d\mu_{j'k} - \sum_h \Gamma_h \, d\mu_h \\
&- \left[(\Gamma_{k'} / v_{j'k'}^-) - \left(\frac{1}{|z_{k'}| |v_{j'k'}^-|} \right) \sum_{j \neq j'} \Gamma_j z_j \right] d\mu_{j'k'}
\end{aligned}
\tag{37}
$$

A similar equation is obtained in the case of an anion-reversible indicator electrode. In that case, which we shall not bother to write down, the only differences are that E^- replaces E^+, the coefficient of $d\mu_{j'k'}$ is that given in expression (36b), and the terms

$$
\left(\frac{q^M s_+^*}{z_{j'} F} \right) \qquad \text{and} \qquad \left(\frac{q^M v_+^*}{z_{j'} F} \right)
$$

are replaced, respectively, by

$$
-\left(\frac{q^M s_-^*}{|z_{k'}| F} \right) \qquad \text{and} \qquad -\left(\frac{q^M v_-^*}{|z_{k'}| F} \right)
$$

b. The Gibbs–Duhem Equations. Equation (37) (or its analog for case *b*) still contains two dependent variables. These may be eliminated through the Gibbs–Duhem equations for the metallic and solution phases. For the metallic phase, the Gibbs–Duhem equation is

$$
-s_M \, dT + v_M \, dp - \sum_i x_i \, d\mu_i = 0
\tag{38}
$$

Here s_M and v_M denote the mean molar entropy and mean molar volume,† respectively, of the homogeneous portion of phase M, and the x_i's are the mole fractions of the metals there. T is the absolute temperature and p is the pressure inside homogeneous phase M.

Select arbitrarily one of the metals i' of phase M and designate it the

---

† The mean molar volume of a phase is equal to the total volume of that phase divided by the total number of moles of all the *independent determinate components* (41). Similarly, the mean molar entropy is the total entropy of the phase divided by the total number of moles.

reference metal. Then solve Eq. (38) for $d\mu_{i'}$ and substitute for it in either form of Eq. (37). Similarly, the Gibbs-Duhem equation for the solution phase is

$$-s_S \, dT + v_S \, dp - x_{j'k'} \, d\mu_{j'k'} - \sum_{j \neq j'} x_{jk} \, d\mu_{jk'}$$
$$- \sum_{k \neq k'} x_{j'k} \, d\mu_{j'k} - \sum_h x_h \, d\mu_h = 0 \tag{39}$$

where s_S and v_S are the mean molar entropy and volume of homogeneous solution phase S and the x's are the mole fractions of the indicated components in the bulk of solution. Select arbitrarily one of the *neutral* components h' as a reference substance in the solution; usually h' is the solvent. Solve Eq. (39) for $d\mu_{h'}$ and substitute for it in either form of Eq. (37). The result of the two substitutions just described is the following equation:

$$d\gamma = -\left\{ \left[\sigma - \left(\frac{s_M}{x_{i'}}\right)\Gamma_{i'} - \left(\frac{s_S}{x_{h'}}\right)\Gamma_{h'} \right] + \left(\frac{q^M s_+^{\cdot}}{z_{j'}F}\right) \right\} dT$$

$$- \left\{ \left[\left(\frac{v_M}{x_{i'}}\right)\Gamma_{i'} + \left(\frac{v_S}{x_{h'}}\right)\Gamma_{h'} \right] - \left(\frac{q^M v_+^{\cdot}}{z_{j'}F}\right) \right\} dp$$

$$- q^M \, dE^+ - \sum_{i \neq i'} \left[\Gamma_i - \left(\frac{x_i}{x_{i'}}\right)\Gamma_{i'} \right] d\mu_i$$

$$- \sum_{j \neq j'} \left[(\Gamma_j/v_{jk}^+) - \left(\frac{x_{jk'}}{x_{h'}}\right)\Gamma_{h'} \right] d\mu_{jk'} \tag{40}$$

$$- \sum_{k \neq k'} \left[(\Gamma_k/v_{j'k}^-) - \left(\frac{x_{j'k}}{x_{h'}}\right)\Gamma_{h'} \right] d\mu_{j'k}$$

$$- \sum_{h \neq h'} \left[\Gamma_h - \left(\frac{x_h}{x_{h'}}\right)\Gamma_{h'} \right] d\mu_h$$

$$- \left\{ \left[(\Gamma_{k'}/v_{j'k'}^-) - \left(\frac{x_{j'k'}}{x_{h'}}\right)\Gamma_{h'} \right] - \left(\frac{1}{|z_{k'}|v_{j'k'}^-}\right) \sum_{j \neq j'} \Gamma_j z_j \right\} d\mu_{j'k'}$$

A similar equation would be obtained for the case of an anion-reversible indicator electrode (case *b*).

Equation (40) is a correct, complete version of the Gibbs adsorption equation for the electrical double layer at a planar ideal polarized electrode. This equation is the *electrocapillary equation*. It is expressed in terms of $m + c + a + b$ independent variables, each of which is an experimentally measurable quantity. One of these variables, $E^{\pm}$, is the potential difference imposed across the terminals of the cell. $E^{\pm}$ is the additional electrical variable mentioned in the introduction (Sec. II.A.1) which characterizes the condition of thermodynamic equilibrium of an ideal polarized electrode. In contrast, a reversible charge-transfer-electrode system having the *same number* of components would have one less degree of freedom, that is, $m + c + a + b - 1$.

4. Relative Surface Excesses

Although Eq. (40) is a correct electrocapillary equation, it appears cumbersome. Its form can be simplified by introducing (*27*) the concept of *relative surface excess* or *relative adsorption* of the components, the *relative surface excess of entropy* and the *relative thickness of the interphase* [cf. Guggenheim (*37*) and Koenig (*28*)].

The adsorption of a neutral molecular species $h \neq h'$ relative to that of the reference component h' is defined by the equation

$$\Gamma_{hh'} = \Gamma_h - \left(\frac{x_h}{x_{h'}}\right)\Gamma_{h'} \tag{41}$$

When the relative surface excess $\Gamma_{hh'}$ of component h is zero, we have

$$\left(\frac{\Gamma_h}{\Gamma_{h'}}\right) = \left(\frac{x_h}{x_{h'}}\right) = \left(\frac{c_h}{c_{h'}}\right) \tag{42}$$

where the c's are molar concentrations in the homogeneous solution. Positive relative surface excess implies

$$\left(\frac{\Gamma_h}{\Gamma_{h'}}\right) > \left(\frac{x_h}{x_{h'}}\right) = \left(\frac{c_h}{c_{h'}}\right) \tag{43}$$

whereas a negative $\Gamma_{hh'}$ implies

$$\left(\frac{\Gamma_h}{\Gamma_{h'}}\right) < \left(\frac{x_h}{x_{h'}}\right) = \left(\frac{c_h}{c_{h'}}\right) \tag{44}$$

Equations (42) to (44) show that the relative surface excess $\Gamma_{hh'}$ is not a direct measure of the amount of component h adsorbed but is rather a measure of the amount by which the adsorption of h exceeds that of the arbitrarily chosen reference component h'. Thus $\Gamma_{hh'} = 0$ does not imply the absence of adsorbed h; rather it implies that components h and h' are both adsorbed in the same ratio as their bulk concentrations. Positive $\Gamma_{hh'}$ implies that the electrical double layer is relatively richer in component h than is the bulk solution; negative $\Gamma_{hh'}$ implies the double layer is relatively poorer in h. Had we chosen another substance, say h'', as the reference component, there would be no assurance whatsoever that $\Gamma_{hh''} = \Gamma_{hh'}$. The relative surface excess of a component is therefore not an invariant, but depends on the arbitrary choice of the reference component whose chemical potential is eliminated via the Gibbs–Duhem equation. Note that the relative surface excess of the reference component $\Gamma_{h'h'} = 0$.

The relative surface excess of the charged components are defined in a similar way. In phase M the adsorption of metallic ions of species i is taken relative to that of ions of the reference species i'. Thus

$$\Gamma_{ii'} = \Gamma_i - (x_i/x_{i'})\Gamma_{i'} \tag{45}$$

The adsorption of the cations and anions of the solution is taken relative to that of neutral reference component h'. Thus

$$\Gamma_{jh'} = \Gamma_j - v_{jk'}^+(x_{jk'}/x_{h'})\Gamma_{h'} \tag{46}$$

and

$$\Gamma_{kh'} = \Gamma_k - v_{j'k}^-(x_{j'k}/x_{h'})\Gamma_{h'} \tag{47}$$

Note that the multiplier of $(\Gamma_{h'}/x_{h'})$ in Eqs. (46) and (47) is not just the mole fraction of the salt furnishing the ion but is the mole fraction of that salt multiplied by the number of moles of ion contained in one formula weight of the salt. In other words, the multiplier is the mole fraction of the ion in the homogeneous solution S. In keeping with this idea, we define the relative surface excess of the ions of the indicator salt by the equations

$$\Gamma_{j'h'} = \Gamma_{j'} - \sum_k v_{j'k}^+(x_{j'k}/x_{h'})\Gamma_{h'} \tag{48}$$

and

$$\Gamma_{k'h'} = \Gamma_{k'} - \sum_j v_{jk'}^-(x_{jk'}/x_{h'})\Gamma_{h'} \tag{49}$$

The summations are required in Eqs. (48) and (49) because cation j' is furnished by all a of the different salts supplying the different anions, whereas the anion k' is supplied by all c of the salts supplying the different cations. Similarly, the relative surface excess of the electrons of the metallic phase may be defined by

$$\Gamma_{ei'} = \Gamma_e - \sum_i z_i(x_i/x_{i'})\Gamma_{i'} \tag{50}$$

The relative surface excess of entropy, $\sigma_{i'h'}$, is defined by the equation

$$\sigma_{i'h'} = \sigma - \left(\frac{s_M}{x_{i'}}\right)\Gamma_{i'} - \left(\frac{s_S}{x_{h'}}\right)\Gamma_{h'} \tag{51}$$

To obtain the relative thickness of the interphase rearrange the coefficient of dp,

$$-\left[\left(\frac{v_M}{x_{i'}}\right)\Gamma_{i'} + \left(\frac{v_S}{x_{h'}}\right)\Gamma_{h'}\right] = -\left[(\Gamma_{i'}/c_{i'}) + (\Gamma_{h'}/c_{h'})\right]$$

where $c_{i'}$ and $c_{h'}$ represent the concentrations of components i' and h' (in moles cm^{-3}) in homogeneous phases M and S, respectively. Let τ be the thickness of the real system, that is, the distance from surface WX to surface ZY in Fig. 5(a). (τ may be called the *thickness of the interphase*.) Let x_D be the distance from $W'X'$ to the dividing surface in the Gibbs model [Fig. 5(b)]. Then,

$$\Gamma_{i'} = \frac{n_{i'}}{A} - c_{i'}x_D$$

$$\Gamma_{h'} = \frac{n_{h'}}{A} - c_{h'}(\tau - x_D)$$

where A is the area of the plane interface.

Hence, the coefficient of dp in Eq. (40) becomes

$$-\left[\left(\frac{\Gamma_{i'}}{c_{i'}}\right) + \left(\frac{\Gamma_{h'}}{c_{h'}}\right)\right] = -\left\{\left[\frac{(n_{i'}/A)}{c_{i'}} + \frac{(n_{h'}/A)}{c_{h'}}\right] - \tau\right\}$$

$$= \left\{\tau - \left[\frac{(n_{i'}/A)}{c_{i'}} + \frac{(n_{h'}/A)}{c_{h'}}\right]\right\}$$

$$= \tau_{i'h'}$$

We shall call $\tau_{i'h'}$ the *relative thickness* of the interphase. It is easy to show that $\tau_{i'h'}$ is invariant over any arbitrary choice of the location of the DS or of the thickness τ of the real system (that is, interphase).

It is important to emphasize again that *the relative surface excesses of entropy and of the components, as well as the relative thickness of the interphase, are completely independent of the arbitrary choice of the dividing surface in the Gibbs model which was made at the beginning of the derivation.* This fact is demonstrated by considering any two explicit choices for the position of the dividing surface (cf., Sec. III.B.1.b). Similarly, one may easily demonstrate that the relative surface excesses are independent of our initial choice of the thickness τ of the real system (that is, the separation of surfaces WX and ZY) provided only that the bounding surfaces WX and ZY lie inside homogeneous regions.

5. The Electrocapillary Equation

Substitution of Eqs. (41), (45) to (49), and (51) into Eq. (40) yields the following final forms of the electrocapillary equation for an ideal polarized electrode.

Case a. Indicator electrode reversible to cation j':

$$
\begin{aligned}
d\gamma = & -\left[\sigma_{i'h'} + \left(\frac{q^M s_+^{\cdot}}{z_{j'}F}\right)\right] dT + \left[\tau_{i'h'} + \left(\frac{q^M v_+^{\cdot}}{z_{j'}F}\right)\right] dp \\
& - q^M dE^+ - \sum_{i \neq i'} \Gamma_{ii'} d\mu_i - \sum_{j \neq j'} (\Gamma_{jh'}/v_{jk}^+) d\mu_{jk'} \\
& - \sum_{k \neq k'} (\Gamma_{kh'}/v_{j'k}^-) d\mu_{j'k} - \sum_{h \neq h'} \Gamma_{hh'} d\mu_h \\
& - \left[(\Gamma_{k'h'}/v_{j'k'}^-) - \left(\frac{1}{|z_{k'}|v_{j'k'}^-}\right) \sum_{j \neq j'} \Gamma_{jh'} z_j\right] d\mu_{j'k'}
\end{aligned}
\tag{52a}
$$

Case b. Indicator electrode reversible to anion k':

$$
\begin{aligned}
d\gamma = & -\left[\sigma_{i'h'} - \left(\frac{q^M s_-^{\cdot}}{|z_{k'}|F}\right)\right] dT + \left[\tau_{i'h'} - \left(\frac{q^M v_-^{\cdot}}{|z_{k'}|F}\right)\right] dp \\
& - q^M dE^- - \sum_{i \neq i'} \Gamma_{ii'} d\mu_i - \sum_{j \neq j'} (\Gamma_{jh'}/v_{jk}^+) d\mu_{jk'} \\
& - \sum_{k \neq k'} (\Gamma_{kh'}/v_{j'k}^-) d\mu_{j'k} - \sum_{h \neq h'} \Gamma_{hh'} d\mu_h \\
& - \left[(\Gamma_{j'h'}/v_{j'k'}^+) - \left(\frac{1}{z_{j'}v_{j'k'}^+}\right) \sum_{k \neq k'} \Gamma_{kh'} |z_k|\right] d\mu_{j'k'}
\end{aligned}
\tag{52b}
$$

E. Physical Implications of the Electrocapillary Equation

Equations (52) are the simplest generally valid forms of the electrocapillary equation for a planar ideal polarized electrode system of arbitrary composition. Either of these equations contains, in principle, all the information obtainable from thermodynamics about the equilibrium properties of the electrical double layer. As is the case for other total differential equations of thermodynamics, this information is derived by considering the various partial differential relationships implicit in the equation.

1. The First Partial Differential Coefficients of γ

a. The Lippmann Equation. It follows from Eq. (52a) that

$$
(\partial\gamma/\partial E^+)_{T,p,\mu} = -q^M
\tag{53a}
$$

Similarly, from Eq. (52b) we obtain

$$(\partial\gamma/\partial E^-)_{T,p,\mu} = -q^M \qquad (53b)$$

In these equations the subscript μ implies that the chemical potentials (activities) of all the components are held constant. Equations (53) state that, along any given electrocapillary curve (Fig. 4), the excess charge density on the metal, at a given value of $E^\pm$, is equal to minus the slope of the curve at that point. Clearly, the maximum of the electrocapillary curve (ECM) corresponds to the condition $q^M = 0$.

Equations (53) define the slope of an electrocapillary curve for an ideal polarized electrode when the potential of the electrode is referred to an indicator electrode. However, for a solution of fixed composition the potential of an indicator electrode versus any ordinary reference electrode (for example, SCE) is a constant, that is,

$$E^\pm = E_{\text{ref}} + \text{const}$$

where E_{ref} is the potential of the ideal polarized electrode versus the ordinary reference electrode including any liquid-junction potential. Thus

$$(\partial\gamma/\partial E_{\text{ref}})_{T,p,\mu} = -q^M \qquad (54)$$

This equation implies that, insofar as our interest is restricted to the study of a fixed-composition ideal polarized electrode system, we can equally well employ either an indicator or an ordinary reference electrode. This general relationship [Eq. (53) or (54)] is known as the *Lippmann equation* (see footnote, p. 252).

b. Relative Surface Excesses of Components. At any constant value of $E^\pm$ (not constant E_{ref}), the following sets of relationships permit the determination of the relative surface excesses of components:

$$\left(\frac{\partial\gamma}{\partial\mu_i}\right)_{T,p,\mu',E^\pm} = -\Gamma_{ii'} \qquad \text{(for } i \neq i') \qquad (55)$$

$$(v_{jk'}^+)\left(\frac{\partial\gamma}{\partial\mu_{jk'}}\right)_{T,p,\mu',E^\pm} = -\Gamma_{jh'} \qquad \text{(for } j \neq j') \qquad (56)$$

$$(v_{j'k}^-)\left(\frac{\partial\gamma}{\partial\mu_{j'k}}\right)_{T,p,\mu',E^\pm} = -\Gamma_{kh'} \qquad \text{(for } k \neq k') \qquad (57)$$

$$\left(\frac{\partial\gamma}{\partial\mu_h}\right)_{T,p,\mu',E^\pm} = -\Gamma_{hh'} \qquad \text{(for } h \neq h') \qquad (58)$$

In Eqs. (55) to (58) the subscript μ' implies that the chemical potentials (activities) of all the components are held constant except that one with respect to which the differentiation is performed. These equations suffice to determine the relative surface excess of all components except the electrons of phase M and cation j' and anion k' of the indicator salt. For direct experimental determination of the relative surface excesses given in Eqs. (55) to (58), one measures the electrocapillary curves at different compositions, varying the activity of one component at a time. Then a plot is made at fixed $E^{\pm}$ of γ versus μ of the component whose composition is being varied. The negative of the slope of this plot gives the relative surface excesses as shown in Eqs. (55) to (58). Although the relative surface excesses of all the components can be determined (see Sec. d, below, for electrons and ions of the indicator salt), the nonrelativized surface excesses (cf. Sec. III.B.1) are not accessible to thermodynamics and therefore to equilibrium measurement. They may, however, be calculated on the basis of a nonthermodynamic model of the double-layer structure.

c. Proof that q^M is Independent of the Choice of Reference Components. We have seen that the values of the relative surface excesses of components depend on the arbitrary choices of the reference components i' in phase M and h' in phase S. However, the original definition of the excess charge densities q^M and q^S on the metal and solution sides of the electrical double layer were in terms of nonrelativized surface excesses of the charged components [Eq. (18)]. It is therefore important to know how these excess charge densities are expressed in terms of relative surface excesses, for it is now clear that only the latter are experimentally accessible. By direct application of the definitions of the relative surface excesses [Eqs. (41), (45)–(50)], it is easy to verify that

$$\left(\sum_i \Gamma_{ii'} z_i - \Gamma_{ei'} \right) = \left(\sum_i \Gamma_i z_i - \Gamma_e \right) = q^M/F \tag{59}$$

and

$$\left(\sum_j \Gamma_{jh'} z_j - \sum_k \Gamma_{kh'} |z_k| \right) = \left(\sum_j \Gamma_j z_j - \sum_k \Gamma_k |z_k| \right) = q^S/F \tag{60}$$

(Recall that $\Gamma_{i'i'} = 0 = \Gamma_{h'h'}$.) Equations (59) and (60) are important for they show that, regardless of the arbitrary choice of reference components i' and h', the equations defining the excess charge density $q^M = -q^S$ remain invariant. Because of this invariance under the arbitrary choice of reference components, q^M, unlike the surface excesses of individual components, is a quantity which can be *measured absolutely*. It has, therefore, among all parameters of the double layer, a special thermodynamic significance. Since the relative surface excesses of components are independent of the choice of the dividing surface in the Gibbs model, it follows that q^M is also invariant over any arbitrary choice of the dividing surface (cf. Sec. III.B.1.b).

d. Relative Surface Excesses of Electrons and Ions of the Indicator Salt. The remaining relative surface excesses of the electrons in phase M and of the ions j' and k' of the indicator salt in phase S are now easily derived. Having determined the relative surface excesses of the metal ions of phase M by Eq. (55), we can obtain the relative surface excess of the electrons by application of Eq. (59). Thus,

$$-\Gamma_{ei'} = (q^M/F) - \sum_{i \neq i'} \Gamma_{ii'} z_i \tag{61}$$

Let the indicator electrode be reversible to cation j' (case a). Then Eq. (56), the partial differential coefficient of the term $d\mu_{j'k'}$ in Eq. (52a), and the definition [Eq. (49)] of the relative surface excess of anion k' give

$$\Gamma_{k'h'} = -\sum_j v_{jk'}^- (\partial\gamma/\partial\mu_{jk'})_{T,p,\mu',E^+} \tag{62a}$$

The relative surface excesses are now determined for all ions in the solution except for the indicator ion j'. Application of Eq. (60) then gives

$$\Gamma_{j'h'} = (1/z_{j'})\left[\sum_k \Gamma_{kh'}|z_k| - \sum_{j \neq j'} \Gamma_{jh'} z_j - (q^M/F)\right] \tag{63a}$$

A similar analysis can be made for the case in which the indicator electrode is reversible to anion k'. In that case we obtain

$$\Gamma_{j'h'} = -\sum_k v_{j'k}^+ (\partial\gamma/\partial\mu_{j'k})_{T,p,\mu',E^-} \tag{62b}$$

and

$$\Gamma_{k'h'} = (1/|z_{k'}|)\left[\sum_{k \neq k'} \Gamma_{kh'}|z_k| - \sum_j \Gamma_{jh'} z_j - (q^M/F)\right] \tag{63b}$$

It is apparent that any information about the electrical double layer which can be obtained by thermodynamic analysis of electrocapillary curves measured with the help of a cation-reversible indicator electrode is equally available from electrocapillary data obtained with an anion-reversible electrode. Therefore, there is no theoretical reason for preferring one type of indicator electrode over another, although there may be cogent experimental reasons for such a preference.

e. The Relative Surface Excess of Entropy. It follows directly from the electrocapillary equation that

$$\left(\frac{\partial\gamma}{\partial T}\right)_{p,E^+,\mu} = -\left[\sigma_{i'h'} + \left(\frac{q^M s_+^{\bullet}}{z_{j'}F}\right)\right] \tag{64a}$$

and

$$\left(\frac{\partial\gamma}{\partial T}\right)_{p,E^-,\mu} = -\left[\sigma_{i'h'} - \left(\frac{q^M s_-^{\bullet}}{|z_{k'}|F}\right)\right] \tag{64b}$$

In Eqs. (64a) and (64b), $\sigma_{i'h'}$ is the relative surface excess of entropy which may be attributed to the existence of the double layer. The terms $(q^M s_+^{\bullet}/z_{j'}F)$ in Eq. (64a) and $(q^M s_-^{\bullet}/|z_{k'}|F)$ in Eq. (64b) may be determined from the measured value of the charge density at the point on the electrocapillary curve under consideration and tabulated values of the entropies of the pure substances comprising the indicator electrode (cf. Appendix B). Equations essentially equivalent to Eqs. (64a) and (64b) were derived by Parsons (*42a*). Some experimental work on the temperature dependence of the electrical double layer was carried out by Anderson and Parsons (*42b*) and by Randles and Whitely (*43*). According to Hansen, et al. (*44*), a temperature variation of $\pm 0.9°C$ will cause fluctuations of only about 0.2 dyne cm^{-1} in the interfacial tension. Barradas (*45*) has also reported only minute variations of γ with T.

f. Relative Thickness of the Interphase. The partial differential coefficient of interfacial tension with respect to pressure is given by

$$\left(\frac{\partial\gamma}{\partial p}\right)_{T,E^+,\mu} = \left[\tau_{i'h'} + \left(\frac{q^M v_+^{\bullet}}{z_{j'}F}\right)\right] \tag{65a}$$

and

$$\left(\frac{\partial\gamma}{\partial p}\right)_{T,E^-,\mu} = \left[\tau_{i'h'} - \left(\frac{q^M v_-^{\bullet}}{|z_{k'}|F}\right)\right] \tag{65b}$$

The terms $(q^M v_+^{\bullet}/z_{j'}F)$ and $(q^M v_-^{\bullet}/|z_{k'}|F)$, which are due to the particular nature of the indicator electrode, may be calculated from the measured value of q^M and from the known densities of the pure substances comprising the indicator electrode. [*Note added in proof:* Hills and Payne (*133*) recently studied the dependence of differential capacity on temperature and pressure.]

g. An Example. We shall illustrate the application of some of the relationships derived above by considering more closely the same set of electrocapillary curves (Fig. 4) for KBr solutions which were discussed earlier from a qualitative standpoint. Each of these curves was obtained at constant temperature and pressure (33) by measuring the interfacial tension of the mercury-solution interface with a capillary electrometer. A mercury-mercurous bromide electrode dipping into the same solution served as indicator electrode. The indicator electrode was reversible to the Br^- anion of the solution (case b). Thus the cell was

$$Cu|Pt|Hg|KBr\ solution|Hg_2Br_2|Hg|Pt|Cu'$$

The measured potential difference was $E^- = \phi^{Cu} - \phi^{Cu'}$.

The metallic phase of the ideal-polarized electrode was pure mercury, and the solution contained only two components, water and KBr. Therefore, mercury and water are the reference components i' and h', respectively. In this case the electrocapillary equation [Eq. (52b)] reduces to

$$d\gamma = -q^M\ dE^- - \Gamma_{K^+,w}\ d\mu_{KBr} \tag{66}$$

where $\Gamma_{K^+,w}$ is the surface excess of K^+ ion, relative to water. Since the solution contains only the indicator salt, the rather complicated coefficient of $d\mu_{j'k'}$ in Eq. (52b) reduces to the relative surface excess of the cation j' of the indicator salt. [If the indicator electrode had been reversible to K^+ ion, we would have used Eq. (52a), which would have reduced to $d\gamma = -q^M\ dE^+ - \Gamma_{Br^-,w}\ d\mu_{KBr}$.] This simplification occurs because (a) the second term (k summation) in that coefficient reduces to zero, and (b) the first term (in square brackets) is equal to the relative surface excess of cation j' [Eq. (50)] when only one salt is present in solution. The last statement is true because the indicator salt KBr alone furnishes K^+ ions to the solution.

Equation (66) implies that, for any constant value of E^-, if $\Gamma_{K^+,w}$ is positive, the interfacial tension γ will decrease with increasing chemical potential, that is, activity of KBr in the solution. The experimental electro-capillary curves A through F (order of increasing KBr concentration and activity) each lie successively lower at all values of E^-. Thus we conclude merely from inspection of the electrocapillary curves that potassium ion is positively adsorbed (relative to water) at all potentials. This fact appears somewhat surprising at first glance, for it would be natural to suppose that, on the left side of the ECM, where q^M is positive according to the Lippmann equation [Eq. (53b)], potassium ions would be repelled away from the interface and that $\Gamma_{K^+,w}$ would be therefore negative. The explanation is to be found in the specific (superequivalent) adsorption of Br^- ion.

By virtue of the invariance of the excess charge density q^M over the choice of the reference components [Eqs. (59)–(60)], we can calculate the relative surface excess of Br^- from the measured values of q^M and $\Gamma_{K^+,w}$ at each point on each curve. Thus, applying Eq. (60) and recalling that $q^M = -q^S$, we have

$$\Gamma_{Br^-,w} = (q^M/F) + \Gamma_{K^+,w} \tag{67}$$

At a given value of E^- on a given electrocapillary curve, q^M is determined by differentiating the curve graphically and applying the Lippmann equation in the form†

$$q^M = -(\partial\gamma/\partial E^-)_{T,p,\mu_{KBr}} \tag{53b}$$

$\Gamma_{K^+,w}$ at the same point on the same curve is determined by differentiating graphically a plot of γ, at constant E^-, versus $\log a_\pm^2$, where $a_\pm$ is the mean activity of KBr in the solution. The value of $\Gamma_{Br^-,w}$ given by Eq. (67) is then the value of the relative surface excess of bromide ion at the same value of E^- and $a_\pm$. Thus at every point along each electrocapillary curve in Fig. 4 we can determine the corresponding values of the excess charge density on the metal and of the relative surface excesses of potassium and of bromide ions.

The experimental results which were obtained by Devanathan and Peries for the 0.1 F KBr solution (curve C in Fig. 4) are shown in Fig. 6. In this figure $+F\Gamma_{K^+,w}$ and $-F\Gamma_{Br^-,w}$ are the ordinates. That is, the ordinates denote the *components of charge* (including sign) on the solution side of the double layer, which are equivalent to the relative adsorption of each ion. The abscissas are the corresponding values of q^M. If the interaction between the ions and the electrode were purely coulombic in nature, we would expect that at the ECM the surface excesses of both ions would be zero. However, we see from Fig. 6 that both ions are positively adsorbed at the ECM (of course, in equal amounts). For positive values of q^M, we would have expected the K^+ ions to be repelled away from the electrode surface and Br^- ions to be attracted toward it. The equivalent component of charge resulting from the adsorption of either ion on the positive side of ECM would then have been negative. The fact is that, for positive q^M, potassium ions are even more strongly adsorbed than at the ECM.

The only explanation which seems capable of accounting for this behavior is specific adsorption of bromide ions. That is, there must be some kind of additional, noncoulombic interaction between bromide ions and the metal surface which brings more bromide ions into the double layer than can be accounted for by simple coulombic attraction. This superequivalent (to q^M) quantity of bromide ion then attracts sufficient additional

---

† Note that in this case the relative surface excess of the electrons in the metallic phase is given by [cf. Eq. (61)] $\Gamma_{e,Hg} = -q^M/F$.

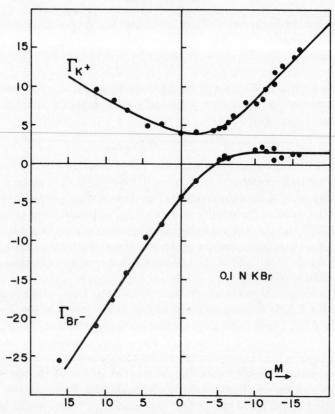

Fig. 6. Components of charge versus q^M for 0.1 F KBr solutions at mercury-aqueous KBr solution interface. Data from (*33*). Curves taken from (*46*). (By permission of Pergamon Press.)

potassium ions (counterions) to maintain the electroneutrality of the entire interfacial region, that is, to make $q^S = -q^M$. As q^M becomes more positive, the amount of specifically adsorbed bromide ions increases, and this increase is reflected in an increased positive adsorption of K^+ counterions. When q^M becomes negative, the specific adsorption of Br^- decreases; eventually, at a sufficiently negative q^M, it reaches zero. Thereafter, as q^M becomes more negative, the relative surface excess of Br^- ion becomes negative, and the corresponding equivalent component of charge in the double layer due to Br^- becomes positive. There is no evidence for specific adsorption of K^+ ions except possibly at extremely negative potentials.

There is an objection which one might be tempted to raise at this point to the foregoing explanation. Since all we can determine experimentally is relative adsorption, one could ask whether the observed values of these relative surface excesses might result not from any real specific attraction of Br^- into the double layer but rather be a consequence of desorption of the reference component, water. In other words, the question is whether we could be dealing with an artifact resulting from our chosen method of relativizing the Γ's. By definition, the relative surface excesses of potassium and bromide ions are given by

$$\Gamma_{K^+,w} = \Gamma_{K^+} - (c_{KBr}/c_w)\Gamma_w$$

and

$$\Gamma_{Br^-,w} = \Gamma_{Br^-} - (c_{KBr}/c_w)\Gamma_w$$

where the c's are the molar concentrations of KBr and water in the solution, and the Γ's on the right side are the nonrelativized surface excesses (cf. Sec. III.B.1). At the point of zero charge (ECM) the relative surface excess of K^+ could be positive even though the nonrelativized surface excess of K^+ reckoned at the OHP, Γ_{K^+}, were itself zero, provided Γ_w were sufficiently large and negative. This would imply that

$$\Gamma_w = -\left(\frac{c_w}{c_{KBr}}\right)\Gamma_{K^+,w}$$

In the case under consideration

$$(c_w/c_{KBr}) = 55.5/0.1 = 555$$

At the ECM the charge equivalent to the relative surface excess of K^+ is about 4 μcoulomb cm^{-2} (Fig. 6). Thus, if $\Gamma_{K^+} = 0$,

$$\Gamma_w \cong -(555 \times 4 \times 10^{-6})/96500 \text{ mole cm}^{-2}$$
$$\cong -2.3 \times 10^{-8} \text{ mole cm}^{-2}$$
$$\cong -1.4 \times 10^{16} \text{ molecules cm}^{-2}$$

But a close packed monolayer of water would contain (47) only about 1.6×10^{15} molecules cm^{-2}. Thus, to account for the observed relative adsorption of K^+ at the ECM solely on the basis of desorption of the reference substance (water), we would have to assume the complete removal of water from at least the first nine layers of solution at the interface! That, clearly, would be an untenable assumption. Thus we are forced to admit that specific attraction of bromide ions to the metal surface is a real and not merely an apparent phenomenon. Of course, it is still true that desorption of water may, and probably does, take place, but it is not capable of accounting for all the observed superequivalent adsorption of the ions.

2. Differential and Integral Capacitance

If electrocapillary curves were perfectly parabolic, a plot of the excess charge density q^M versus $E^\pm$ (or E_{ref}) would be a straight line. In fact, such plots are always curved. Figure 7 shows a graph of q^M versus E_{NCE} (normal calomel electrode) for the mercury electrode in 0.1 F KBr solution, that is, the same ideal polarized electrode system we have just considered (curve C in Figs. 4 and 6). The slope of this curve at any point is known as the *differential capacitance* of the electrical double layer and is denoted by C. Thus,

$$C = (\partial q^M / \partial E)_{T,p,\mu} \tag{68}$$

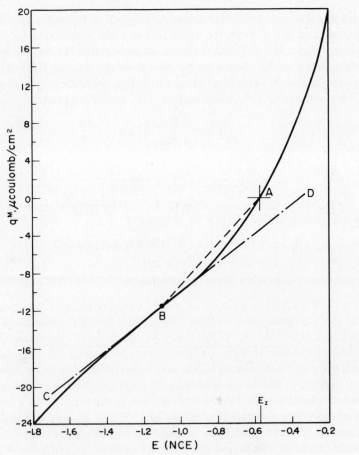

Fig. 7. q^M versus E_{NCE} for 0.1 F KBr. Data from (*22*).

Here E may be either $E^{\pm}$ or E_{ref}. The units of C are μcoulomb cm^{-2} volt^{-1} or μfarad cm^{-2}. Applying the Lippmann equation [Eq. (54)], we see that C is also minus the second derivitive of γ with respect to E. Thus

$$C = -(\partial^2\gamma/\partial E^2)_{T,p,\mu} \tag{69}$$

C may therefore be obtained by performing a second graphical differentiation on the electrocapillary curve. Such a procedure is not very satisfactory, owing to the inherent inaccuracy of repeated graphical differentiation. Fortunately, the differential capacity C can be measured directly using an ac impedance bridge. In fact, direct measurement of C is more accurate than is direct measurement of γ. Much of the detailed knowledge about the structure of the double layer has been obtained by measuring C directly and then calculating the other double-layer parameters by means of Eq. (69) and the electrocapillary equation.

It follows from Eq. (69) that, at any value of E, the corresponding value of the excess charge density on the metal is given by

$$q^M = \int_{E_z}^{E} C \, dE \tag{70}$$

where E_z denotes the value of $E^{\pm}$ (or E_{ref}) at the ECM. Figure 8 shows the differential capacitance of ideal polarized mercury electrodes in several different 0.1 F solutions of potassium salts, including KBr. These $C - E$ curves were measured by Grahame et al. (22) using an ac impedance bridge. By graphically integrating these $C - E$ curves he was able to determine q^M to a high degree of precision. The plot of q^M for 0.1 F KBr solution shown in Fig. 7 was constructed from Grahame's table of data.

Integration of differential capacity curves between any two values of E, say E_1 and E_2, gives the difference of the excess charge density, q^M, between these two potentials, that is,

$$q^M(E_2) - q^M(E_1) = \int_{E_1}^{E_2} C \, dE \tag{71}$$

However, to determine the absolute magnitude of q^M at a given E, it is necessary to know independently the value of E_z. For liquid metal electrodes, E_z may be determined directly from electrocapillary curves or with a streaming electrode (48,49). For solid electrodes it is not possible to measure γ directly (at least not very accurately), and the accurate determination of E_z for solid electrodes remains one of the major unsolved problems in electrochemistry. Still, it is possible to measure C for solid electrodes, and much of what is known about the electrical double layer at

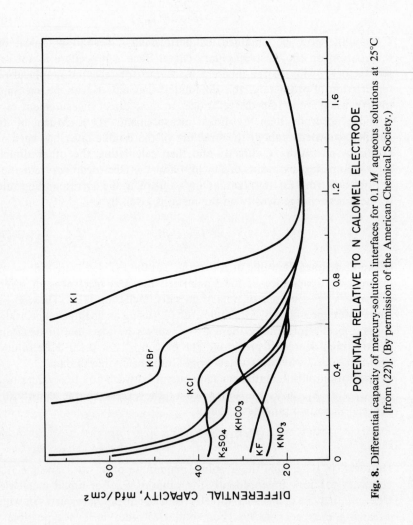

Fig. 8. Differential capacity of mercury-solution interfaces for 0.1 M aqueous solutions at 25°C [from (22)]. (By permission of the American Chemical Society.)

solid electrodes has been derived from such measurements [cf. Delahay (5), pp. 129–134].

It also follows from Eq. (69) that a second integration of a $C - E$ curve will yield the electrocapillary curve itself. Thus

$$\gamma = \gamma_{ECM} - \int_E^{E_z} \int_E^{E_z} C \, dE \, dE \tag{72}$$

To determine the value of γ at any E from a $C - E$ curve, it is necessary to know independently two limits of integration (E_z) and the value of the interfacial tension at the electrocapillary maximum (γ_{ECM}). However, even if γ_{ECM} is not known, provided E_z is known, one can still determine γ to within an additive constant by double integration of the $C - E$ curve.

Closely related to the differential capacitance is the *integral capacitance* K, of the electrical double layer. K is defined by the equation

$$K = q^M/(E - E_z) \tag{73}$$

For any point on a $q^M - E$ curve, K is equal to the slope of a chord drawn from the point of zero charge to the point in question. For example, in Fig. 7 the integral capacity of the electrical double layer at point B is the slope of the dashed line $A - B$ drawn from the point of zero charge (A) to point B. In this particular example, point B corresponds to $q^M = -11.6$ μcoulomb cm^{-2} and $E_{NCE} = -1.10$ volts. Also $E_z = -0.574$ volt on the NCE scale. Thus

$$K = -11.6/(-1.10 + 0.57) \ \mu\text{farad cm}^{-2}$$

$$= 22 \ \mu\text{farad cm}^{-2}$$

On the other hand, the differential capacitance at the same point is the slope of the tangent to the $q^M - E$ curve at that point, that is, the slope of the dash-dot line CBD. In this case the differential capacitance, $C = 16$ μfarad cm^{-2}. However, as E approaches E_z, it is clear that the value of the integral capacitance approaches that of the differential capacitance:

$$\lim_{E \to E_z} K = C(E_z) \tag{74}$$

In general, C is a more useful quantity for elucidating the structure of the electrical double layer than is K.

3. Second Cross-Partial-Differential Coefficients of γ

The electrocapillary equation [Eq. (52)] is a total differential equation, and the interfacial tension γ possesses continuous derivitives of all orders.

Thus the second cross-partial-differential coefficients of γ exist, and their values are independent of the order of differentiation. Applying this idea to the independent variables $E^{\pm}$ and μ_h, for example, we obtain (for $h \neq h'$)

$$\left[\frac{\partial}{\partial \mu_h}\left(\frac{\partial \gamma}{\partial E^{\pm}}\right)_{T,p,\mu}\right]_{T,p,\mu',E^{\pm}} = \left[\frac{\partial}{\partial E^{\pm}}\left(\frac{\partial \gamma}{\partial \mu_h}\right)_{T,p,E^{\pm},\mu'}\right]_{T,p,\mu}$$

or

$$\left(\frac{\partial q^M}{\partial \mu_h}\right)_{T,p,E^{\pm},\mu'} = \left(\frac{\partial \Gamma_{hh'}}{\partial E^{\pm}}\right)_{T,p,\mu} \tag{75}$$

Here the subscript μ' means that all chemical potentials are held constant, except the one with respect to which the differentiation is performed. The subscript μ denotes constant composition.

Equation (75) implies that, at corresponding points, the slope of a plot of the excess charge density q^M versus the chemical potential (logarithm of activity) of neutral component h, at constant $E^{\pm}$, is equal to the slope of a plot of the relative surface excess $\Gamma_{hh'}$ of component h versus $E^{\pm}$ at constant composition. Such a cross-differentiation relationship can provide a valuable cross-check to ensure that different kinds of experimental data about the electrical double layer are actually self-consistent.

As an illustration of such a consistency check, we shall consider briefly the adsorption of n-butyl alcohol on an ideal polarized mercury electrode in HCl solution. This study was reported by Blomgren and co-workers (50). Electrocapillary curves for aqueous HCl solutions containing n-butyl alcohol were measured. The indicator electrode was a reversible hydrogen electrode in the same solution (case a). The relevant form of the electrocapillary, equation [Eq. (52a)] was

$$d\gamma = -q^M \, dE^+ - \Gamma_{Cl^-,w} \, d\mu_{HCl} - \Gamma_{Bu,w} \, d\mu_{Bu}$$

where the subscript Bu refers to n-butyl alcohol, and w indicates that water is the reference substance in solution. In one series of experiments the concentration (activity) of HCl was held constant (0.1 F), and the concentration (activity) of alcohol was varied. The relative surface excess $\Gamma_{Bu,w}$ of the alcohol was determined from the electrocapillary curves by application of Eq. (58), and the excess charge density at corresponding points was determined by means of the Lippmann equation [Eq. (53a)]. In Fig. 9, $\Gamma_{Bu,w}$ is plotted as a function of the potential E^+ of the ideal polarized mercury electrode with respect to the hydrogen indicator electrode. (This potential is denoted E_H in the figure.) Each curve refers to a different

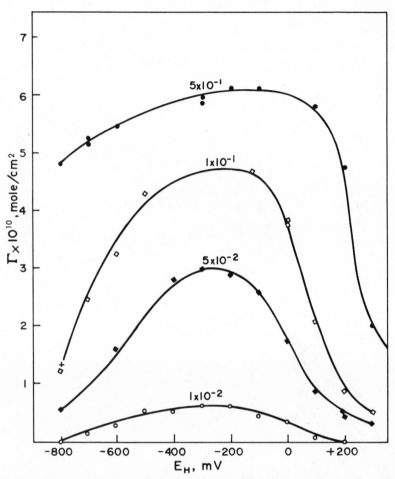

Fig. 9. Relative surface excess of *n*-butyl alcohol in 0.1 *F* HCl versus E^+ [from (*50*)]. (By permission of the American Chemical Society.)

concentration of alcohol. The corresponding graphs of $q^S = -q^M$ versus E_H are shown in Fig. 10. For this system, Eq. (75) becomes

$$-\left(\frac{\partial q^S}{\partial \mu_{Bu}}\right)_{E_H, \mu_{HCl}} = \left(\frac{\partial \Gamma_{Bu,w}}{\partial E_H}\right)_\mu \qquad (76)$$

In particular, Eq. (76) implies that whenever the $\Gamma_{Bu,w}$ versus E_H curves (adsorption isotherms) show a maximum, the corresponding $q^S - E_H$ curves must intersect. Inspection of Figs. 9 and 10 shows that this is indeed true. The potential of the point of intersection is equal to the potential of

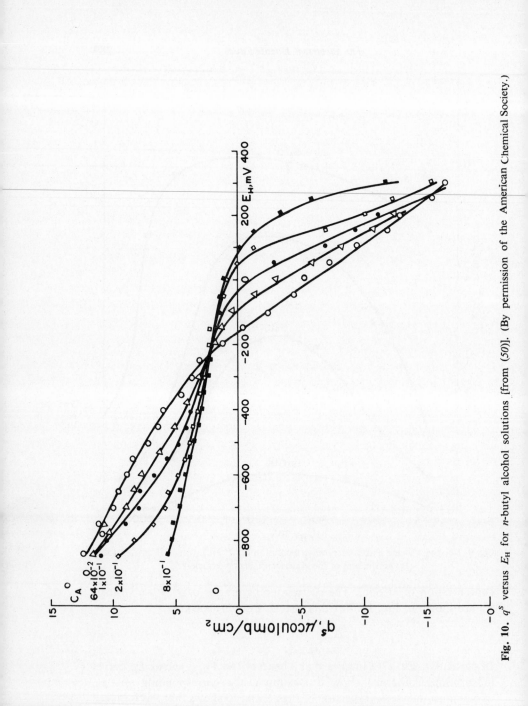

Fig. 10. q^s versus E_H for n-butyl alcohol solutions [from (50)]. (By permission of the American Chemical Society.)

the maxima. As a matter of fact, it is possible to determine the position of the maxima of the adsorption isotherms more precisely from the crossing points of the charge-density curves than from the isotherms themselves. This illustrates another advantage of such a cross-differentiation relationship. Closer inspection of Figs. 9 and 10 shows that Eq. (76) is verified at other points, too.

Many other useful cross-partial-differentiation relationships may be derived from the electrocapillary equation in a similar manner. An example is

$$\left(\frac{1}{v_{jk'}^+}\right)\left(\frac{\partial \Gamma_{jh'}}{\partial \mu_{j'k}}\right)_{E^\pm,\mu'} = \left(\frac{1}{v_{j'k}^-}\right)\left(\frac{\partial \Gamma_{kh'}}{\partial \mu_{jk'}}\right)_{E^\pm,\mu'} \tag{77}$$

for $j \neq j'$ and $k \neq k'$.

4. Esin and Markov Coefficients

Thus far we have considered the independent electrical variable to be the potential $E^\pm$ of the ideal polarized electrode with respect to the indicator electrode. For some purposes, however, the excess charge density q^M appears to be a more convenient parameter for describing the electrical characteristics of the double layer. As we have seen, q^M is invariant over any arbitrary choice of the reference components or of the indicator salt. Thus in a certain sense, q^M may be considered a "more fundamental" electrical variable than the potential difference across the cell terminals. [This point of view has been favored especially by Parsons (40,51–54) for the consideration of adsorption isotherms, and it has been used also in a theoretical treatment of adsorption kinetics (55) and in a study of the effect of specific adsorption in the kinetics of charge-transfer processes (56). Recently, Damaskin (57) has offered arguments against this point of view, at least insofar as the treatment of the adsorption isotherms for organic compounds at an ideal polarized electrode is concerned.]

A modification of the basic electrocapillary equation in which q^M appears as the independent electrical variable is easily derived by considering Parsons' auxiliary function $\xi^\pm$ (40) which is defined by

$$\xi^\pm = \gamma + q^M E^\pm \tag{78}$$

Differentiation of Eq. (78) yields

$$d\gamma = d\xi^\pm - q^M \, dE^\pm - E^\pm \, dq^M \tag{79}$$

Substitution of Eq. (79) into the left side of the electrocapillary equation [Eq. (52a) or (52b)] yields:

Case a. (indicator electrode reversible to cation j'):

$$d\xi^+ = +E^+ \, dq^M - \cdots \tag{80a}$$

Case b. (indicator electrode reversible to anion k'):

$$d\xi^- = +E^- \, dq^M - \cdots \tag{80b}$$

Except for the term $E^\pm \, dq^M$, all the other terms on the right side of Eqs. (80a) and (80b) are identical including sign to the corresponding terms in Eqs. (52a) and (52b).

Equations (80) are not so useful themselves as are a particular set of cross-differentiation relationships which are easily derived from them. These cross-differentiation relationships define the so-called Esin and Markov coefficients.† We define the Esin and Markov coefficient of a neutral component h of the solution as the slope of a plot of $E^\pm$ versus the chemical potential of component h at constant temperature, pressure, excess charge density q^M, and chemical potentials of all the other components. It follows directly from Eqs. (80) that

$$\left(\frac{\partial E^\pm}{\partial \mu_h}\right)_{T,p,q^M,\mu'} = -\left(\frac{\partial \Gamma_{hh'}}{\partial q^M}\right)_{T,p,\mu} \qquad \text{(for } h \neq h') \tag{81}$$

Thus the Esin and Markov coefficient is a measure of the rate of change of the relative surface excess of a component with q^M at constant composition [cf. Eq. (75)]. Similarly, we can derive Esin and Markov coefficients for the charged components. Thus

$$\left(\frac{\partial E^\pm}{\partial \mu_i}\right)_{T,p,q^M,\mu'} = -\left(\frac{\partial \Gamma_{ii'}}{\partial q^M}\right)_{T,p,\mu} \qquad \text{(for } i \neq i') \tag{82}$$

$$\left(\frac{\partial E^\pm}{\partial \mu_{jk'}}\right)_{T,p,\mu',q^M} = -\left(\frac{1}{v_{jk'}^+}\right)\left(\frac{\partial \Gamma_{jh'}}{\partial q^M}\right)_{T,p,\mu} \qquad \text{(for } j \neq j') \tag{83}$$

$$\left(\frac{\partial E^\pm}{\partial \mu_{j'k}}\right)_{T,p,q^M,\mu'} = -\left(\frac{1}{v_{j'k}^-}\right)\left(\frac{\partial \Gamma_{kh'}}{\partial q^M}\right)_{T,p,\mu} \qquad \text{(for } k \neq k') \tag{84}$$

---

† Esin and Markov (*58*) showed in 1939 that the potential of the ECM versus a "constant-potential" reference electrode was a linear function of the logarithm of the concentration for several aqueous electrolyte solutions. Grahame (*7*), in 1955, suggested that this phenomenon be called the *Esin and Markov Effect*, and Parsons (*51*) generalized the concept in 1957.

In Eqs. (81) to (84), as before, the subscript μ' means that the chemical potentials (activities) of all components are held constant except for the component with respect to which the differentiation is being performed. The subscript μ implies constant composition.

We shall refer to plots of $E^{\pm}$ versus either μ or the logarithm of the activity of the component in question as *Esin and Markov plots*. Similarly, both $(\partial E^{\pm}/\partial \mu)$ and $(\partial E^{\pm}/\partial \ln a) = RT(\partial E^{\pm}/\partial \mu)$ will be referred to as the *Esin and Markov coefficient*. This slight imprecision in language is justified by convenience in discussions. In any particular case it will always be clear which definition of the Esin and Markov coefficient is being employed.

Like other cross-differentiation relationships, the Esin and Markov coefficients can provide valuable self-consistency checks for experimental data. In addition, these coefficients for charged components have been shown to provide another criterion for specific adsorption (*51*) (see Sec. III.D.3).

F. The Electrocapillary Equation for Reversible Charge-Transfer Electrodes

From the standpoint of the thermodynamic theory of electrocapillarity, the only essential difference between an ideal polarized electrode and an electrode at which a reversible charge-transfer (faradaic) process takes place is that the latter electrode has one less degree of freedom than it would have were it ideally polarizable. Thus the electrocapillary equation for a reversible charge-transfer electrode must have one less independent variable than the electrocapillary equation would have for a system containing the same components but for which the charge-transfer reaction is prohibited. The theory of the electrocapillary curve for reversible charge-transfer electrodes was treated first by Grahame and Whitney (*26*), and it was considered later in a somewhat more detailed way by Mohilner (*59*). A generalized version of the latter treatment will be given in this section. Little experimental data on this aspect of electrocapillarity has been reported yet. Koenig et al. (*60*) have reported recently on the mercury-mercurous ion electrode.

Let the electrode system have the same composition as the generalized ideal polarized electrode system considered in Sec. II.C and II.D. If this system were ideally polarized, the electrocapillary equation would be Eqs. (52) (a or b, depending on whether the indicator electrode is chosen reversible to cation j' or anion k'). In the present case this equation is not directly applicable because a reversible charge-transfer reaction takes place at the metal-solution interface. In general this charge-transfer reaction may be represented by the half-reaction

$$O + ne = R \tag{85}$$

Here O represents the oxidized form of the couple, n is the number of electrons involved, and R represent the reduced form of the couple. Since this reaction is an over-all electrode reaction no components are common to both sides.

In general O and R stand for sets of reactant species (oxidized forms of couple) and product species (reduced forms of couple), whereas for simple cases O and R are single species. In either case the condition of charge-transfer equilibrium for the half-reaction is given by an equation expressing the equality of the electrochemical potentials of reactants and products [cf. Van Rysselberghe (*19*)]. Thus

$$\bar{\mu}_O + n\bar{\mu}_e = \bar{\mu}_R \tag{86}$$

For complicated half-reactions the symbols $\bar{\mu}_O$ and $\bar{\mu}_R$ stand for linear combinations of the electrochemical potentials of reactant species and product species, some of which may belong to phase M with others belonging to phase S. In a simple case substance O might be a metallic ion in solution, and substance R could be either a metal in phase M (metal deposition) or an ion or a neutral substance in phase S (simple redox reaction). In all cases, the electrons belong to phase M and n is the net number of electrons lost by phase M when half-reaction (85) goes once (advances by one dedonder) to the right.

Equation (86), which shows that not all the electrochemical potentials of the substances involved in the half-reaction are independent, provides the additional relationship required to obtain the electrocapillary equation for the reversible charge-transfer electrode. In essence, one substitutes Eq. (86) into Eq. (52). Consideration of a simple redox reaction will serve to illustrate the method of derivation. The cell is

$$Cu \,|\, M \,|\, S \,|\, IN \,|\, Cu'$$

and Eqs. (24) to (30) and (32) still apply.

Let both substances O and R be cations of the solution. The O and R are each represented by a member of the set of subscripts $\{j\}$. (For convenience, assume that the O-R couple does not involve ion j' of the indicator salt.) In keeping with the previous definition of the salt components of the solution, O and R are considered furnished to the solution by two salts designated by the subscripts Ok' and Rk', where O and R are the members of the set $\{j\}$ corresponding to cations O and R, respectively. Thus

$$\mu_{Ok'} = v_{Ok'}^{+}\bar{\mu}_O + v_{Ok'}^{-}\bar{\mu}_{k'}$$

and   (87)

$$\mu_{Rk'} = v_{Rk'}^{+}\bar{\mu}_R + v_{Rk'}^{-}\bar{\mu}_{k'}$$

Substituting Eqs. (87) into Eq. (86) and taking differentials, we obtain

$$\left(\frac{1}{v_{Ok'}^{+}}\right) d\mu_{Ok'} - \left(\frac{v_{Ok'}^{-}}{v_{Ok'}^{+}}\right) d\bar{\mu}_{k'} + nd\bar{\mu}_e = \left(\frac{1}{v_{Rk'}^{+}}\right) d\mu_{Rk'} - \left(\frac{v_{Rk'}^{-}}{v_{Rk'}^{+}}\right) d\bar{\mu}_{k'} \quad (88)$$

Since the metallic phase M of the reversible charge-transfer electrode is in contact with the terminal Cu, we have

$$d\bar{\mu}_e = d\bar{\mu}_e^{Cu} \tag{29}$$

For simplicity let IN be reversible to anion k' of the solution and let the temperature and pressure be constant. Then (cf. Appendix B)

$$d\bar{\mu}_{k'} = |z_{k'}| \, d\bar{\mu}_e^{Cu'} \tag{89}$$

Owing to the electroneutrality of the salts [Eq. (49)], we have

$$(v_{Ok'}^{-}/v_{Ok'}^{+}) = (z_O/|z_{k'}|)$$

and   (90)

$$(v_{Rk'}^{-}/v_{Rk'}^{+}) = (z_R/|z_{k'}|)$$

where z_O and z_R represent the (cationic) ionic charges of substances O and R, respectively. Moreover, by the half-reaction (85), we have

$$z_O - z_R = n \tag{91}$$

Substituting Eqs. (29), (89), (90), and (91) in Eq. (88), we obtain

$$\left(\frac{1}{v_{Ok'}^{+}}\right) d\mu_{Ok'} - n(d\bar{\mu}_e^{Cu'} - d\bar{\mu}_e^{Cu}) = \left(\frac{1}{v_{Rk'}^{+}}\right) d\mu_{Rk'} \tag{92}$$

Recall that the definition of the potential $E^{\pm}$ of the electrode M with respect to the indicator electrode is given by

$$FE^{\pm} = (\bar{\mu}_e^{Cu'} - \bar{\mu}_e^{Cu}) \tag{28}$$

Thus Eq. (92) reduces to

$$\left(\frac{1}{v_{Ok'}^{+}}\right) d\mu_{Ok'} - nF \, dE^{-} = \left(\frac{1}{v_{Rk'}^{+}}\right) d\mu_{Rk'} \tag{93}$$

This equation provides the relationship required to convert the electro-capillary equation for an ideal polarized electrode into a correct electro-capillary equation for a charge-transfer electrode. The electrochemical potential of either salt Ok' or salt Rk' in Eq. (52b) may be eliminated by means of Eq. (93). Substitute for the quantity $(1/v_{Rk'}^+)\,d\mu_{Rk'}$ in Eq. (52b) to obtain the electrocapillary equation:

$$dy = -(q^M - nF\Gamma_{Rh'})\,dE^- - \left(\frac{1}{v_{Ok'}^+}\right)(\Gamma_{Oh'} + \Gamma_{Rh'})\,d\mu_{Ok'}$$

$$- \sum_{i \neq i'} \Gamma_{ii'}\,d\mu_i - \sum_{j \neq j',O,R} (1/v_{jk'}^+)\Gamma_{jh'}\,d\mu_{jk'}$$

$$- \sum_{k \neq k'} (1/v_{j'k}^-)\Gamma_{kh'}\,d\mu_{j'k} - \sum_{h \neq h'} \Gamma_{hh'}\,d\mu_h$$

$$- \left\{ \left[(\Gamma_{j'}/v_{j'k'}^+) - \left(\frac{x_{j'k'}}{x_{h'}}\right)\Gamma_{h'} \right] - \left(\frac{1}{z_{j'}v_{j'k'}^+}\right) \sum_{k \neq k'} \Gamma_k|z_k| \right\} d\mu_{j'k'} \quad (94)$$

Equation (94) differs from its ideal polarized analog in the following important ways:

(a) The number of independent variables is one less. (The term involving $d\mu_{Rk'}$ has been eliminated.) To each solution composition corresponds one and only one value of E^- and consequently one and only one value of y.

(b) The term involving dE^- is no longer simply $-q^M\,dE^-$ but rather $-(q^M - nF\Gamma_{Rh'})\,dE^-$. Thus, the slope of the electrocapillary curve is no longer given by the classical Lippmann equation [Eq. (53)]. Rather, for the electrocapillary curve of a reversible charge-transfer electrode, the slope is

$$(\partial y/\partial E^-)_{T,p,\mu} = -(q^M - nF\Gamma_{Rh'}) \quad (95)$$

As before, the subscripts T and p denote constant temperature and pressure. The subscript μ in Eq. (95) implies that the activities of all the substances whose chemical potentials *appear* in Eq. (94), that is, all substances except salt Rk', are held constant. E^- is then changed in accordance with the Nernst equation by changing $\mu_{Rk'}$ in the solution.

(c) The coefficient of $d\mu_{Ok'}$ now involves the sum of the relative surface excesses of both members of the couple.

(d) Terms in the j summation for $j = O$, and $j = R$ are now deleted.

If we had chosen instead to eliminate the chemical potential of the salt Ok', the electrocapillary equation would have been the same except for the third and fourth terms in Eq. (94). These terms would have been replaced by

$$-(q^M + nF\Gamma_{Oh'})\, dE^- \qquad \text{and} \qquad -\left(\frac{1}{v_{Rk'}^+}\right)(\Gamma_{Ok'} + \Gamma_{Rk'})\, d\mu_{Rk'}$$

respectively. Had the indicator electrode been chosen reversible to cation j' instead of anion k', an additional term involving $d\mu_{j'k'}$ would have been obtained.

To illustrate the preceeding argument with a concrete example, consider the case of a Hg electrode dipping into an aqueous solution containing VSO_4, $V_2(SO_4)_3$, and H_2SO_4. Let the indicator electrode be a Hg, Hg_2SO_4 electrode in contact with the same solution.† The indicator ion is, therefore, sulfate. The "binary salts" from which the solution is made up are then H_2SO_4, indicator salt, VSO_4, and $V_2(SO_4)_3$. The Hg electrode M is not ideal polarized but, of course, is involved in the reversible V^{3+}/V^{2+} charge-transfer reaction. If the charge-transfer reaction were prohibited, the electrocapillary equation for this system would be, at constant temperature and pressure,

$$d\gamma = -q^M\, dE^- - \Gamma_{V^{2+},w}\, d\mu_{VSO_4} - \tfrac{1}{2}\Gamma_{V^{3+},w}\, d\mu_{V_2(SO_4)_3} - \tfrac{1}{2}\Gamma_{H^+,w}\, d\mu_{H_2SO_4} \tag{96}$$

However, since the V^{3+}/V^{2+} reaction is a reversible charge-transfer reaction which can occur at the Hg electrode, this equation does not apply. Rather, by Eq. (86), we have

$$\bar{\mu}_{V^{3+}} + \bar{\mu}_e = \bar{\mu}_{V^{2+}}$$

Thus, Eq. (93) becomes

$$\tfrac{1}{2}\, d\mu_{V_2(SO_4)_3} - F\, dE^- = d\mu_{VSO_4} \tag{97}$$

Substituting for the chemical potential of VSO_4 in the electrocapillary equation [Eq. (96)] by means of Eq. (97), one obtains the following electrocapillary equation for the reversible V^{3+}/V^{2+} electrode at constant temperature and pressure:

$$d\gamma = -(q^M - \Gamma_{V^{2+},w}F)\, dE^- - \tfrac{1}{2}(\Gamma_{V^{2+},w} + \Gamma_{V^{3+},w})\, d\mu_{V_2(SO_4)_3}$$
$$- \tfrac{1}{2}\Gamma_{H^+,w}\, d\mu_{H_2SO_4} \tag{98}$$

† In a laboratory experiment one could not use such an indicator electrode in direct contact with the solution containing the V^{3+}/V^{2+} couple because this redox reaction would proceed spontaneously at the surface of the indicator electrode also. Instead one would employ an ordinary reference electrode (for example, SCE) connected with the solution by a salt bridge, and then calculate from E_{ref} what the potential E^- of the electrode of interest would have been.

Comparing Eq. (98) with Eq. (96), one sees that the primary difference is, as stated above, that Eq. (98) has three independent variables, whereas the ideal polarized analog, Eq. (96), had four.

The electrocapillary equation for any other type of charge-transfer electrode can be derived in a similar way. The basic idea is to eliminate the chemical potential of a neutral substance furnishing one of the components involved in the charge-transfer reaction. Changes required to convert the ideal polarized electrocapillary equation to one for a charge-transfer electrode for several different types of charge-transfer reactions are summarized in Table II.

III. ADSORPTION AND DOUBLE-LAYER MODELS

A. Introduction

The thermodynamic theory of electrocapillarity (Section II) provides a set of general, unequivocal relationships among the measurable, macroscopic equilibrium properties of the electrical double layer. These relationships, which are all derivable from the basic electrocapillary equation [Eq. (52) or (94)], will remain valid regardless of how concepts of the molecular structure of the double layer may change, since their derivation was without resort to any molecular conceptions whatever. As a result, the thermodynamic theory by itself provides no information about double-layer structure. For interpretation of double-layer effects in electrode kinetics, structural information is needed. Thus it is necessary to resort to a nonthermodynamic molecular model of double-layer structure. The merit of a model may be assessed both by its agreement with the thermodynamic theory of electrocapillarity and by its ability to provide reasonable correlations between electrode kinetics and double-layer properties. Clearly no model can be correct if it predicts behavior which is at variance with the thermodynamic theory. It is likely that a more stringent test for a model will be based on its ability to correlate electrode kinetics and double-layer structure.

Thus far, the structural model of the electrical double layer which has been most successful is the one proposed by Grahame (6) (see Fig. 1). Grahame's model is essentially a refinement of an earlier one first proposed by Stern (18) in 1924. (Stern was awarded the Nobel Prize in physics in 1943 for his work on molecular beams.) The latter was, in turn, really a combination of two still earlier models: the compact- or rigid-layer model of Helmholtz (61) and Quincke (62), and the diffuse-layer model of Gouy (65) and Chapman (66). [For a brief historical account of double-layer

TABLE II

CHANGES IN ELECTROCAPILLARY EQUATION TO ACCOUNT FOR REVERSIBLE CHARGE-TRANSFER REACTION, $O + ne \rightleftharpoons R$

Nature of O	Nature of R	Chemical potential term eliminated	Ion to which indicator electrode is reversible	Analog of the Lippmann equation	New term involving chemical potential of couple member not eliminated	Additional terms required for indicator salt	Changed summation index ranges
Cation in S	Metal in M	$\frac{1}{v_{O_{k'}}^+} d\mu_{O_{k'}}$	cation j'	$\frac{\partial \gamma}{\partial E+} = -(q^M + nF\Gamma_{O_{h'}})$	$-(\Gamma_{O_{h'}} + \Gamma_{R_{i'}}) d\mu_R$	$-\dfrac{n}{|z_{k'}||v_{j'_{k'}}^-|} \Gamma_{O_{h'}} d\mu_{j'_{k'}}$	$i \neq i'$, R $j \neq j'$, O
Cation in S	Metal in M	$\frac{1}{v_{O_{k'}}^+} d\mu_{O_{k'}}$	anion k'	$\frac{\partial \gamma}{\partial E-} = -(q^M + nF\Gamma_{O_{h'}})$	$-(\Gamma_{O_{h'}} + \Gamma_{R_{i'}}) d\mu_R$	none	$i \neq i'$, R $j \neq j'$, O
Cation in S	Metal in M	$d\mu_R$	cation j'	$\frac{\partial \gamma}{\partial E+} = -(q^M - nF\Gamma_{R_{i'}})$	$-\frac{1}{v_{O_{k'}}^+}(\Gamma_{O_{h'}} + \Gamma_{R_{i'}}) d\mu_{O_{k'}}$	$\dfrac{n}{|z_{k'}||v_{j'_{k'}}^-|} \Gamma_{R_{i'}} d\mu_{j'_{k'}}$	$i \neq i'$, R $j \neq j'$, O
Cation in S	Metal in M	$d\mu_R$	anion k'	$\frac{\partial \gamma}{\partial E-} = -(q^M - nF\Gamma_{R_{i'}})$	$-\frac{1}{v_{O_{k'}}^+}(\Gamma_{O_{h'}} + \Gamma_{R_{i'}}) d\mu_{O_{k'}}$	none	$i \neq i'$, R $j \neq j'$, O
Cation in S	Cation in S	$\frac{1}{v_{O_{k'}}^+} d\mu_{O_{k'}}$	cation j'	$\frac{\partial \gamma}{\partial E+} = -(q^M + nF\Gamma_{O_{h'}})$	$-\frac{1}{v_{R_{k'}}^+}(\Gamma_{O_{h'}} + \Gamma_{R_{h'}}) d\mu_{R_{k'}}$	$-\dfrac{n}{|z_{k'}||v_{j'_{k'}}^-|} \Gamma_{O_{h'}} d\mu_{j'_{k'}}$	$\neq j'$, O, R
Cation in S	Cation in S	$\frac{1}{v_{O_{k'}}^+} d\mu_{O_{k'}}$	anion k'	$\frac{\partial \gamma}{\partial E-} = -(q^M + nF\Gamma_{O_{h'}})$	$-\frac{1}{v_{R_{k'}}^+}(\Gamma_{O_{h'}} + \Gamma_{R_{h'}}) d\mu_{R_{k'}}$	none	$j \neq j'$, O, R
Cation in S	Cation in S	$\frac{1}{v_{R_{k'}}^+} d\mu_{R_{k'}}$	cation j'	$\frac{\partial \gamma}{\partial E+} = -(q^M - nF\Gamma_{R_{h'}})$	$-\frac{1}{v_{O_{k'}}^+}(\Gamma_{O_{h'}} + \Gamma_{R_{h'}}) d\mu_{O_{k'}}$	$\dfrac{n}{|z_{k'}||v_{j'_{k'}}^-|} \Gamma_{R_{h'}} d\mu_{j'_{k'}}$	$j \neq j'$, O, R
Cation in S	Cation in S	$\frac{1}{v_{R_{k'}}^+} d\mu_{R_{k'}}$	anion k'	$\frac{\partial \gamma}{\partial E-} = -(q^M - nF\Gamma_{R_{h'}})$	$-\frac{1}{v_{O_{k'}}^+}(\Gamma_{O_{h'}} + \Gamma_{R_{h'}}) d\mu_{O_{k'}}$	none	$j \neq j'$, O, R
Cation in S	Neutral species in S	$\frac{1}{v_{O_{k'}}^+} d\mu_{O_{k'}}$	cation j'	$\frac{\partial \gamma}{\partial E+} = -(q^M + nF\Gamma_{O_{h'}})$	$-(\Gamma_{O_{h'}} + \Gamma_{R_{h'}}) d\mu_R$	$-\dfrac{n}{|z_{k'}||v_{j'_{k'}}^-|} \Gamma_{O_{h'}} d\mu_{j'_{k'}}$	$j \neq j'$, O $h \neq h'$, R
Cation in S	Anion in S	$\frac{1}{v_{O_{k'}}^+} d\mu_{O_{k'}}$	cation j'	$\frac{\partial \gamma}{\partial E+} = -(q^M + nF\Gamma_{O_{h'}})$	$-\frac{1}{v_{j'_R}^-}(\Gamma_{O_{h'}} + \Gamma_{R_{h'}}) d\mu_{j'_R}$	$-\dfrac{n}{|z_{k'}||v_{j'_{k'}}^-|} \Gamma_{O_{h'}} d\mu_{j'_{k'}}$	$j \neq j'$, O $k \neq k'$, R
Neutral species in S	Anion in S	$d\mu_O$	cation j'	$\frac{\partial \gamma}{\partial E+} = -(q^M + nF\Gamma_{O_{h'}})$	$-\frac{1}{v_{j'_R}^-}(\Gamma_{O_{h'}} + \Gamma_{R_{h'}}) d\mu_{j'_R}$	none	$h \neq h'$, O $k \neq k'$, R

concepts see Butler's book (*63*).] Helmholtz regarded the double layer as a condenser of constant capacity, that is, he thought of the metal-solution interface as two sheets of charge, one on the metal, the other on the solution, separated by a constant distance. If this model were correct, then according to the thermodynamic theory [Eqs. (68) and (69)] the electrocapillary curve should be a perfect parabola. The careful electrocapillary experiments of Gouy (*64*) early in this century demonstrated that real electrocapillary curves are not strictly parabolic. To explain the shape of real electrocapillary curves, Gouy (*65*), in 1910, and, independently, Chapman (*66*), in 1913, introduced the idea that the solution "plate" of the Helmholtz condenser ought not be considered a rigid sheet of charge located at a constant distance from the metal. Instead, because of the thermal motion of the ions, this solution plate would have a diffuse structure. The diffuse-layer theory of Gouy and Chapman, although correctly predicting a nonparabolic shape for the electrocapillary curve, failed badly to account even qualitatively for experimentally determined electrocapillary behavior. In effect, the original Gouy–Chapman theory overcorrected the Helmholtz compact-layer theory.

Stern's model combined the two ideas. His model has a rigid, inner layer next to the electrode surface containing solvent molecules and an outer diffuse layer. The latter obeys the equations of Gouy and Chapman. The Stern theory actually corrected the Gouy–Chapman theory in two ways. First, Stern took explicit account of the fact that, owing to their finite size,† the ions of the diffuse layer cannot approach the electrode surface closer than a certain minimum distance x_2 (Fig. 1). Second, Stern recognized that some ions might be held to the electrode in a rigid monolayer through the operation of close range forces (specific adsorption). Stern's model correctly predicted the main features of electrocapillary curves and, more significantly, of differential capacitance curves, which, being second derivatives of electrocapillary curves [cf. Eq. (69)], provide a more sensitive test. However, very precise measurements of differential capacitance by Grahame (*6,16,68,69*) led him to modify Stern's model in an important way. Grahame introduced the idea that, when there is a layer of specifically adsorbed ions, these ions (usual anions) lie closer to the electrode surface than the plane of closest approach of nonspecifically adsorbed ions of the diffuse layer. (Stern had hinted at the possible necessity of considering different distances of closest approach for cations and anions, but he did not develop this idea.) Grahame denoted the distance of closest approach

† Gouy (*67*) was aware that the finite size of ions might have to be considered, but he did not develop the idea in explicit form as did Stern.

of the specifically adsorbed layer of ions by x_1 (Fig. 1), and he called the plane of closest approach of the electrical centers of these ions the *inner Helmholtz plane*, IHP. The plane of closest approach of the nonspecifically adsorbed ions, located at a distance $x_2 > x_1$ from the electrode surface, was called the *outer Helmholtz plane*, OHP. Thus, as was indicated in Sec. I, the OHP forms the boundary between the inner and the diffuse layer. When specific adsorption of ions occurs, it takes place inside the inner layer.

We shall develop in this section some of the elementary ideas of the structural theory of the double layer with emphasis on the theory of the diffuse layer of Gouy and Chapman as modified by Stern, for that theory currently occupies a central role in the correlation of double layer effects in electrode kinetics. We shall also consider some aspects of inner-layer structure. For a detailed discussion of inner-layer structure, the reader is referred to the recent critical review of Maçdonald and Barlow (*15*). Before beginning the discussion of double-layer structure, it will be useful to attempt to classify the various measures of adsorption and also the types of adsorption which are encountered in structural discussions.

B. Classification of Measures and Types of Adsorption

1. Measures of Adsorption

In Sec. II it was shown that the only quantities expressing the amount of adsorption which are thermodynamically accessible are the relative surface excesses (relative adsorption) of the components. Therefore only relative surface excesses are available from equilibrium measurements. On the other hand, relative surface excesses do not result from calculations based on structural models. Rather, the surface excesses which one obtains from structural model calculations, for example, Gouy–Chapman–Stern calculations, are nonrelativized. Since experimental tests of structural theories involve comparisons of calculated nonrelativized surface excesses with experimentally determined relativized surface excesses, it is important to understand the relation between these two classes of adsorption measure. We shall illustrate the difference between the two types of surface excesses by considering the example of nonspecifically adsorbed ionic species.

a. Comparison of Relative Surface Excesses and Surface Excesses Reckoned at the OHP. The diagram in Fig. 11(*a*) represents the concentration profile of a nonspecifically adsorbed cationic species j at an ideal polarized electrode. The origin of the x axis is taken at the metal surface.

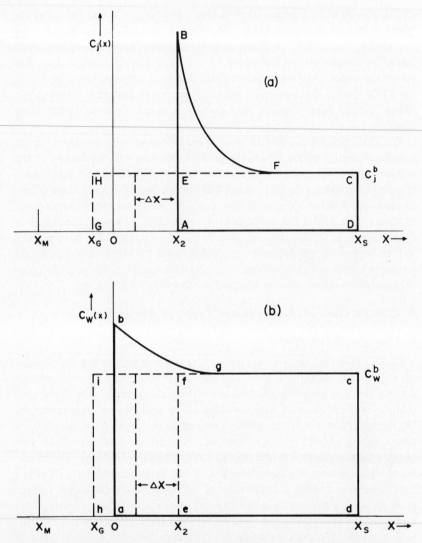

Fig. 11. (a) Hypothetical concentration-distance profile for nonspecifically adsorbed ion. (b) Hypothetical concentration-distance profile for water.

x_2 denotes the location of the outer Helmholtz plane, that is, the plane of closest approach of the electrical centers of the cations of species j. Assume that the cation j is attracted toward the electrode surface. Then the concentration $c_j(x)$ (in moles cm^{-3}), expressed as a function of the distance x

from the electrode surface, will rise from the value c_j^b in the homogeneous bulk of the solution to $c_j(x_2)$ at the OHP, and it will then drop abruptly to zero, since no cations of this type are inside the inner layer. The positions x_M and x_S in Fig. 11(a) represent the locations of the two bounding surfaces WX and ZY in Fig. 5(a) which were used to delimit the real system, that is, interphase, discussed in the derivation of the Gibbs adsorption equation. Recall that the position of these surfaces was entirely arbitrary except for the restriction that WX be located inside homogeneous metal phase M and that ZY be located inside homogeneous solution phase S. For convenience, consider an electrode of unit area, that is, $A = 1$ cm^2 [cf. Eq. (13)]. Let the total number of moles of cation j per unit area in the interphase be n_j. Then n_j is given by the area under the solid curve representing $c_j(x)$, that is, the area $ABCD$.

Consider now a Gibbs model whose dividing surface is located at x_2. The total number of moles n_{jG}^{OHP} of cation j per unit area in the Gibbs model is given by the area of the rectangle $AECD$. The surface excess of cation j, which is calculated from this particular model will be called the *surface excess of cation j reckoned at the OHP* and denoted by the symbol Γ_j^{OHP}, which is given by the area EBF in Fig. 11(a), that is,

$$\Gamma_j^{OHP} = n_j - n_{jG}^{OHP} \tag{99}$$

This quantity may be calculated on the basis of diffuse-layer theory. However, Γ_j^{OHP} cannot be measured; only $\Gamma_{jh'}$ can be measured.

Figure 11(b) represents the corresponding concentration profile of the water, the reference component h', versus distance from the electrode surface. Since water is also inside the inner layer $c_w(x)$ falls to zero at about $x = 0$. [We are assuming in Fig. 11(b) that the water concentration rises somewhat from its bulk value near the electrode surface, but this point is not essential to the argument. $c_w(x)$ could equally well fall, or remain constant.] The total number of moles of water, n_w, per unit area of electrode surface in the interphase (that is, real system) is given by the area $abcd$. The number of moles of water n_{wG}^{OHP} in the corresponding Gibbs model with dividing surface at x_2 is represented by the area $efcd$. The surface excess of water reckoned at the OHP (Γ_w^{OHP}) is given by the area $abgfe$:

$$\Gamma_w^{OHP} = n_w - n_{wG}^{OHP} \tag{100}$$

The relative surface excess of cation j is given by (cf. Section II.E.1.b)

$$\Gamma_{jw} = \Gamma_j^{OHP} - \left(\frac{x_j^b}{x_w^b}\right)\Gamma_w^{OHP} \tag{101}$$

where x_j^b and x_w^b are the mole fractions of cation j and of water in the homogeneous bulk of the solution. Since the ratio of mole fractions is the same as the ratio of volume concentrations (moles cm^{-3})

$$\Gamma_{jw} = \Gamma_j^{OHP} - \left(\frac{c_j^b}{c_w^b}\right)\Gamma_w^{OHP} \tag{102}$$

b. Proof That Relative Surface Excess Is Independent of the Location of the Dividing Surface. The position of the dividing surface in the Gibbs model was left arbitrary in the thermodynamic derivation (Sec. II). Thus the experimentally accessible, relative surface excesses appearing in the thermodynamic electrocapillary equation must also be independent of the location of the dividing surface. This fact may be demonstrated by arbitrarily choosing any other dividing surface displaced, say, a distance Δx closer to the metal surface than the OHP. The number of moles of cation j in this new Gibbs model with dividing surface at $(x_2 - \Delta x)$ is

$$n_{jG}^{x_2 - \Delta x} = c_j^b[x_S - (x_2 - \Delta x)] = c_j^b(x_S - x_2) + c_j^b \Delta x = n_{jG}^{OHP} + c_j^b \Delta x$$

Thus the surface excess of cation j reckoned at the position $(x_2 - \Delta x)$ is

$$\Gamma_j^{(x_2 - \Delta x)} = n_j - n_{jG}^{(x_2 - \Delta x)} = n_j - n_{jG}^{OHP} - c_j^b \Delta x$$

or

$$\Gamma_j^{(x_2 - \Delta x)} = \Gamma_j^{OHP} - c_j^b \Delta_x$$

Similarly, the surface excess of water reckoned at the new dividing surface is

$$\Gamma_w^{(x_2 - \Delta x)} = \Gamma_w^{OHP} - c_w^b \Delta x$$

Thus the relative surface excess of cation j defined in terms of the surface excesses reckoned at $(x_2 - \Delta x)$ is

$$\Gamma_{jw} = \Gamma_j^{(x_2 - \Delta x)} - \left(\frac{c_j^b}{c_w^b}\right)\Gamma_w^{(x_2 - \Delta x)}$$

$$= \Gamma_j^{OHP} - c_j^b \Delta x - \left(\frac{c_j^b}{c_w^b}\right)(\Gamma_w^{OHP} - c_w^b \Delta x) \tag{103}$$

$$= \Gamma_j^{OHP} - \left(\frac{c_j^b}{c^w}\right)\Gamma_w^{OHP}$$

Equation (103) is clearly identical to Eq. (102). Thus we have proved that the relative surface excess of any component is independent of the location of the dividing surface. [This proof that the relative surface excesses of components are invariant for any arbitrary choice of the location of the dividing surface in the Gibbs model implies that the value of the excess charge density q^M is likewise invariant over such choice (cf. Section II.E.1.c).]

c. Significance of the Gibbs Surface. The *Gibbs surface* is defined as a dividing surface located at a position x_G in the Gibbs model such that the number of moles of reference component, for example, h', in this particular Gibbs model is precisely equal to the number of moles of that component in the real system. Thus, in an aqueous solution, the surface excess of water reckoned at the Gibbs surface is zero, that is,

$$\Gamma_w^{x_G} = 0$$

The relative surface excess of cation j defined in terms of surface excesses reckoned at the Gibbs surface is

$$\Gamma_{jw} = \Gamma_j^{x_G} - (c_j^b/c_w^b)\Gamma_w^{x_G} = \Gamma_j^{x_G} \qquad (104)$$

It is apparent from Eq. (104) that the Gibbs surface has a certain thermodynamic significance and one can, in fact, derive the electrocapillary equation via surface excesses reckoned at the Gibbs surface (*26*). However, this method has no advantage or disadvantage over the method employing the Gibbs-Duhem equation. The real significance of the Gibbs surface is its usefullness for comparing surface excesses computed from models of the solution side of the double layer with measured, that is, relative surface excesses.

Grahame and Parsons (*70*) have pointed out that diffuse-layer theory permits the calculation only of the surface excesses of nonspecifically adsorbed ions *reckoned at the OHP and not at the x_G dividing surface*. It follows, therefore, that one should not expect perfect agreement between measured values of the relative surface excesses Γ^{x_G} and the values of Γ^{OHP} calculated on the basis of structural theory. In Fig. 11(*a*) Γ^{OHP} is represented by the area *EBF* and Γ^{x_G} is represented by difference in area *EBF–GHEA*. The inherent relative error in assuming that the measured surface excess Γ^{x_G} is identical with the surface excess calculated on the basis of diffuse-layer theory, Γ^{OHP} is

$$\frac{|\Gamma^{OHP} - \Gamma^{x_G}|}{|\Gamma^{OHP}|}$$

From Fig. 11(*a*) it is clear that this relative error is given by the ratio of area *HEAG* to area *EBF*. This error will become less important as the amount of adsorbed ion (*EBF*) increases and as the bulk concentration c_j^b

decreases. We shall consider the problem of the comparison between measured and calculated surface excesses in more detail later (Sec. III.C.7.c).

It is occasionally stated (46) that the discrepancy between measured surface excesses and "true" (nonrelativized) surface excesses should always be negligible for aqueous solutions since the ratio c_j^b/c_w^b is low in aqueous solution. However, this is not necessarily the case as can be seen immediately from Fig. 11. Even if the water concentration is perfectly uniform up to the metal surface (that is, x_G located at the metal surface) there will still be a discrepancy. Moreover, this discrepancy will become more and more serious as c_j^b increases and as Γ_j^{OHP} diminishes.

2. Types of Adsorption

In developing the structural theory of the double layer, it is convenient to consider separately the adsorption of ionic and neutral components of the solution. In the case of ionic components it is useful to treat separately adsorption in the inner layer (specific adsorption) and adsorption in the diffuse layer (nonspecific adsorption). Whether there is specific adsorption there will be nonspecific adsorption of ions in the diffuse layer. Thus the classes of adsorption we shall consider are (1) nonspecific adsorption of ions in the diffuse layer in the absence of specific adsorption; (2) nonspecific adsorption of ions in the diffuse layer in the presence of specific adsorption; (3) specific adsorption of ions in the inner layer; and (4) adsorption of neutral molecules, especially organic molecules.

C. The Diffuse Layer

The basic equations of the Gouy–Chapman–Stern (GCS) theory of the diffuse layer will be derived. These equations and some of their consequences will be discussed.

1. Derivation of the Basic Equation

The GCS theory considers the diffuse layer at a plane electrode. It is assumed that the diffuse layer is in electrostatic and osmotic equilibrium and that the ions in the layer obey a Boltzmann distribution law in which the only work terms required are purely electrostatic. It is further assumed that the electrical potential at any point in the diffuse layer depends only on the distance x from the electrode surface. [Any actual variation of electrical potential (*micropotential*) in planes parallel to the electrode surface is neglected. The electrical potential ϕ considered in the theory is, therefore, a smeared out, or average, *macropotential*.] The ions of the diffuse layer are assumed to interact with the electric field there as point charges according to the laws of classical electrostatics. Thus the work required to transport an ion from the homogeneous bulk of the solution to a position x inside the diffuse layer is given by the product of the ionic charge times the difference in potential between that position and the bulk of the solution. The electrical work required to transport one mole of ions of species i of charge z_i (including sign) from the bulk of solution to position x is $z_i F(\phi(x) - \phi^S)$, where ϕ^S is the inner potential of the homogeneous interior of the solution phase.† Denoting the ionic concentration (moles cm^{-3}) in the bulk of solution by c_i^b, the concentration at distance x is

$$
c_i(x) = c_i^b \exp\left\{ -\frac{z_i F}{RT} \left[\phi(x) - \phi^S \right] \right\}
$$

$$
= c_i^b \exp\left(-\frac{z_i F}{RT} \phi \right)
$$

(105)

Here, for simplicity in notation, we let $\phi = \phi(x) - \phi^S$. The excess charge density per unit volume $\rho(x)$ at any distance x is defined by

$$
\rho(x) = \sum_i z_i F c_i(x)
$$

(106)

Substituting Eq. (105) into Eq. (106), one obtains

$$
\rho(x) = \sum_i z_i F c_i^b \exp\left(-\frac{z_i F}{RT} \phi \right)
$$

(107)

According to classical electrostatic theory (*71,72*), the relation between the electrical potential ϕ at any point in an isotropic dielectric medium and the excess charge density per unit volume ρ at that location is given by Poisson's equation

$$
\mathbf{V} \cdot \varepsilon \, \mathbf{V} \phi = -\rho
$$

Here ε is the *permittivity* (or specific inductive capacity) of the dielectric at the point. ε has dimensions of capacitance per unit length (see Appendix C). Under the assumption that ϕ depends only on x, the Poisson equation reduces to

$$
\frac{d}{dx}\left(\varepsilon \frac{d\phi}{dx} \right) = -\rho(x)
$$

(108)

---

† The inner potential of the solution ϕ^S is not a function of x. Strictly,
$$
\phi^S = \lim_{x \to \infty} \phi(x)
$$

In general, the permittivity of a dielectric medium depends on the gradient of potential, that is, the electric field strength $\mathscr{E}$. ($\mathscr{E} = -(d\phi/dx)$ in the one-dimensional case.) At ordinary field strengths, ε is a constant, but as $\mathscr{E}$ increases ε begins to decrease toward a limiting (low) value (dielectric saturation). At the field strengths actually prevailing near the OHP (up to several million volts cm^{-1}) there is certainly some dielectric saturation which, strictly speaking, should be taken into account in the derivation of the diffuse-layer equations. However, in the GCS theory this effect is neglected. Under this assumption, the permittivity at any point in the diffuse layer is assigned the same value as the permittivity of the pure solvent at low field strengths. Thus the Poisson equation becomes simply

$$\frac{d^2\phi}{dx^2} = -\frac{\rho(x)}{\varepsilon} \tag{109}$$

where ε is now considered a constant. Substitution of the expression for $\rho(x)$ [Eq. (107)] into Eq. (109) gives

$$\frac{d^2\phi}{dx^2} = -\frac{F}{\varepsilon} \sum_i z_i c_i^b \exp\left(-\frac{z_i F}{RT}\phi\right) \tag{110}$$

The mathematical problem of the diffuse layer is to solve this second-order differential equation (Poisson–Boltzmann equation) subject to the boundary conditions of the GCS model.

Equation (110) may be integrated directly if one recalls that

$$\frac{d}{dx}\left(\frac{d\phi}{dx}\right)^2 = 2\left(\frac{d\phi}{dx}\right)\frac{d^2\phi}{dx^2}$$

Thus the quantity $2(d\phi/dx)$ is an integrating factor for Eq. (110). Multiplying both sides of Eq. (110) by this factor and performing an indefinite integration, we obtain

$$\left(\frac{d\phi}{dx}\right)^2 = \left(\frac{2RT}{\varepsilon}\right) \sum_i c_i^b \exp\left(-\frac{z_i F}{RT}\phi\right) + K \tag{111}$$

where K is an integration constant. This constant may be evaluated by applying the boundary condition

$$\lim_{x \to \infty}\left(\frac{d\phi}{dx}\right) = 0 \tag{112}$$

This condition simply expresses the fact that the electric field is zero in the homogeneous bulk of solution, that is, the inner potential in the bulk of the solution is a constant ϕ^S, and thus $\phi = \phi(x) - \phi^S \rightarrow 0$. Taking the limit of both sides of Eq. (111) as $x \rightarrow \infty$ and applying Eq. (112) to evaluate the integration constant, one obtains the equation of the electric field strength $\mathscr{E}(x)$ at any point x inside the diffuse layer:

$$-\mathscr{E}(x) = \frac{d\phi}{dx} = \pm \left(\frac{2RT}{\varepsilon}\right)^{\frac{1}{2}} \left\{\sum_i c_i^b \left[\exp\left(-\frac{z_i F}{RT} \phi\right) - 1\right]\right\}^{\frac{1}{2}} \quad (113)$$

As is usual in cases in which square roots arise in chemical problems, the sign of the square root in Eq. (113) must be chosen on the basis of physical reasoning. The basic physical criterion for choosing the sign in this case is the electroneutrality of the entire interfacial region. Two cases arise depending upon whether there is specific adsorption: (a) In the absence of specific adsorption the excess charge density on the metal q^M will be balanced solely by nonspecific adsorption of ions of opposite sign in the diffuse layer. Cations will tend to migrate from regions of higher toward regions of lower electrical potential, that is, in the direction of the field, and anions will tend to migrate in the opposite direction. For example, if q^M is negative, then $d\phi/dx$ must be positive so that cations will be attracted toward, and anions repelled from, the electrode surface (Fig. 12). Thus in the absence of specific adsorption the sign on the right side of Eq. (113) must be chosen opposite to that of q^M. (b) If there is specific adsorption, Eq. (113) may still be applied to the diffuse layer. However, in this case [cf. Eq. (105)] the excess charge per unit area which is balanced by nonspecific adsorption in the diffuse layer is not q^M but rather the algebraic sum, $(q^M + q^i)$, where q^i is the excess charge density resulting from specific adsorption of ions at the IHP (Fig. 13). Thus in the presence of specific adsorption the sign of the right side of Eq. (113) must be chosen opposite to that of the sum $(q^M + q^i)$.

If we take the limit of both sides of Eq. (113) as $x \rightarrow x_2$ (the OHP), we obtain†

$$-\mathscr{E}(x_2) = \frac{d\phi}{dx}\bigg|_{x=x_2} = \pm \left(\frac{2RT}{\varepsilon}\right)^{\frac{1}{2}} \left\{\sum_i c_i^b \left[\exp\left(-\frac{z_i F}{RT} \phi_2\right) - 1\right]\right\}^{\frac{1}{2}} \quad (114)$$

This equation gives the electric field strength at the OHP as a function of ϕ_2, the potential difference across the diffuse layer.

---

† The main difference between the original Gouy-Chapman theory and the Gouy-Chapman-Stern theory is that in the former theory it was assumed that this limit could be taken as $x \rightarrow 0$, that is, the field strength at the electrode surface would be given by an equation of the form of (114). In that case the potential ϕ_2 would be replaced by $(\phi^M - \phi^S)$, where ϕ^M is the inner potential of the metal.

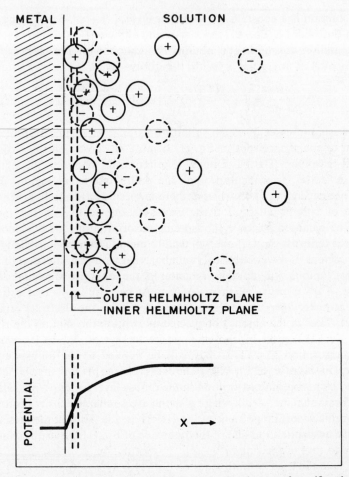

Fig. 12. Schematic diagram of the double layer in the absence of specific adsorption. $q^M < 0$. Dotted circles represent " ghosts " of anions which have been repelled away from electrode surface. The lower figure is a schematic of the corresponding potential profile [from (6)]. (By permission of the American Chemical Society.)

The field strength at the OHP may also be calculated by means of Gauss' law provided there are no free charges located between the metal surface and the OHP, that is, provided there is no specific adsorption. Under these conditions the inner layer consists of a thin layer of dielectric (solvent) separating the diffuse layer from the metal surface. For such a "parallel-plate-condenser" situation, Gauss' law states that the number of lines of

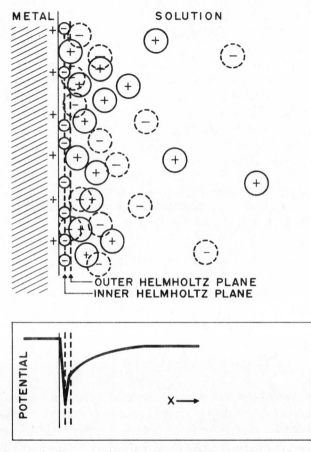

Fig. 13. Schematic diagram of the electrical double layer with specific adsorption of anions. $q^M > 0$. Dotted circles represents "ghosts" of anions repelled in diffuse layer. Lower diagram schematic of the corresponding potential profile [from (6)]. (By permission of the American Chemical Society.)

electric displacement crossing the dielectric layer per unit area of metal surface is just equal to the charge density on the metal surface q^M. The number of lines of displacement per unit area is the magnitude D of the electric displacement vector. Now, for an isotropic dielectric, the magnitude of the electric displacement D is related to the magnitude of the electric field strength $\mathscr{E}$ by the equation

$$D = \varepsilon\mathscr{E} \qquad (115)$$

where ε is the permittivity. Since there are no free charges inside the inner layer in the absence of specific adsorption,† the value of D will remain constant all the way across the inner layer. Owing to dielectric saturation, the permittivity is certainly lower in the first layer of solvent, next to the metal, than in the layer of solvent at the OHP (cf. Fig. 2). Thus it does not follow that just because the displacement D remains constant across the inner layer that the field strength $\mathscr{E}$ does also. (The latter would imply that ϕ varied linearly across the inner layer.) However, it does follow that, if one knows the value of ε at the OHP, one can then calculate the electric field strength there from a knowledge of q^M by means of Gauss' law. According to the assumptions of the GCS theory, the permittivity ε has reached its low-field (bulk) value at the OHP. This is undoubtedly not so. However, as we shall see below, dielectric saturation does not cause serious errors insofar as diffuse-layer calculations are concerned. Under this assumption, Gauss' law and Eq. (115) give

$$q^M = D(x_2) = \varepsilon \mathscr{E}(x_2) = -\varepsilon \frac{d\phi}{dx}\bigg|_{x=x_2} \tag{116}$$

where ε is the value of the permittivity of the solvent at low field strengths. The combination of Eqs. (116) and (114) gives the equation relating the measurable excess charge density q^M to the potential ϕ_2 at the OHP:

$$q^M = \pm(2RT\varepsilon)^{\frac{1}{2}}\left\{\sum_i c_i^b\left[\exp\left(-\frac{z_iF}{RT}\phi_2\right) - 1\right]\right\}^{\frac{1}{2}} \tag{117}$$

This equation is valid only in the absence of specific adsorption, for it is only in that case that Eq. (116) may be used to calculate the magnitude of the displacement D and the field strength $\mathscr{E}$ at the OHP. However, the modification in Eq. (117) required to account for specific adsorption may be derived simply. When there is specific adsorption, the number of lines of displacement per unit area passing between the IHP and the OHP will be given, according to Gauss' law, by

$$D(x) = (q^M + q^i) \qquad \text{(for } x_1 < x \leq x_2) \tag{118}$$

Thus the electric field at the OHP will be given by

$$(q^M + q^i) = D(x_2) = \varepsilon \mathscr{E}(x_2) = -\varepsilon \frac{d\phi}{dx}\bigg|_{x=x_2} \tag{119}$$

† There are induced charges because of polarization of the dielectric, but these are accounted for by ε.

Hence the potential ϕ_2 at the OHP in the presence of specific adsorption is

$$(q^M + q^i) = \pm(2RT\varepsilon)^{\frac{1}{2}}\left\{\sum_i c_i^b\left[\exp\left(-\frac{z_iF}{RT}\phi_2\right) - 1\right]\right\}^{\frac{1}{2}} \quad (120)$$

When the solution contains only a single z-z electrolyte, Eq. (117) or Eq. (120) takes a simpler form. Let c^b stand for the bulk electrolyte concentration and let $z = z_+ = |z_-|$ be the ionic charge without regard to sign. In this case Eq. (117) becomes

$$q^M = \pm(2RT\varepsilon)^{\frac{1}{2}}(c^b)^{\frac{1}{2}}\left[\exp\left(\frac{zF}{RT}\phi_2\right) + \exp\left(-\frac{zF}{RT}\phi_2\right) - 2\right]^{\frac{1}{2}} \quad (121)$$

The last term on the right side of Eq. (121) is a perfect square, which may be rewritten as follows:

$$\left[\exp\left(\frac{zF}{RT}\phi_2\right) + \exp\left(-\frac{zF}{RT}\phi_2\right) - 2\right]^{\frac{1}{2}}$$

$$= \left\{\left[\exp\left(\frac{zF}{2RT}\phi_2\right) - \exp\left(-\frac{zF\phi_2}{2RT}\right)\right]\right\}^{\frac{1}{2}}$$

$$= \left[4\sinh^2\left(\frac{zF\phi_2}{2RT}\right)\right]^{\frac{1}{2}} = 2\sinh\left(\frac{zF\phi_2}{2RT}\right)$$

Therefore, the $q^M - \phi_2$ relationship for a z-z electrolyte solution becomes

$$q^M = +(8RT\varepsilon)^{\frac{1}{2}}(c^b)^{\frac{1}{2}}\sinh\left(\frac{zF\phi_2}{2RT}\right) \quad (122)$$

The plus sign is selected so that the sign of q^M will be the same as the sign of ϕ_2.

To use Eq. (117) or Eq. (122) in practical calculations, it is necessary to substitute the appropriate value of the permittivity ε of the solvent at the temperature of interest into the equation. The permittivity (cf. Appendix C) is equal to the product of the (dimensionless) *dielectric coefficient* K_D (often called the *dielectric constant*) multiplied by the *permittivity of free space* ε_0. The latter, a universal constant, has a value of 8.849×10^{-14} farad cm^{-1}. At 25°C, the dielectric coefficient of water is 78.49. Thus, for water at 25°C, we have

$$\varepsilon = K_D\varepsilon_0 = 78.49(8.849)(10)^{-14} \text{ farad cm}^{-1}$$

$$= 6.946 \times 10^{-12} \text{ farad cm}^{-1}$$

If the concentration of a z-z electrolyte c^b is expressed in moles per cubic centimeter, and ϕ_2 is given in volts, Eq. (122) for aqueous solutions at 25°C becomes

$$q^M = 371.1(c^b)^{\frac{1}{2}} \sinh (19.46 \, z\phi_2) \quad \mu \text{ coulomb cm}^{-2} \qquad (123)$$

If, instead, c^b is given in moles per liter, one obtains

$$q^M = 11.72(c^b)^{\frac{1}{2}} \sinh (19.46 \, z\phi_2) \quad \mu \text{ coulomb cm}^{-2} \qquad (124)$$

In the presence of specific adsorption the same equations apply except that q^M is replaced by $(q^M + q^i)$.

2. $\phi_2 - E$ Relation

Equation (122) may be rearranged to give ϕ_2 as an explicit function of q^M:

$$\phi_2 = \left(\frac{2RT}{zF}\right) \sinh^{-1} \left[\frac{q_M}{(8RT\varepsilon c^b)^{\frac{1}{2}}}\right] \qquad (125)$$

[In the presence of specific adsorption, q^M in Eq. (125) is replaced by $q^M + q^i$.]

To obtain the dependence of ϕ_2 on the measured electrode potential E for a real electrode, it is necessary to determine experimentally the dependence of q^M on E. For an ideal polarized electrode, this dependence may be obtained either from electrocapillary curves by application of the Lippmann equation [Eq. (53)] or from differential capacity curves by application of Eq. (70). After the $q^M - E$ relationship is known from experiment, one obtains a graph of the $\phi_2 - E$ dependence by calculating ϕ_2 for each value of q^M [Eq. (125)] and plotting each ϕ_2 versus the value of E corresponding to the same q^M. Grahame found that NaF exhibits essentially no specific adsorption at mercury, except, perhaps, at the most positive potentials within the range of ideal polarizability [cf. Payne (73)]. The ϕ_2 versus E graphs for the ideal polarized mercury electrode in several concentrations of aqueous NaF are shown in Fig. 14. Extensive tables of ϕ_2 versus E for NaF solutions have been calculated by Russell (76).

To determine the $\phi_2 - E$ relationship for any electrolyte solution, other than z-z, at an ideal polarized electrode, one uses the procedure just described except that the $\phi_2 - q^M$ relation is determined by application of Eq. (117).

3. Differential Capacitance of the Diffuse Layer

If the q^M versus E graphs were linear, the differential capacitance (C versus E curves) would be constant [cf. Eq. (68)] and the double layer

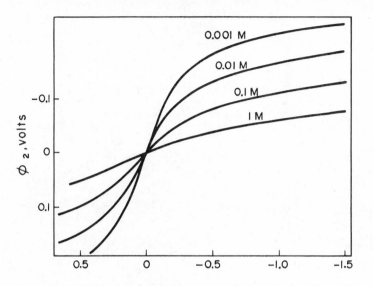

Fig. 14. Potential drop across the diffuse layer for NaF solutions calculated from the data of Grahame (*74,75*). Potential scale is $E - E_z$. Vertical scale is ϕ_2 [from (*9*)]. (By permission of Wiley.)

would behave as if it were a simple electrical capacitor (Helmholtz model). In fact the differential capacity curves always show marked dependence on electrode potential. Figure 15(*b*) shows C versus $E - E_z$ curves obtained by Grahame for the NaF solutions referred to above. It will be noted that, as the concentration of NaF decreases, a pronounced minimum appears in the differential capacities at the point of zero charge, that is, at $E = E_z$. This phenomenon, which is quite general for dilute solutions in the absence of specific adsorption, may be easily understood in terms of Stern's model of the double layer and the GCS theory. Let ϕ^M be the inner potential of the metal electrode. Then the total (*Galvani*, or *absolute*) potential drop from metal to solution, $\Delta\phi = \phi^M - \phi^S$, may be expressed as the sum of the potential differences across the inner layer and across the diffuse layer; that is,

$$\Delta\phi = (\phi^M - \phi^S) = [\phi^M - \phi(x_2)] + [\phi(x_2) - \phi^S] = [\phi^M - \phi(x_2)] + \phi_2$$

Differentiating with respect to q^M, we have

$$\frac{\partial \Delta\phi}{\partial q^M} = \frac{\partial [\phi^M - \phi(x_2)]}{\partial q^M} + \frac{\partial \phi_2}{\partial q^M} \tag{126}$$

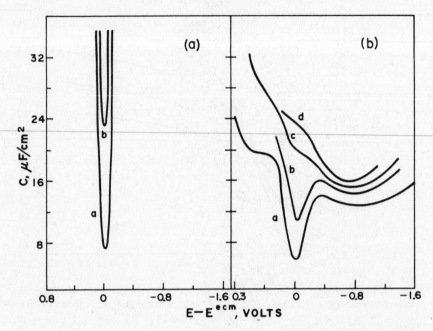

Fig. 15. Differential capacity for an electrode in the absence of specific adsorption. (*a*) capacity of the diffuse layer, (*b*) measured total differential capacity for NaF solutions from the data of Grahame (*6*). Curves: (*a*) 0.001 *F*, (*b*) 0.01 *F*, (*c*) 0.1 *F*, (*d*) 1.0 *F* NaF [from (*8*)]. (By permission of Butterworths.)

Experimentally, the electrical variable which can be changed is $E^{\pm}$ or E_{ref}. $E^{\pm}$ may be expanded as [cf. Eq. (19)]

$$E^{\pm} = (\phi^{\text{Cu}} - \phi^{M}) + (\phi^{M} - \phi^{S}) + (\phi^{S} - \phi^{\text{IN}}) + (\phi^{\text{IN}} - \phi^{\text{Cu}'})$$

Note that $\phi^{M} - \phi^{S}$ is the only nonconstant term in this expression if the solution composition is fixed. (For E_{ref} the same situation exists, except there is an additional constant, the liquid junction potential.) Thus, for a given solution any change in E, the potential of the ideal polarized electrode with respect to the indicator electrode or an ordinary reference electrode, must result entirely from a change in the Galvani potential drop $\Delta\phi = \phi^{M} - \phi^{S}$. Hence,

$$\frac{\partial\Delta\phi}{\partial q^{M}} = \frac{\partial E}{\partial q^{M}} = \frac{1}{C}$$

The left side of Eq. (126) is therefore equal to the reciprocal of the measured differential capacitance [cf. Eqs. (68) and (69)]. Each term on the right side of Eq. (126) also has the dimensions of capacitance, so we may rewrite Eq. (126) in the form

$$\frac{1}{C} = \frac{1}{C^i} + \frac{1}{C^d} \tag{127}$$

C^i is the *differential capacitance of the inner layer*, and C^d is the *differential capacitance of the diffuse layer*. Equation (127) has the same form as the equation from classical electrical theory for the equivalent capacitance of a circuit consisting of two capacitances, C^i and C^d, in series. Thus the measured (total) differential capacitance C of an ideal polarized electrode in the absence of specific adsorption may be represented by an equivalent electrical circuit consisting of a series combination of the differential capacitances of the inner and diffuse layers. The circuit diagram of the electrical double layer at such an electrode is

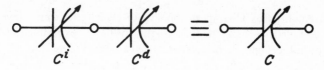

Both the inner and diffuse layers are represented as variable capacitors to emphasize that their values C^i and C^d are not constants.

By differentiating Eq. (122), it is seen that the differential capacitance of the diffuse layer (for a z-z electrolyte) has a hyperbolic cosine dependence on ϕ_2:

$$C^d = \frac{\partial q^M}{\partial \phi_2} = \left(\frac{2z^2 F^2 \varepsilon c^b}{RT} \right)^{\frac{1}{2}} \cosh \left(\frac{zF\phi_2}{2RT} \right) \tag{128}$$

For aqueous solutions at 25°C, Eq. (128) becomes

$$C^d = 228.5z(c^b)^{\frac{1}{2}} \cosh (19.46z\phi_2) \tag{129}$$

where C^d is in microfarad cm^{-2}, ϕ_2 is in volts, and c^b is the salt concentration in moles liter^{-1}.

Graphs of C^d versus $E - E_z$ for the two lowest concentrations of NaF are shown in Fig. 15(*a*).

In view of Fig. 15 and Eq. (127), it is easy to understand qualitatively why the minimum in the measured differential capacity C appears at the point of zero charge for dilute solutions. Since C^d has a minimum at the ECM, its contribution to the measured differential capacitance will be largest in the neighborhood of that point. As E deviates either positively or negatively from E_z, the magnitude of C^d increases rapidly [Fig. 15(*a*)] and, thus, its contribution to C diminishes. Since C^d varies linearly with the square root of the bulk electrolyte concentration, the contribution of C^d to C near E_z will become more and more important as c^b is lowered. Thus in 10^{-3} F NaF [curve a, Fig. 15(b)] the measured differential capacitance is influenced strongly by the diffuse layer near the ECM, and a pronounced minimum appears in the $C - E$ curve. In 10^{-2} F NaF (curve b), the contribution of C^d to C near E_z, is still apparent, but the minimum is not so pronounced. In 10^{-1} F NaF (curve c), the contribution of C^d to C is small even at E_z, and only a shoulder appears. For any concentration, the contribution of C^d to C at potentials far from E_z will be negligible [cf. Eq. (127)]. The fact that the measured differential capacitance C of an ideal polarized electrode seldom rises above 40 to 50 μF cm^{-2} implies that the differential capacitance C^i of the inner layer must remain finite, since for large C^d,

$$C = C^i$$

The original Gouy-Chapman theory, in which no inner layer was considered, failed to account even qualitatively for experimental differential capacitance curves because C was assumed equal to C^d at all potentials [cf. Fig. 15(a)]. The introduction by Stern of the inner-layer capacitor in series with the diffuse layer provided a model which was obviously in qualitative agreement with experiment. Quantitative verification of the GCS theory required knowledge of the value of the inner layer capacitance C^i and of its variation with electrode potential and concentration. Work reported by Frumkin (77) in 1940 indicated that semiquantitative agreement between theory and experiment could be obtained by assuming C^i to have one constant value at negative q^M and another (higher) constant value at positive q^M [cf. Butler (63)]. However, the very accurate measurements of differential capacitance by Grahame (6) showed conclusively that C^i actually varies with electrode potential over the entire potential range. In the absence of specific adsorption (NaF), Grahame found that the inner-layer capacitance C^i, in contrast to the diffuse layer capacitance C^d, is essentially independent of electrolyte concentration. However, C^i was found to depend strongly on the excess charge density of the metal q^M and on the temperature. For discussion of the inner layer in the absence of specific adsorption see Delahay (5), Parsons (9), and Macdonald and Barlow (15).

4. Potential Profile and Effective Thickness of the Diffuse Layer

The potential-distance ($\phi - x$) profile of the diffuse layer may be determined by integrating the equation for the electric field, Eq. (113), using the boundary condition that $\phi = \phi_2$ at the OHP. Knowing the $\phi - x$ relationship, one may then calculate concentration-distance profiles for each ionic species by application of Eq. (105). We shall illustrate the procedure by considering the case of a z-z electrolyte for which Eq. (113) becomes

$$\frac{d\phi}{dx} = -\left(\frac{8RT}{\varepsilon}\right)^{\frac{1}{2}} (c^b)^{\frac{1}{2}} \sinh\left(\frac{zF\phi}{2RT}\right) \tag{130}$$

This equation is valid everywhere inside the diffuse layer, that is, for all $x \geq x_2$. Note that the minus sign is chosen on the right side since q^M and $d\phi/dx$ must have opposite signs [cf. Eq. (122)]. Let us define $x' = x - x_2$. Then $dx' = dx$. Substituting for x in terms of x', the distance from the OHP, and separating variables in Eq. (130), we obtain

$$\int_0^{x'} dx' = -\left(\frac{\varepsilon}{8RTc^b}\right)^{\frac{1}{2}} \int_{\phi_2}^{\phi(x')} \text{csch}\left(\frac{zF\phi}{2RT}\right) d\phi \tag{131}$$

The integration on the right side may be performed by applying the formula (6)

$$\int \text{csch}\,\theta\, d\theta = \ln\left(\tanh\left|\frac{\theta}{2}\right|\right) + \text{const} \tag{132}$$

Using Eq. (132), we see that Eq. (131) becomes

$$x' = -\left(\frac{\varepsilon RT}{2z^2F^2c^b}\right)^{\frac{1}{2}} \left[\ln\left(\tanh\left|\frac{zF\phi}{4RT}\right|\right)\right]_{\phi_2}^{\phi(x')}$$

or

$$x' = -\left(\frac{\varepsilon RT}{2z^2F^2c^b}\right)^{\frac{1}{2}} \ln\left[\tanh\left|\frac{zF\phi(x')}{4RT}\right| \Big/ \tanh\left|\frac{zF\phi_2}{4RT}\right|\right] \tag{133}$$

The factor

$$\left(\frac{\varepsilon RT}{2z^2F^2c^b}\right)^{\frac{1}{2}} = \frac{1}{\kappa}$$

where κ is the so-called Debye reciprocal length, which occurs in the Debye-Hückel theory of electrolytes (78). Inserting this parameter in Eq. (133) and rearranging, we obtain

$$\tanh\left|\frac{zF\phi(x')}{4RT}\right| = \left(\tanh\left|\frac{zF\phi_2}{4RT}\right|\right) e^{-\kappa x'}$$

or

$$\phi(x') = \pm\left(\frac{4RT}{zF}\right) \tanh^{-1}\left\{\left(\tanh\left|\frac{zF\phi_2}{4RT}\right|\right) e^{-\kappa x'}\right\} \tag{134}$$

Equation (134) gives the electrical potential $\phi(x')$ inside the diffuse layer at any distance x' from the OHP as a function of x', and of the potential at the OHP, ϕ_2. The latter may be calculated by means of Eqs. (117) or (120), depending on whether there is specific adsorption. The sign of the right side of Eq. (134) will be that of q^M in the absence of specific adsorption, or of $(q^M + q^i)$ in the presence of specific adsorption.

Since the maximum value of $\tanh|zF\phi_2/4RT|$ is unity, it follows that the quantity in braces on the right side of Eq. (134) will always be less than or equal to $e^{-\kappa x'}$, that is,

$$\left(\tanh\left|\frac{zF\phi_2}{4RT}\right|\right)e^{-\kappa x'} \le e^{-\kappa x'}$$

For example, if $x' = 9.2/\kappa$,

$$\left(\tanh\left|\frac{zF\phi_2}{4RT}\right|\right)e^{-\kappa x'} \le e^{-9.2} = 10^{-4}$$

Now for $\theta \ll 1$, we have

$$\tanh^{-1}\theta \to \theta$$

Thus for large distances from the OHP, Eq. (134) reduces to

$$\phi(x') = \pm\left(\frac{4RT}{zF}\right)\left(\tanh\left|\frac{zF\phi_2}{4RT}\right|\right)e^{-\kappa x'} \tag{135}$$

Equation (135) implies that, for large distances from the OHP, the electrical potential falls off exponentially with increasing distance. For small values of ϕ_2, Eq. (135) reduces further to

$$\phi(x') = \phi_2\, e^{-\kappa x'} \tag{136}$$

Equation (136) is convenient for estimating the effective thickness of the diffuse layer. For example, the thickness $x' = \tau_{99.99}$ within which 99.99 per cent of the diffuse-layer potential drop occurs is calculated as follows:

$$\frac{\phi(\tau_{99.99})}{\phi_2} = 0.0001 = e^{-\kappa\tau_{99.99}}$$

or

$$\tau_{99.99} = \frac{\ln 0.0001}{-\kappa} = \frac{9.2}{\kappa}$$

At 25°C in aqueous solutions of z-z electrolytes,

$$\kappa = 3.28(10)^7 z(c^b)^{\frac{1}{2}}$$

where c^b is the electrolyte concentration in moles per liter. Thus

$$\tau_{99.99} = \frac{2.8(10)^{-7}}{z(c^b)^{\frac{1}{2}}} \quad cm \tag{137}$$

The effective thicknesses given in Table I were calculated by means of this formula. From the standpoint of correlating double-layer structure with electrode kinetics, the important conclusion to be drawn from such calculations is that the effective thickness of the diffuse layer is almost always negligibly small in comparison with the thickness of the diffusion layer. Therefore, no significant error is involved in neglecting the existence of the double layer in the solution of boundary value problems for mass transfer. [Exceptions to this rule occur only in the case of extremely fast electrode reactions (*10*).]

5. Surface Excesses of Nonspecifically Adsorbed Ions

a. General. The volume concentration of any ionic species i at any position $x \geq x_2$ in the diffuse layer is given by Eq. (105). Therefore, the excess number of moles per cubic centimeter of ionic species i at any distance x' ($x' = x - x_2$) from the OHP is

$$c_i(x') - c_i^b = c_i^b \left[\exp\left(-\frac{z_i E \phi(x')}{RT} \right) - 1 \right] \tag{138}$$

If the ion is not also specifically adsorbed, its nonrelativized surface excess reckoned at the OHP (Γ_i^{OHP}) may be calculated by integrating Eq. (138) from the OHP ($x' = 0$) to the homogeneous bulk of solution ($x \to \infty$), that is,

$$^d\Gamma_i^{OHP} = c_i^b \int_0^\infty \left[\exp\left(-\frac{z_i F \phi(x')}{RT} \right) - 1 \right] dx' \tag{139}$$

The symbol $^d\Gamma_i^{OHP}$ represents the contribution of the diffuse layer to the total surface excess Γ_i^{OHP}. In the absence of specific adsorption these two quantities are, of course, equal. Equation (139) gives the excess number of moles of the ionic species in a column of solution of 1 cm^2 cross-sectional area and extending from the OHP into the homogeneous bulk of the solution. If the ionic species is specifically adsorbed, then Eq. (139) gives only the contribution of the diffuse layer to the surface excess of species i. The total surface excess Γ_i^{OHP} of species i (reckoned at the OHP) would then be

$$\Gamma_i^{OHP} = {}^i\Gamma_i + {}^d\Gamma_i^{OHP} \tag{140}$$

where $^i\Gamma_i$ is the number of moles per square centimeter of species i specifically adsorbed at the IHP. Unlike Γ_i^{OHP}, and the diffuse-layer contribution $^d\Gamma_i^{OHP}$, the quantity $^i\Gamma_i$ is not a surface excess. Rather $^i\Gamma_i$ is a *surface concentration*, that is, it is the actual number of moles of species i adsorbed per square centimeter of electrode at the IHP.

The contribution of each species of ion to the total excess charge density in the diffuse layer is given by [cf. Eq. (139)]

$$q_i^d = z_i F \, {}^d\Gamma_i^{OHP}$$

$$= z_i F c_i^b \int_0^\infty \left[\exp\left(-\frac{z_i F \phi(x')}{RT} \right) - 1 \right] dx' \tag{141}$$

The total excess charge density q^d in the diffuse layer† is

$$q^d = \sum_i z_i F \, {}^d\Gamma_i^{OHP} = \sum_i q_i^d \tag{142}$$

In the absence of specific adsorption, all the excess charge density on the solution side of the double layer q^S will be due to nonspecific adsorption in the diffuse layer. Therefore, in that case we have

$$-q^M = q^S = q^d = \sum_i z_i F \, {}^d\Gamma_i^{OHP} \tag{143}$$

Note that, although the ionic surface excesses reckoned at the OHP, which are calculated on the basis of the GCS theory, are not the same as the corresponding relative surface excesses (cf. Sec. III.B.1), their sum does equal the measured charge density q^M. This illustrates again the invariance of q^M over the placement of the dividing surface in the Gibbs model.

b. Solution of a Single Symmetric Electrolyte. In the case of a single z-z electrolyte of bulk concentration c^b, Eq. (139) may be integrated in closed form. For the cation we have

$$ {}^d\Gamma_+^{OHP} = c^b \int_0^\infty \left[\exp\left(-\frac{zF\phi(x')}{RT} \right) - 1 \right] dx' \tag{144}$$

where $z = z_+ = |z_-|$.

The integrand in Eq. (144) may be rewritten as follows:

$$\left[\exp\left(-\frac{zF\phi(x')}{RT} \right) - 1 \right]$$

$$= \left[\exp\left(-\frac{2zF\phi(x')}{RT} \right) - 2 \exp\left(-\frac{zF\phi(x')}{RT} \right) + 1 \right]^{\frac{1}{2}}$$

$$= \left[\exp\left(\frac{zF\phi(x')}{RT} \right) - 2 + \exp\left(-\frac{zF\phi(x')}{RT} \right) \right]^{\frac{1}{2}} \exp\left(-\frac{zF\phi(x')}{2RT} \right)$$

$$= \pm \left[2 \sinh\left(\frac{zF\phi(x')}{2RT} \right) \right] \exp\left(-\frac{zF\phi(x')}{2RT} \right)$$

† Grahame (6) used the symbol η^d for this quantity.

Thus

$$^d\Gamma_+^{OHP} = -2c^b \int_0^\infty \left[\sinh\left(\frac{zF\phi(x')}{2RT}\right) \right] \exp\left(-\frac{zF\phi(x')}{2RT}\right) dx' \quad (145)$$

[The minus sign in Eq. (145) is chosen so that $^d\Gamma_+^{OHP}$ will be positive when $\phi(x')$ is negative.] From the equation for the electric field strength in the diffuse layer [Eq. (113)], we deduce that, for a z-z electrolyte,

$$dx' = dx = -\left(\frac{\varepsilon}{8RTc^b}\right)^{\frac{1}{2}} \operatorname{csch}\left(\frac{zF\phi}{2RT}\right) d\phi \quad (146)$$

Substituting for dx' in Eq. (145) by means of Eq. (146) and recalling that $\phi \to \phi_2$ as $x' \to 0$ and $\phi \to 0$ as $x' \to \infty$, we obtain

$$^d\Gamma_+^{OHP} = +\left(\frac{\varepsilon c^b}{2RT}\right)^{\frac{1}{2}} \int_{\phi_2}^0 e^{-zF\phi/2RT} \, d\phi \quad (147)$$

Integrating Eq. (147), we obtain

$$^d\Gamma_+^{OHP} = (zF)^{-1}(2RT\varepsilon c^b)^{\frac{1}{2}}(e^{-zF\phi_2/2RT} - 1) \quad (148)$$

Similarly, for the anion

$$^d\Gamma_-^{OHP} = (zF)^{-1}(2RT\varepsilon c^b)^{\frac{1}{2}}(e^{zF\phi_2/2RT} - 1) \quad (149)$$

Grahame (6) denoted the quantity $(2RT\varepsilon c^b)^{\frac{1}{2}}$ appearing in Eqs. (148) and (149) by A. With this notation, the contributions of the cation and anion to the total excess charge in the diffuse layer are

$$q_+^d = zF \, ^d\Gamma_+^{OHP} = A\left[\exp\left(-\frac{zF\phi_2}{2RT}\right) - 1 \right] \quad (150)$$

and

$$q_-^d = -zF \, ^d\Gamma_-^{OHP} = -A\left[\exp\left(\frac{zF\phi_2}{2RT}\right) - 1 \right] \quad (151)$$

where

$$q_+^d + q_-^d = q^d \quad (152)$$

Equations (148) to (152) are valid for a solution of a single z-z electrolyte either in the presence or absence of specific adsorption. In the absence of specific adsorption one has, in addition,

$$q_+^d + q_-^d = q^d = q^S = -q^M \quad (153)$$

c. Solutions of Unsymmetric Electrolytes and Mixed Electrolyte Solutions.
The simple formulas just derived for the surface excesses of cation and
anion in the diffuse layer are restricted to solutions of a single z-z
electrolyte. In the case of a solution of a single unsymmetric electrolyte, or
of a mixed electrolyte solution, the general equation [Eq. (139)] is again the
starting point. For a solution of a single 2:1 electrolyte (for example,
Na_2SO_4), or of a single 1:2 electrolyte ($BaCl_2$), Grahame (79) derived
formulas analogous to Eqs. (148) to (151). For 2:1 electrolytes, he obtained

$$q_+^d = A[e^{-F\phi_2/RT}(1 + 2e^{F\phi_2/RT})^{\frac{1}{2}} - 3^{\frac{1}{2}}] \tag{154}$$

and

$$q_-^d = -A[(1 + 2e^{F\phi_2/RT})^{\frac{1}{2}} - 3^{\frac{1}{2}}] \tag{155}$$

whereas, for 1:2 electrolytes,

$$q_+^d = A[(1 + 2e^{-F\phi_2/RT})^{\frac{1}{2}} - 3^{\frac{1}{2}}] \tag{156}$$

and

$$q_-^d = -A[e^{F\phi_2/RT}(1 + 2e^{-F\phi_2/RT})^{\frac{1}{2}} - 3^{\frac{1}{2}}] \tag{157}$$

In Eqs. (154) to (157),

$$A = (2RT\varepsilon c^b)^{\frac{1}{2}} \tag{158}$$

where c^b is the *salt concentration*, that is, the concentration of the divalent
ion.

For solutions of single salts of other unsymmetric valence types, and
for mixed electrolyte solutions, particular formulas might also be derived.
However, with the availability of modern digital computers, it is a relatively
simple matter to carry out a numerical integration of the general equation,
Eq. (139). Numerical integrations were performed by Joshi and Parsons
(80) for mixtures of HCl and $BaCl_2$. For any ionic species i in any mixture
of electrolytes whatever, Eq. (139) for the surface excess and Eq. (113) for
the electric field strength both hold. Thus the surface excess of that ion in
the diffuse layer (reckoned at the OHP) is obtained by substituting for dx'
in Eq. (139) by means of Eq. (113). The result is

$$^d\Gamma_i^{OHP} = \pm\left(\frac{\varepsilon}{2RT}\right)^{1/2} \int_{\phi_2}^0 \frac{c_i^b[\exp(-z_iF\phi/RT) - 1]}{\{\sum_i c_i^b[\exp(-z_iF\phi/RT) - 1]\}^{1/2}} \, d\phi \tag{159}$$

The sign in Eq. (159) is selected according to the same criterion as in the
case of Eq. (113). For example, when the ionic species considered is
cationic, then, if $\phi_2 > 0$, the minus sign is selected; if $\phi_2 < 0$, the plus sign
is selected. For anions, the opposite sign rule holds.

Numerical integration of Eq. (159) gives $^d\Gamma_i^{OHP}$ as a function of ϕ_2. Then, in the absence of specific adsorption, the value of ϕ_2 is simply calculated by means of Eq. (117) and substituted into the numerically integrated result. If there is specific adsorption, Eq. (120) must be used to calculate ϕ_2.

6. Ionic Concentrations at the OHP

The concentration of ions at the OHP may be calculated using the Boltzmann distribution postulate [Eq. (105)] of the GCS theory. Thus the concentration c_i^{OHP} of ionic species i at the OHP is given by

$$c_i^{OHP} = c_i^b \exp\left(-\frac{z_i F \phi_2}{RT}\right) \qquad (160)$$

Here c_i^b is the bulk concentration of the ionic species. The potential ϕ_2 in Eq. (160) is determined by Eq. (117) in the absence of specific adsorption and by Eq. (120) in the presence of specific adsorption.

Equation (160) makes it clear that the concentration of an ionic species at the OHP may differ greatly from the bulk concentration. For example, at 25°C, if ϕ_2 is -0.059 volt (a not unreasonably high value; cf. Fig. 14), Eq. (160) takes the following form for a divalent cation

$$\log\left(\frac{c_i^{OHP}}{c_i^b}\right) = \frac{2(0.059)}{0.059} = 2$$

or

$$c_i^{OHP} = 100 c_i^b$$

Thus, if the divalent cation in question is present in only millimolar concentration in the bulk of the solution, its concentration at the OHP will be 0.1 M! This kind of result illustrates in a striking way that one should expect the double layer to have an important effect on the kinetics of faradaic processes, for it is the concentration of an electroactive species at the electrode surface which must be considered in the calculation of the rates of electrode processes.

7. Validity of the GCS Theory

The GCS theory of the diffuse layer which has been developed in broad outline in the preceding paragraphs strongly resembles in its formalism the Debye and Hückel theory (78) of electrolyte solutions. (Historically, it is interesting to note that the initial work of Gouy and Chapman on the theory of the diffuse layer preceded by a full decade the work of Debye and Hückel on electrolyte solutions.) As in the case of electrolyte theory, the GCS theory has proved a very successful first approximation, but significant progress beyond this first approximation has been difficult. That the GCS theory has been used successfully in the treatment of the diffuse layer even in the case of relatively concentrated electrolytes ($>1\ M$) is at first glance quite surprising, in view of the low range of concentrations over which the simple Debye-Hückel theory is valid. However, when one recalls that the Debye-Hückel theory employs a linearized approximation to the exponential in the Poisson-Boltzmann equation, whereas the GCS theory uses the full exponential [cf. Eq. (110)], there is less reason for surprise. If a comparison is to be made between diffuse-layer theory and electrolyte theory it would be better to compare the GCS theory with the extended Debye-Hückel theory of Gronwall and colleagues (*78*), which employs the complete Taylor series expansion of the exponential in the Poisson-Boltzmann equation. The latter theory is valid over a considerably greater concentration range than is the simple Debye-Hückel theory. Nevertheless, direct comparisons between diffuse-layer theory and bulk-electrolyte theory may not be a fruitful approach, anyway, in spite of the similarities in the formalisms of the two theories. As Parsons (*9*) has pointed out, the reasons for the difficulties in making advancement beyond the very successful first approximations in the two cases appear to be different.

The shortcomings of the GCS theory naturally stem from defects in its fundamental postulates which, in turn, result from weaknesses in the simple physical model on which the theory is based. A partial list of these weaknesses would include (a) the Boltzmann distribution postulate, (b) neglect of dielectric saturation, (c) the assumption that the plane of closest approach (OHP) is the same for all ions regardless of their different radii, (d) neglect of possible effect of ion-pair formation.

a. The Boltzmann Distribution Postulate. The Boltzmann distribution postulate appears defective because it implies that the only work required to move an ion from the bulk of solution to a given position inside the diffuse layer is the simple electrostatic work of interaction between the charge of the ion and the electric field—the term $z_iF[\phi(x) - \phi^S]$. This oversimplification was the result of the assumption in the GCS model that the ions of the diffuse layer interact with the electric field as classical point charges. In reality, the ions of the diffuse layer occupy finite volumes, and they are polarizable, that is, they may be distorted by the field. These effects are not included in the simple, point charge interaction term. A number of workers have considered such effects (*81–86*). As Delahay (*5*) has pointed out recently, the true thermodynamic criterion for electrochemical equilibrium in the diffuse layer is not the simple Boltzmann expression [Eq. (105)]; rather it should be the uniformity of the electrochemical potentials of each ionic species throughout the diffuse layer, that is,

$$\bar{\mu}_i^b = \bar{\mu}_i(x)$$

where $\bar{\mu}_i^b$ is the bulk value of the electrochemical potential of species i and $\bar{\mu}_i(x)$ is the value of the electrochemical potential of that same species at a distance x from the electrode surface inside the diffuse layer.

To obtain useful expressions for this condition, it is not sufficient simply to expand the electrochemical potentials in the usual manner. Rather, explicit formulations for ion polarization and other interaction terms for the ion with the electric field would have to be taken into account. A recent, and possibly very fruitful, approach based on the theory of local thermodynamic balance of Prigogine et al. (*87*) has been published by Hurwitz and colleagues (*88*). These workers considered not only the previously mentioned effects but, also, the effects of dielectric saturation and ion-pair formation and have presented estimates of the possible magnitude of the defects of the simple GCS theory on double-layer corrections in electrode kinetics. It would appear from their work that, in most cases, these corrections would result in a change in the rate constant of less than an order of magnitude.

b. Dielectric Saturation. In Sec. III.C.1, it was pointed out that the GCS theory assumes that ε, the dielectric permittivity, remains constant throughout the diffuse layer. It can be readily shown that the electric field at the OHP can reach values of several million volts per centimeter. Calculations by Grahame (*89*) show that ε changes markedly at such field strengths. However, Grahame stated: ". . . dielectric saturation does not seriously affect the validity of calculations of observable quantities made with the assumption of a constant dielectric coefficient" (*79*). Macdonald and Barlow (*15*), Hurwitz et al. (*88*), and Joshi and Parsons (*80*) have reached similar conclusions. For further references concerning this subject, consult Delahay's book (*5*).

c. Ionic Radii and the Location of the OHP. One of the most obvious defects in the simple GCS theory is its assumption that the same OHP is valid for all ions regardless of their different radii. In fact, a smaller solvated ion will be able to approach the surface of the electrode more closely than a larger solvated ion. Thus, an obvious way to improve the GCS model would be to consider an OHP for the smaller ion which lies closer to the surface of the electrode than the OHP for the larger ion. In contrast to the two previous classes of corrections to the GCS theory (cf. Secs. a and b above), whose significance still remains rather obscure, the importance of considering different planes of closest approach for nonspecifically adsorbed ions of different radii has been strongly indicated both by the electrocapillary measurements on mixed electrolyte solutions by Joshi and Parsons *(80)†* and by a recent study of double-layer effects in electrode kinetics by Aramata and Delahay *(90,91)*.

Joshi and Parsons measured the electrocapillary curves at a mercury electrode for aqueous mixed electrolyte solutions containing HCl and $BaCl_2$.‡ The surface excesses of each of the three ions of this solution can be calculated according to the classical GCS theory by application of the general equation [Eq. (159)]. Application of this equation requires a knowledge of the value of the potential ϕ_2 at the OHP, which in turn is calculated from the general equation relating the potential at the OHP to the measurable charge density q^M, Eq. (117). However, since the ions have different radii, it is clear that the value of ϕ_2 calculated from Eq. (117) will be a kind of weighted average of the values of the electrical potential actually prevailing at the different planes of closest approach (OHP's) for the different ions. Thus, the location of the OHP referred to in the classical theory x_2 is actually a kind of " average " value of the distances of closest approach of the different ions.

Joshi and Parsons considered electrocapillary curves in the region of negative charge densities, where essentially complete expulsion of nonspecifically adsorbed anions from the first layers inside the diffuse layer could be assumed. Thus, they could be concerned, justifiably, with considering ony two OHP's. One, for the hydrogen ion, is located at a distance

---

† Actually Joshi and Parsons considered separately three different possible modifications to the GCS theory, dielectric saturation, variation of location of OHP, and ionic association (cf. Sec. d. below) and tested whether any of them would bring the theoretical curves for surface excesses (cf. Fig. 16) into closer coincidence with the experimental points. The effect of a dielectric-saturation correction alone was shown, by substituting a lower value for the permittivity, to cause the theoretical curve to move further away from the experimental points.

‡ Three-component solutions (water and two salts with a common ion) constitute the principal class of mixed electrolyte solutions for which activity coefficients for both solutes are available. Activity coefficients of the salts are, of course, required for precise thermodynamic analysis of electrocapillary data [cf. Eq. (52)]. In most systems it has been possible to determine the activity coefficient of only one electrolyte solute. Whenever one of the solutes (for example, HCl) is known to obey Harned's rule [p. 459 in *(78)*], the activity coefficient of the other solute (for example, $BaCl_2$, $SrCl_2$, $AlCl_3$, $CeCl_3$) may be calculated without further assumptions *(92)*.

x_2' from the electrode surface; the other, for the barium ion, is located at a distance $x_2'' > x_2'$ from the electrode surface.

The experimentally determined relative surface excesses of Ba^{2+} and H^+ [expressed as components of charge in the diffuse layer; cf. Eq. (141)] and the nonrelativized surface excesses reckoned at the "average" OHP calculated theoretically from the classical GCS equation [Eq. (159)] are shown in Fig. 16. Figure 17 shows the effect on the theoretical surface excesses by considering that the distance of closest approach to the electrode is 2.15 A greater for Ba^{2+} than for H^+. It will be noted that this modification in the GCS theory does move the theoretical curves

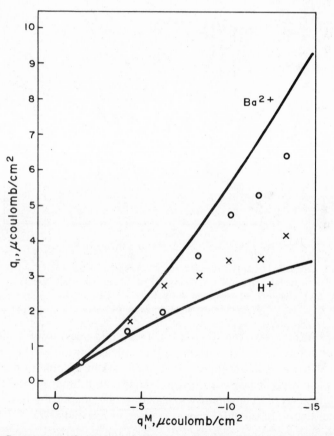

Fig. 16. Components of charge equivalent to the relative surface excess due to Ba^{2+} and H^+ in solution, 0.2 F HCl + 0.05 F BaCl$_2$. Dots are for Ba^{2+}, crosses for H^+. Solid lines calculated according to the GCS theory [from (80)]. (By permission of Pergamon Press.)

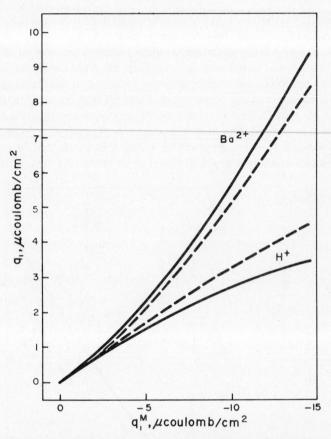

Fig. 17. Calculated results for components of charge in 0.2 F HCl + 0.05 F BaCl$_2$. Solid line calculated from GCS theory. Dotted line calculated assuming OHP for H$^+$ is 2.15 A closer to electrode surface than OHP for Ba^{2+} [from (80)]. (By permission of Pergamon Press.)

(dotted lines) away from the GCS curves (solid lines) toward the experimental points (cf. Fig. 16). Details of the calculations may be found in the original article. Thus, it appears that simply taking account of the different sizes of nonspecifically adsorbed ions can, indeed, improve the theory of the diffuse layer. A qualitative explanation of some of the basic ideas in this correction to the GCS theory is given in Appendix D.

d. Ionic Association. To the extent that ion pairs can form in the solution or in the diffuse layer, the results of the classical GCS calculations will be modified for, obviously, an ion pair such as BaCl$^+$, will be attracted or repelled less in the diffuse layer than a doubly charged ion. Several workers who have studied double-layer effects in electrode kinetics have pointed to ion-pair formation as a possible explanation for discrepancies. Joshi and Parsons (*80*), in the same paper referred to above, attempted to estimate the effect of ion-pair formation by considering several values for the dissociation constant of the ion pair $BaCl^+$. These workers found that they could obtain a shift of the calculated nonrelativized surface excesses at the "average" OHP toward the relative surface excesses provided that a value of 0.18 mole/liter for the dissociation constant was used. They quote McDougall and Davies (*93*) as giving the measured dissociation constant for this ion pair as 1.35 moles/liter. They concluded that the effect of ionic association in the bulk of the solution could not produce the desired correction [cf. Delahay and Aramata (*94*)]. However, a combination of dielectric saturation and ion-pair formation inside the diffuse layer could provide a partial explanation of the discrepancy between theory and experiment.

e. Statistical Mechanical Theories of the Diffuse Layer. Recently, two groups of workers, Buff and Stillinger (*95*) and Levich and Krylov (*96,97*), have applied statistical mechanics to the problem of the diffuse layer. This fundamentally different approach to the problem has not yet been tested experimentally. Delahay (*5*) has discussed these theories briefly and has indicated that this new approach may actually be more fruitful in the long run than the various piece-meal corrections to the GCS theory which have been outlined above.

D. Specific Ionic Adsorption

1. Definition

Specific, or superequivalent, ionic adsorption may be defined either phenomenologically in terms of measurable quantities or on the basis of a structural model of the double layer. In the introduction (Sec. I.B) a definition in terms of the structural model of Grahame was given. According to this definition, specific ionic adsorption occurs when ions can interact so strongly with the metal that they lose their solvation, at least in the direction toward the metal, penetrate into the inner layer, and become fixed to the metal with their electrical centers at the IHP.

A phenomenological definition of specific ionic adsorption is most conveniently given in three parts. At the *point of zero charge* one says *there is specific adsorption if the measured* (that is, relative) *surface excess of any ionic species is positive.* At potentials more *positive* than the point of zero charge ($q^M > 0$), one says there is specific adsorption if the relative surface excess of any *cation* is positive. At potentials more *negative* than the point of zero charge ($q^M < 0$), one says there is specific adsorption if the relative surface excess of any *anion* is greater than zero.

Since $q^S = -q^M = 0$ at the point of zero charge, the sum of the measured surface excesses of all the ions (in equivalents) must add up to zero. For this reason it is frequently said that an electrolyte (salt) is specifically adsorbed. In the example in Sec. II.E.1.g the relative surface excesses of K^+ and Br^- at the point of zero charge are equal and positive; consequently, one says there is specific adsorption and that the electrolyte KBr is specifically adsorbed. It is believed, however, that usually only one ion, ordinarily the anion, is actually adsorbed into the inner layer, and that the extra ions of opposite sign are adsorbed nonspecifically in the diffuse layer merely as counterions to maintain the electroneutrality of the interphase. Thus a positive relative surface excess of a cation does not necessarily imply that the cation itself is adsorbed in the inner layer; rather, it usually implies that the anion adsorbed in the inner layer attracts sufficient cationic counterions into the diffuse layer to maintain the electroneutrality of the interphase.

It is important to realize that the phenomenological definition of specific adsorption does not include, strictly speaking, any statement about which of the ions is in the inner layer. [In fact, one need not even believe in the existence of an inner layer to believe in specific adsorption. Unequivocal experimental evidence from electrocapillary measurements which would meet the above phenomenological criterion of specific adsorption were obtained by Gouy (*65*) about a decade before Stern's concept of an inner layer was published (*18*). Early work with simple inorganic salts showed that changing the cation had very little effect on electrocapillary behavior but changing the anion could have marked effect. Therefore, it became common to speak of certain anions, which today we would say are specifically adsorbed, as "surface active," a term which persists in the literature. The same term was also applied to many organic substances which adsorbed at the metal-solution interface.]

In addition, a quasi-phenomenological definition of specific adsorption may be given. According to this definition, one says there is specific adsorption if the experimental data cannot be explained by the theory of the diffuse layer.

Although the last definition may appear less satisfying in view of the defects in the theory of the diffuse layer (Sec. III.B.7), it is, in fact, the most important definition of specific adsorption. The pure model definition is, of course, nonoperational and, therefore, cannot provide an experimental criterion for specific adsorption in terms of a molecular model. The last definition which does lead to interpretation based on models is the one most widely used in practice.

Whenever according to the phenomenological definition there is specific adsorption, there will necessarily also be specific adsorption in the sense of the quasi-phenomenological (GCS) definition. However, the converse is not necessarily true. Specific adsorption may exist in the quasi-phenomenological sense, although no specific adsorption would be said to be present according to the simple phenomenological definition given above. For example, according to the GCS theory [cf. Eq. (148)] the surface excess of a cation reckoned at the OHP, $^d\Gamma_+^{OHP}$, approaches an asymtotic value as q^M becomes increasingly positive. In the case of a single $z - z$ electrolyte, the equivalent charge, q^d_+ [cf. Eq. (150)], approaches an asymtotic value of about -2 μcoulombs/cm^2. Thus for $q^M > 0$, one would say that there is specific adsorption in the quasi-phenomenological (GCS) sense whenever $-2 < zF\Gamma_{+h'}$, where $\Gamma_{+h'}$ is the measured surface excess of the cation (compound h' is the reference component in solution). An example is the case of aqueous potassium acetate solution at an ideal polarized mercury electrode; cf. Fig. 23. A similar idea applies to $^d\Gamma_-^{OHP}$ [cf. Eq. (149)]. q^d_- [cf. Eq. (151)] approaches about 2 μcoulombs/cm^2 as q^M becomes increasingly negative. Hence there is specific adsorption in the sense of the GCS theory whenever $q^M < 0$ and $2 > -zF\Gamma_{-h'}$, where $\Gamma_{-h'}$ is the relative surface excess of the anion. Specific adsorption in the quasi-phenomenological sense thus includes cases which would not be termed specific adsorption in the simple phenomenological sense.†

Various criteria may be applied to experimental data to determine whether an electrolyte is specifically adsorbed at an ideal polarized electrode. Several of these criteria will be discussed below in terms of the definitions just given.

2. Criteria for Detecting the Presence of Specific Ionic Adsorption at Ideal Polarized Electrodes

The criteria for detection of specific adsorption at ideal polarized electrodes may be divided into two categories, depending upon whether the experimental data are derived directly from electrocapillary curves or from differential capacity curves.

---

† The author is indebted to Professor Paul Delahay for calling this fact to his attention.

a. Based on Electrocapillary Curves

(1) *Relative Surface Excesses.* The example of KBr solutions at mercury electrodes given in Section II illustrated the application of a criterion
based on relative surface excesses. For a simple electrolyte, one measures
electrocapillary curves for a series of solutions and then makes plots
according to Eq. (62b):

$$\frac{1}{RT}\left(\frac{\partial \gamma}{\partial \ln a_{KBr}}\right)_{T,p,E^-} = -\Gamma_{K^+,w} \tag{161}$$

to determine the relative surface excess of the anion or cation, depending
on the nature of the indicator electrode employed. Then, applying the
phenomenological definition, one can decide whether there is superequivalent adsorption of the electrolyte. (Insofar as one is concerned with
the phenomenological definition, the term *superequivalent adsorption* is
probably preferable to the term *specific adsorption*, for the latter term
usually connotes some concept of double-layer structure.)

If a similar series of measurements were made using, for example, all the
potassium halides, one would discover that in the case of KF the relative
surface excess of each ion was zero at the potential of the ECM, that K^+
was positively adsorbed and F^- negatively adsorbed at potentials more
negative than E_z, and vice versa for potentials more positive than E_z.
Thus, one would conclude that KF was not specifically adsorbed (except
possibly at the very most positive potentials). Comparing the results
obtained for KF with the results obtained for KCl, KBr, and KI, one would
observe an increasing amount of specific adsorption with changing anion
and conclude that the anion was the chief cause of the specific adsorption.
A series of similar experiments using the same anions but varying the
cations over the alkali metals would indicate no changes from the K^+
series (except possibly in the case of Cs^+). Invoking Grahame's model of
specific adsorption one would conclude that the anions enter the inner
layer, but the cations do not, with the possible exception of Cs^+. Cesium,
however, is not the only inorganic cation which exhibits specific adsorption.
Frumkin (98) has cited evidence based on electrocapillary curves for Tl^+
and Pb^{2+} solutions at mercury electrodes which indicates that Tl^+ and
Pb^{2+} are also specifically adsorbed. In addition, he has obtained other
kinds of evidence which indicate that Cd^{2+}, Zn^{2+}, and Cs^+ may also be
specifically adsorbed. He states: "We must conclude that the frequently
quoted statement that inorganic cations are never specifically adsorbed is
incorrect."

(2) *Esin and Markov Effect.* One of the earliest criteria for specific adsorption (8) was the shift with changing electrolyte concentration of the potential of the ECM measured with respect to a "constant-potential" reference electrode. This shifting of the point of zero charge is an example of the Esin and Markov effect. Parsons, in 1957 (51), derived the general behavior of the Esin and Markov coefficient in the presence and absence of specific adsorption using the quasi-phenomenological definition of specific adsorption. The main ideas involved in Parsons' analysis will now be given. For simplicity, we shall consider only the case of an ideal polarized electrode dipping into a solution of a single z-z electrolyte. Extension to the case of mixed electrolytes follows the same route, but the mathematical formulas are more complicated. Little experimental work on mixed electrolyte solutions has yet been done.

The Esin and Markov coefficient was defined in Sec. II.E.4 as the partial derivative of $E^{\pm}$ with respect to μ of either a neutral component or salt at constant q^M. In that general derivation the particular form of the Esin and Markov coefficient was not given for the case in which the partial differentiation is made with respect to the chemical potential $\mu_{j'k'}$ of the indicator salt. In general $(\partial E^{\pm}/\partial \mu_{j'k'})$ is cumbersome. However, for the z-z electrolyte, this Esin and Markov coefficient has a simple form. For the z-z electrolyte solution at constant temperature and pressure, the electrocapillary equation is

$$dy = -q^M \, dE^{\pm} - \Gamma_{\mp,w} \, d\mu_{\text{salt}} \tag{162}$$

where $\Gamma_{\mp,w}$ is the relative surface excess of either the anion or the cation of the solution. When the indicator electrode is reversible to the cation, the relative surface excess appearing in Eq. (162) is that of the anion, and vice versa (cf. Sec. II.E.1.g). Using Parsons' auxiliary function $\xi^{\pm}$, one converts to an equation in which the excess charge density is the independent variable

$$d\xi^{\pm} = E^{\pm} \, dq^M - \Gamma_{\mp,w} \, d\mu_{\text{salt}} \tag{163}$$

From the second cross-partial-differentiation relationship implicit in Eq. (163), one obtains the form of the Esin and Markov coefficient for a solution of a single electrolyte:

$$\left(\frac{\partial E^{\pm}}{\partial \mu_{\text{salt}}}\right)_{T,p,q^M} = -\left(\frac{\partial \Gamma_{\mp,w}}{\partial q^M}\right)_{T,p,\mu_{\text{salt}}} \tag{164}$$

For practical purposes, it is more convenient to express the coefficient on the right side of Eq. (164) in terms of the equivalent components of charge on the solution side of the double layer. As before, let

$$z = z_+ = |z_-|$$

where z_+ and z_- are the charges, including sign, of the cation and anion, respectively. Then, define q_+, the contribution of the cation to q^s, by

$$q_+ = zF\Gamma_{+,w} \tag{165}$$

and, similarly, for the anion,

$$q_- = -zF\Gamma_{-,w} \tag{166}$$

Owing to the electroneutrality of the interphase,

$$q^M = -(q_+ + q_-) = -q^s \tag{167}$$

It is also convenient to employ the activity of the salt in solution (a_{salt}) instead of the chemical potential.

$$\mu_{\text{salt}} = \mu_{\text{salt}}^0 + RT \ln a_{\text{salt}} = \mu_{\text{salt}}^0 + RT \ln a_\pm^2 \tag{168}$$

where μ_{salt} is the standard chemical potential of the salt and $a_\pm$ is the mean ionic activity. Thus,

$$d\mu_{\text{salt}} = RT\, d \ln a_{\text{salt}} = RT\, d \ln a_\pm^2 \tag{169}$$

Using Eqs. (164), (165), and (169), at constant T and p,

$$\left(\frac{\partial E^\pm}{\partial \mu_{\text{salt}}}\right)_{q^M} = \frac{1}{RT}\left(\frac{\partial E^\pm}{\partial \ln a_{\text{salt}}}\right)_{q^M} = \frac{1}{RT}\left(\frac{\partial E^\pm}{\partial \ln a_\pm^2}\right)_{q^M} = \pm\frac{1}{zF}\left(\frac{\partial q_\mp}{\partial q^M}\right)_{a_\pm} \tag{170}$$

Actually it is unnecessary to discuss both Esin and Markov coefficients for the cation and anion reversible indicator electrode since these are simply related† by the formula

$$\left(\frac{\partial E^+}{\partial \ln a_\pm^2}\right)_{q^M} + \left(\frac{\partial E^-}{\partial \ln a_\pm^2}\right)_{q^M} = \frac{RT}{zF} \tag{171}$$

Henceforth, we shall discuss just the case of the cation reversible indicator electrode.

---

† By Eq. (167), $-q_+ = q^M + q_-$. Thus, Eq. (170) becomes

$$\left(\frac{\partial E^-}{\partial \ln a_\pm^2}\right)_{q^M} = +\frac{RT}{zF}\left[1 + \left(\frac{\partial q_-}{\partial q^M}\right)_{a_\pm}\right]$$

Substituting Eq. (170) into the above gives Eq. (171).

Equation (170) is a thermodynamic equation derived from the general electrocapillary equation. The quantities q_+ and q_- were defined by Eqs. (165) and (166) in terms of the relative surface excesses of the anions and the cations. To use the Esin and Markov coefficient to test structural models of the electrical double layer, we must first assume that replacing the relative surface excesses by surface excesses reckoned at the OHP is justifiable. It is actually at this point that the quasi-phenomenological definition of specific adsorption enters. That definition implies

$$q_+ \equiv zF\Gamma_{+,w} = zF\Gamma_+^{OHP} \equiv q_+^{OHP} \equiv q_+^d + q_+^i \tag{172}$$

and

$$q_- \equiv -zF\Gamma_{-,w} = -zF\Gamma_-^{OHP} \equiv q_-^{OHP} \equiv q_-^d + q_-^i \tag{173}$$

Although it is true that

$$q_+^{OHP} + q_-^{OHP} = q_+ + q_- = q^S = -q^M \tag{174}$$

because the invariance of q^M over the choice of the dividing surface, it is only an assumption that Eqs. (172) and (173) individually are true (cf. Sec. III.B.1). The last identity signs in Eqs. (172) and (173) are written because it is by definition implicit in the model that the amount of charge on the solution side of the double layer is split into two portions, an inner-layer part and a diffuse-layer part. Using the quasi-phenomenological definition of specific adsorption and assuming that only the anion is specifically adsorbed, we may then write the Esin and Markov coefficient

$$\left(\frac{\partial E^+}{\partial \ln a_\pm^2}\right)_{q^M} = \frac{RT}{zF}\left(\frac{\partial q_-^d}{\partial q^M}\right)_{a_\pm} + \frac{RT}{zF}\left(\frac{\partial q_-^i}{\partial q^M}\right)_{a_\pm} \tag{175}$$

We can show, best, how the Esin and Markov coefficient may be used as a criterion for detecting the presence of adsorption by deriving first its form when there is no specific adsorption, that is, by assuming $q_-^i = 0$. According to the GCS theory,

$$q^M = 2A \sinh\left(\frac{zF\phi_2}{2RT}\right) \tag{176}$$

where A is given, as before, by

$$A = (2RT\varepsilon c^b)^{\frac{1}{2}} \tag{158}$$

Then q^d is given by

$$q^d = -A(e^{zF\phi_2/2RT} - 1) \tag{151}$$

Substituting for ϕ_2 in Eq. (151) by means of Eq. (176) and then differentiating with respect to q^M, one obtains

$$\left(\frac{\partial q_-^d}{\partial q^M}\right)_{a_\pm} = -A \exp\left[\sinh^{-1}\left(\frac{q^M}{2A}\right)\right]\left(\frac{\partial \sinh^{-1}(q^M/2A)}{\partial q^M}\right)_{a_\pm} \qquad (177)$$

Noting that (99)

$$\frac{d \sinh^{-1} u}{dx} = \frac{1}{(u^2 + 1)^{\frac{1}{2}}}\left(\frac{du}{dx}\right) \qquad (178)$$

we see that Eq. (177) becomes

$$\left(\frac{\partial q_-^d}{\partial q^M}\right)_{a_\pm} = -\frac{1}{2} \exp\left[\sinh^{-1}\left(\frac{q^M}{2A}\right)\right]\left[1 + \left(\frac{q^M}{2A}\right)^2\right]^{\frac{1}{2}} \qquad (179)$$

Since (100)

$$\sinh^{-1} x = \ln[x + (x^2 + 1)^{\frac{1}{2}}] \qquad (180)$$

Eq. (179) becomes

$$\left(\frac{\partial q_-^d}{\partial q^M}\right)_{a_\pm} = -\frac{1}{2} - \frac{1}{2}\left\{\frac{(q^M/2A)}{[1 + (q^M/2A)^2]^{\frac{1}{2}}}\right\} \qquad (181)$$

Substituting Eq. (181) into Eq. (175), assuming no specific adsorption, one obtains for the Esin and Markov coefficient

$$\left(\frac{\partial E^+}{\partial \ln a_\pm^2}\right)_{T,p,q^M} = -\frac{RT}{2zF}\left\{1 + \frac{(q^M/2A)}{[1 + (q^M/2A)^2]^{\frac{1}{2}}}\right\} \qquad (182)$$

Parsons tested Eq. (182) using the experimental data of Grahame obtained for NaF solutions at an ideal polarized mercury electrode. The results are shown in Fig. 18. It will be noted from this figure that the slope of the plot of E^+ versus $\ln a_\pm$ is zero for fairly negative charges and that it appears to approach a constant negative value for large positive charges; it can also be seen from this graph that the limiting slope for positive charges is approximately twice the slope of the zero charge line. From Eq. (182) we can deduce that this is precisely the behavior one should expect if there is no specific adsorption in NaF solutions and if there are no serious errors in the assumptions used in deriving Eq. (182). When $q^M \gg 1$, Eq. (182) reduces to

$$\left(\frac{\partial E^+}{\partial \ln a_\pm^2}\right)_{T,p,q^M} = -\frac{RT}{zF}$$

Thus for high positive charges the limiting slope of the Esin and Markov coefficient in the absence of specific adsorption should approach $-RT/zF$

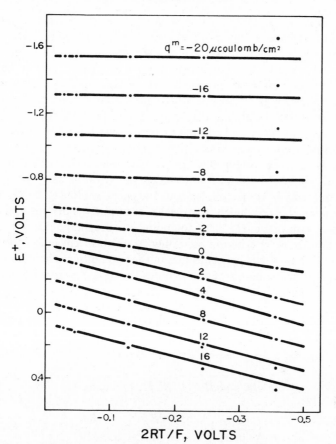

Fig. 18. Esin and Markov plots for NaF solutions calculated from the data of Grahame [from (*51*)]. (By permission of Butterworths.)

and this is, in fact, the slope obtained by Parsons. For $q^M = 0$ we obtain

$$\left(\frac{\partial E^+}{\partial \ln a_\pm^2}\right)_{q^M = 0} = -\frac{RT}{2zF} \tag{183}$$

and, in fact, Parsons' graphs show that the slope for the zero charge line is exactly half the limiting slope at positive charges. Finally, when $q^M/2A \ll -1$, one deduces that

$$\left(\frac{\partial E^+}{\partial \ln a_\pm^2}\right)_{q^M} = 0$$

which is the result found.

Thus, assuming that there is no specific adsorption of F^- at the mercury-solution interface, one finds good agreement between the GCS theory and experimental Esin and Markov plots. According to Delahay (5), this agreement is most convincing around the point of zero charge; at potentials far from the point of zero charge, the diffuse layer actually plays a minor role in determining the properties of the metal-solution interface.

Thus, it appears that one may use the Esin and Markov coefficient as a criterion for specific adsorption in the following way: If a given ideal polarized electrode system containing a single z-z electrolyte gives agreement with Eq. (182) or Fig. 18, one deduces that, to the extent that the Esin and Markov coefficient is a sufficiently sensitive test, there is no specific adsorption. If one obtains Esin and Markov plots at variance with this behavior, it is concluded that specific adsorption is present. Using the particular form of Eq. (182) for the point of zero charge [Eq. (183)], we can explain the older criterion according to which specific adsorption was indicated by a shift with concentration of the point of zero charge when the electrocapillary maximum was determined using a "constant-potential" reference electrode.

Applying the Nernst equation to a galvanic cell consisting of the indicator electrode and the reference electrode dipping into the solution under consideration, neglecting the liquid junction potential, assuming that single ion activities may be replaced by mean ionic activities, one deduces that

$$dE_{\text{ref}} = dE^{\pm} + (RT/2zF) \, d \ln a_{\pm}^2$$

Hence,

$$\left(\frac{\partial E_{\text{ref}}}{\partial \ln a_{\pm}^2} \right)_{q^M} = \left(\frac{\partial E^{\pm}}{\partial \ln a_{\pm}^2} \right)_{q^M} + \frac{RT}{2zF}$$

Now, applying Eq. (183), the form of Eq. (182) for the point of zero charge, one deduces that

$$\left(\frac{\partial E_{\text{ref}}}{\partial \ln a_{\pm}^2} \right)_{q^M = 0} = \frac{RT}{2zF} - \frac{RT}{2zF} = 0$$

Hence the potential of the point of zero charge measured with respect to any constant-potential reference electrode should not shift if there is no specific adsorption. It will be shown in Sec. III.D.3 that, when there is strong specific adsorption, the Esin and Markov coefficient is essentially independent of the charge density. Thus, it follows that, if there is specific adsorption, the point of zero charge measured with respect to a constant-potential reference electrode will shift with changing electrolyte concentration.

b. Based on Differential Capacity Curves. Differential capacity curves provide an important means of studying the electrical double layer at ideal polarized electrodes because the differential capacitance measured with an ac impedance bridge can be determined very accurately. Grahame and Soderberg (*101*) devised a criterion for detecting the presence of specific adsorption which is simple to apply, although Eq. (193), on which it is based, appears somewhat formidable. The method is based on Grahame's concept of the *components of the differential capacity* C_+ and C_-, attributed to the ions of the solution. As before, we shall consider only the particular case of a single z-z electrolyte. Grahame also worked out the method for unsymmetric electrolytes (*79*).

The components of charge in the solution side of the double layer, q_+ and q_- are related to the charge density on the metal by the requirement of the electroneutrality of the interphase:

$$q^M = -q_+ - q_- \tag{184}$$

Differentiating this equation with respect to E^- at constant μ, one obtains

$$\left(\frac{\partial q^M}{\partial E^-}\right)_\mu = -\left(\frac{\partial q_+}{\partial E^-}\right)_\mu - \left(\frac{\partial q_-}{\partial E^-}\right)_\mu \tag{185}$$

The quantity on the left side is the measurable differential capacity C, defined by Eq. (68). The two quantities on the right side are the components of the differential capacity attributable to the cations and the anions, respectively. They are defined by the equations

$$C_+ = -\left(\frac{\partial q_+}{\partial E^-}\right)_\mu = -zF\left(\frac{\partial \Gamma_{+,w}}{\partial E^-}\right)_\mu \tag{186}$$

and

$$C_- = -\left(\frac{\partial q_-}{\partial E^-}\right)_\mu = zF\left(\frac{\partial \Gamma_{-,w}}{\partial E^-}\right)_\mu \tag{187}$$

The signs of C_+ and C_- have been chosen so that $C_+ + C_-$ will always be positive, that is, have the same sign as C. Thus,

$$C = C_+ + C_- \tag{188}$$

In this section we shall refer only to an anion-reversible indicator electrode. Applying the cross-differentiation relationship, Eq. (75), we have

$$\left(\frac{\partial q^M}{\partial \mu}\right)_{E^-} = \left(\frac{\partial \Gamma_{+,w}}{\partial E^-}\right)_\mu \tag{189}$$

Differentiating again with respect to E^- and recalling the definitions of C and C_+, we obtain

$$zF\left(\frac{\partial C}{\partial \mu}\right)_{E^-} = -\left(\frac{\partial C_+}{\partial E^-}\right)_{\mu} \tag{190}$$

Now assume that there is no specific adsorption and that the GCS theory is valid. Then, as before,

$$q_+ = q_+^d = A\left[\exp\left(-\frac{zF\phi_2}{2RT}\right) - 1\right] \tag{191}$$

and

$$q^M = 2A \sinh\left(\frac{zF\phi_2}{2RT}\right) \tag{176}$$

Solving, as in the previous section, for ϕ_2 in terms of q^M, one writes

$$q_+^d = A\left\{\exp\left[-\sinh^{-1}\left(\frac{q^M}{2A}\right)\right] - 1\right\} \tag{192}$$

An equation for $(\partial C_+^d/\partial E^-)_{\mu}$ is next derived by differentiating Eq. (192) twice with respect to E^-. Equations (178) and (180) are used, and after algebraic rearrangement, one obtains Grahame's equation

$$\left(\frac{\partial C_+^d}{\partial E^-}\right)_{\mu} = \frac{C}{2[(q^M/2A) + 1]^{\frac{1}{2}}}\left(\left\{\left[\left(\frac{q^M}{2A}\right)^2 + 1\right]^{\frac{1}{2}} - \left(\frac{q^M}{2A}\right)\right\}\left(\frac{1}{C}\right)\left(\frac{\partial C}{\partial E^-}\right)_{\mu} - \frac{C}{2A[(q^M/2A)^2 + 1]}\right) \tag{193}$$

In this equation C is the total measured differential capacity of the double layer assuming no specific adsorption. Equation (193) is a theoretical equation which shows how the partial differential coefficient $(\partial C_+/\partial E^-)_{\mu}$ should behave if there is no specific adsorption. From Eq. (190) one can calculate the actual value of this partial-differential coefficient from measured differential capacity data. The criterion for detecting the presence of specific adsorption is to compare the theoretical curve from Eq. (193) and the actual curve from experimental measurements. If the two curves coincide, one concludes there is no specific adsorption; if they do not coincide one concludes there is specific adsorption.

Figure 19 shows a plot from Grahame and Soderberg illustrating the use of this criterion. The lower curve, obtained for KF agrees very well over the entire potential range with the theoretical curve, and one concludes that KF exhibits essentially no specific adsorption except possibly at the most positive potentials [cf. Payne (73)].

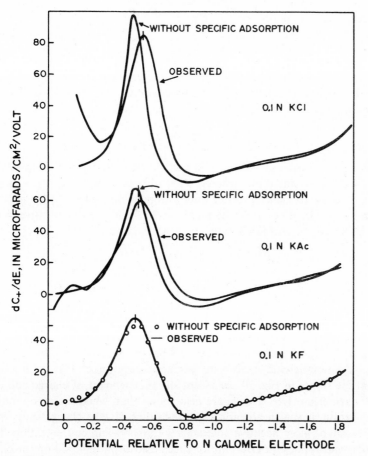

Fig. 19. Observed and calculated values of dC_+/dE illustrating Grahame and Soderberg's (*101*) criterion for specific adsorption. (By permission of the American Institute of Physics.)

The other two curves in Fig. 19, for potassium chloride and potassium acetate, exhibit the behavior expected for the presence of specific adsorption in that there is a considerable range of potentials over which the theoretical and experimental curves do not coincide.

3. The Esin and Markov Coefficient in the Presence of Specific Ionic Adsorption

In Sec. III.C.2.a(2) we derived the general equation for the Esin and Markov coefficient when there is specific adsorption of the anion, Eq. (175).

That equation may be expanded to give

$$\left(\frac{\partial E^+}{\partial \ln a_{\pm}^2}\right)_{q^M} = \frac{RT}{zF}\left\{\left(\frac{\partial q_-^i}{\partial q^M}\right)_{a_{\pm}} + \left(\frac{\partial q_-^d}{\partial q^d}\right)_{a_{\pm}}\left(\frac{\partial q^d}{\partial q^M}\right)_{a_{\pm}}\right\} \qquad (194)$$

where q^d is the total charge in the diffuse layer and q_-^i is the charge resulting from the layer of specifically adsorbed anions at the IHP. Applying the condition of electroneutrality, the last term in Eq. (194) can be written

$$\left(\frac{\partial q^d}{\partial q^M}\right)_{a_{\pm}} = -\left[1 + \left(\frac{\partial q_-^i}{\partial q^M}\right)_{a_{\pm}}\right] \qquad (195)$$

Obviously, evaluation of Eq. (194) would be quite tedious. However, as Parsons (*51*) pointed out, when specific adsorption is strong, there can be a large range of concentration and charge within which $(q^M + q_-^i)/2A$ is large and negative, and, therefore, $q^d/2A$ is large and positive. In the presence of specific adsorption, Eq. (181) becomes

$$-\left(\frac{\partial q_-^d}{\partial q^d}\right)_{a_{\pm}} = -\tfrac{1}{2} + \tfrac{1}{2}\left\{\frac{(q^d/2A)}{[1 + (q^d/2A)^2]^{\frac{1}{2}}}\right\} \qquad (196)$$

Thus, when $q^d/2A$ is large and positive the right side of Eq. (193) approaches zero. Therefore, in the region where $q^d/2A$ is large and positive, the second term in Eq. (191) vanishes, and the measured Esin and Markov coefficient, Eq. (191), will be due essentially to the specifically adsorbed ions. Figure 20 shows the results for KCl, KBr, and KI solutions. In contrast to the situation in which there is no specific adsorption (cf. Fig. 18), it is found that the slopes in Fig. 20 are essentially independent of charge and differ for each electrolyte. This figure also shows that, when the anion, but not the cation, is specifically adsorbed, the point of zero charge shifts in the direction of more negative potentials with increasing concentration. Thus, when one applies the criterion for specific adsorption based on the change of the point of zero charge, one can also deduce something about which ion is specifically adsorbed, provided that only one ion is specifically adsorbed. If the shift with increasing concentration is toward more negative potentials, the anion is specifically adsorbed; toward more positive potentials, the cation.

By analysis of the Esin and Markov plots for specifically adsorbed ions, one can deduce information about the nature of the adsorption isotherm which the specifically adsorbed ions obey. To construct a detailed theory of specific adsorption, as Parsons has pointed out (*102*), it is necessary to know what type of isotherm is obeyed to separate ion-electrode interactions from ion-ion interactions. For a discussion of isotherms see Sec. III.D.1.d.

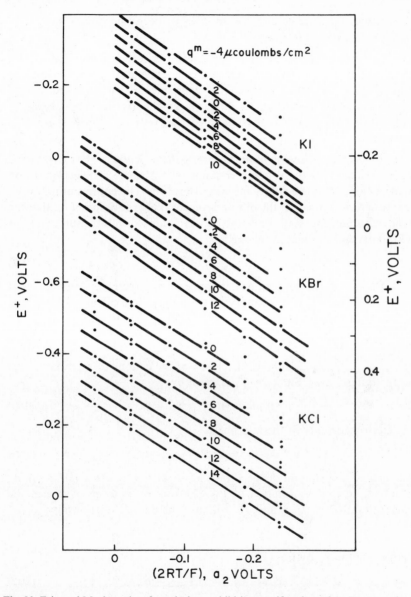

Fig. 20. Esin and Markov plots for solutions exhibiting specific adsorption. Dots, results of Devanathan and Peries (*33*); crosses, calculated from differential capacity data of Grahame [from (*51*)]. (By permission of Butterworths.)

4. Methods for Determining Amount of Specific Adsorption at an Ideal Polarized Electrode

To determine the amount of specifically adsorbed ions, it is always necessary to begin with the quasi-phenomenological definition and to calculate the difference between the experimental surface excesses and those computed according to diffuse-layer theory. This calculation is straightforward, provided it is assumed that only one ion is specifically adsorbed.

a. Calculated Directly from Electrocapillary Curves. For simplicity, again assume that the solution contains only a single z-z electrolyte and that the anion alone exhibits specific adsorption. The entire surface excess of the cation is, then, in the diffuse layer. Making the assumption that the relative surface excess of the cation may be confused with the surface excess of the cation reckoned at the OHP,

$$\Gamma_{+,w} = {}^d\Gamma_+^{OHP} \tag{197}$$

The quantity on the left can be determined from electrocapillary curves as described in Sec. II. The quantity on the right can be calculated according to the GCS theory. Thus,

$$zF\Gamma_{+,w} = q_+^d = A(e^{-zF\phi_2/2RT} - 1) \tag{198}$$

where q_+^d, as before, is the contribution of the cations to the total charge in the diffuse layer. Similarly q_-^d is given by

$$q_-^d = -A(e^{zF\phi_2/2RT} - 1) \tag{199}$$

By electroneutrality of the interphase,

$$q^M = -q^S = -q_+^d - q_-^d - q_-^i \tag{200}$$

where q_-^i is the amount of specifically adsorbed charge at the IHP resulting from the anions. A solution of Eq. (200) for q_-^i in terms of measurable quantities, that is; $\Gamma_{+,w}$ and q^M, will give the desired relationship:

$$q_-^i = -q^M - q_+^d - q_-^d \tag{201}$$

Substitute for q_+^d by Eq. (198):

$$q_-^i = -q^M - zF\Gamma_{+,w} - q_-^d \tag{202}$$

Also, by Eq. (198),

$$\frac{zF\phi_2}{2RT} = -\ln\left(\frac{zF\Gamma_{+,w}}{A} + 1\right) \tag{203}$$

Substituting Eq. (203) into Eq. (199), rearranging, and substituting the result for q_-^d into Eq. (202), we obtain

$$q_-^i = -q^M - zF\Gamma_{+,w} - A\left(\frac{zF\Gamma_{+,w}}{zF\Gamma_{+,w} + A}\right) \tag{204}$$

b. Calculated from Differential Capacity Curves. Provided only one ion is specifically adsorbed, the amount specifically adsorbed can also be determined from analysis of differential capacity curves following the method due to Grahame and Soderberg (*101*). Assume that the anion is specifically adsorbed. From Eq. (190), we have

$$C_+ = -zF \int \left(\frac{\partial C}{\partial \mu}\right)_{E^-} dE + k \tag{205}$$

where k is an integration constant. Integrating once more and applying the definition of C_+, Eq. (186), we obtain

$$q_+ = -\int C_+ \, dE + k' \tag{206}$$

where k' is a second integration constant. Substituting Eq. (205) into Eq. (206), one obtains

$$q_+ = zF \iint \left(\frac{\partial C}{\partial \mu}\right)_{E^-} dE \, dE - kE + k' \tag{207}$$

Thus, the procedure is to integrate the measured $(\partial C/\partial \mu)_{E^-}$ curves to obtain the amount of cation in the double layer. Once q_+ is known at a particular value of E, the corresponding value of q_-^i can be determined using the method in the preceding section. The Grahame-Soderberg method requires knowledge of the value of the integration constants k and k', which, in turn, depend upon where one chooses to begin the graphical integration. The principle of the method is to begin the integration of the differential capacity curves at a sufficiently negative potential that it can be assumed that the anion is entirely repelled from the inner layer so that the electrical double layer obeys the diffuse-layer theory. At the most negative limit of integration C_+ may be determined by the theory of the diffuse layer. We have

$$C_+ = -\left(\frac{\partial q_+}{\partial q^M}\right)_\mu \left(\frac{\partial q^M}{\partial E^-}\right)_\mu \tag{208}$$

Applying the GCS theory for a z-z electrolyte and using the identities shown in Sec. III.D.3, one obtains

$$\left(\frac{\partial q_+}{\partial q^M}\right)_\mu = -\frac{1}{2}\left\{1 - \frac{(q^M/2A)}{[(q^M/2A)^2 + 1]^{\frac{1}{2}}}\right\} \tag{209}$$

Hence C_+ is given by

$$C_+ = \frac{C}{2}\left\{1 - \frac{(q^M/2A)}{[(q^M/2A)^2 + 1]^{\frac{1}{2}}}\right\} \tag{210}$$

Here C is the value of the measured differential capacity at the negative limit of integration. Thus k can be evaluated. The second integration constant k', appearing in Eqs. (206) and (207), is simply the value of q_+ at the lower limit of integration, which is given by

$$q_+ = A\{\exp[-\sinh^{-1}(q^M/2A)] - 1\}$$

$$= A\left\{\frac{1}{(q^M/2A) + \left[\left(\dfrac{q^M}{2A}\right)^2 + 1\right]^{\frac{1}{2}}} - 1\right\} \tag{211}$$

The value of q^M is determined by integration of the differential capacity curve from the point of zero charge to the potential in question. To select the "best" values of the integration constants, Grahame and Soderberg also evaluated the integration constants by another method which involved application of the value of the Esin and Markov coefficient at the point of zero charge and actually used a combination of approaches in their work.

This method can, of course, be extended to the case of other than a z-z electrolyte and the appropriate formulas were given in the original paper. Figure 21 shows a plot of the amount of specifically adsorbed ion q_-^i for several aqueous electrolyte solutions. One observes, in accordance with expectations, that KF shows no specific adsorption and that the most strongly adsorbed anions, Br^- in this figure, remain adsorbed at potentials much more negative than the more weakly adsorbed anions, for example, acetate. Parsons recently recalculated the amount of specific adsorption from Grahame's original data and these recalculated curves were reported in a paper by Mott and Watts-Tobin (*103*).

Devanathan and Canagaratna (*46*) have recently criticized the method of Grahame and Soderberg and have indicated that, instead, a method originated by Devanathan (*20*) using single differential capacitance curves be employed. They base their argument on a comparison of surface excesses calculated via the method of Grahame and Soderberg with those determined by the electrocapillary measurements of Devanathan and Peries (*33*). However, it appears that this criticism is open to question. A better explanation of the deviations is probably the one offered originally by Grahame and Parsons (*70*) and which has been discussed in Sec. III.B.1, namely, the difference in the locations of the OHP and the Gibbs surface.

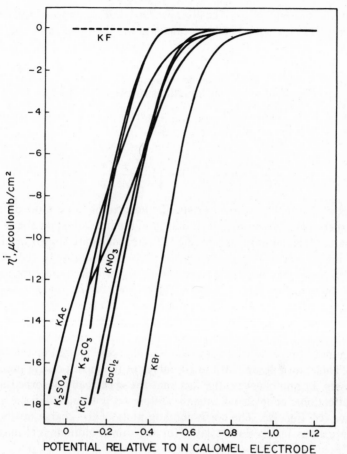

Fig. 21. Component of charge attributable to anions specifically adsorbed at mercury surface, 25°C, 0.1 N aqueous solutions. Potential relative to normal calomel electrode [from (*101*)]. (By permission of the American Institute of Physics.)

5. Effect of Specific Adsorption on ϕ_2

To consider double-layer corrections in electrode kinetics it is necessary to know the effect on ϕ_2 of the specific adsorption of an ion which does not itself undergo oxidation or reduction. In the absence of specific adsorption the shape of the ϕ_2 versus potential curves is that of a *symmetric odd function* of difference in potential, $E - E_z$ (cf. Fig. 14). When only one ion, for example, the anion, is specifically adsorbed the above methods permit the amount of specifically adsorbed ion at any potential to be determined. It is then possible to calculate the total charge in the diffuse layer

$$-q^d = q^M + q^i \tag{212}$$

For a z-z electrolyte, using the equations of the GCS theory,

$$
\begin{aligned}
\phi_2 &= \frac{2RT}{zF} \sinh^{-1}\left(\frac{q^M + q^i}{2A}\right) \\
&= \frac{2RT}{zF} \ln\left\{\left(\frac{q^M + q^i}{2A}\right) + \left[\left(\frac{q^M + q^i}{2A}\right)^2 + 1\right]^{\frac{1}{2}}\right\}
\end{aligned}
\tag{213}
$$

Values of ϕ_2 in the presence of specific adsorption were plotted against the experimental electrode potential by Grahame and Soderberg for a number of the solutions they studied. Their results are shown in Fig. 22. Each line is the $\phi_2 - E_{\text{NCE}}$ curve for a 0.1 N solution. In the case of KF, which exhibits no specific adsorption, ϕ_2 is, indeed, an odd function of the difference of potential $E - E_z$. However, for the solutions in which there is specific adsorption of the anions, there is a reversal of the ϕ_2 curve at the more positive potentials. Thus, in the case of specific anionic adsorption ϕ_2 exhibits a maximum. (This maximum appears to be a minimum in Fig. 22 because negative potentials are plotted up the vertical axis.) This "turning up" of the ϕ_2 curves at the most positive potentials expresses, in another way, the fact that the specifically adsorbed anions attract cationic counterions into the diffuse layer to maintain the charge balance (cf. Fig. 12). The corresponding plots of the charge attributable to the cations in the double layer for the same solutions are shown in Fig. 23.

Parry and Parsons (*104*) have shown in the case of specifically adsorbed aromatic ions, such as benzene-m-disulfonate ion, that ϕ_2 can exhibit a minimum as well as a maximum. The minimum occurs because a saturation value of the amount of specifically adsorbed organic ion is attained.

6. Other Topics

We have considered in this section a few aspects of the behavior of the electrical double layer at ideal polarized electrodes with emphasis on the practical analysis of experimental data. Much of the scientific importance of specific adsorption data lies in their application to the problem of deciding the correct molecular model for the inner layer. Such discussion

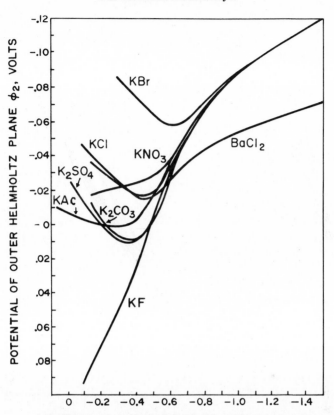

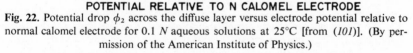

Fig. 22. Potential drop ϕ_2 across the diffuse layer versus electrode potential relative to normal calomel electrode for 0.1 N aqueous solutions at 25°C [from (*101*)]. (By permission of the American Institute of Physics.)

includes the question of the actual potential distribution in the inner layer, the detailed nature of the forces which cause specific adsorption, and the isotherms for specific adsorption. For information on these more advanced aspects of specific adsorption, one may consult Delahay's book (*5*) for an over-all view and the reviews of Parsons (*8,9*) and of Macdonald and Barlow (*15*) for many details. Copious references to the literature of specific adsorption are given in the works cited.

A simplified treatment of adsorption isotherms is given in the next section in connection with the adsorption of neutral substances and some concepts developed there can be applied to the specific adsorption isotherms of ions as well.

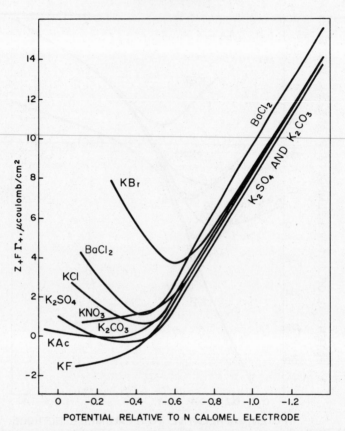

Fig. 23. Component of charge attributable to cations in diffuse layer. 0.1 N aqueous solutions, 25°C at mercury versus potential relative to normal calomel electrode [from (*101*)]. (By permission of the American Institute of Physics.)

E. Adsorption of Neutral Compounds

1. Electrocapillary Curves

The presence in solution of neutral organic compounds can affect, markedly, the properties of the electrical double layer. The basic thermodynamic relationship giving the effect of the adsorption of a neutral compound on an electrocapillary curve is given by Eq. (58):

$$(\partial\gamma/\partial\mu_h)_{T,p,E^\pm,\mu} = -\Gamma_{hh'} \tag{214}$$

Figures 24 and 25 illustrate the marked effect which the presence of a neutral organic substance can have on the electrocapillary curve of an

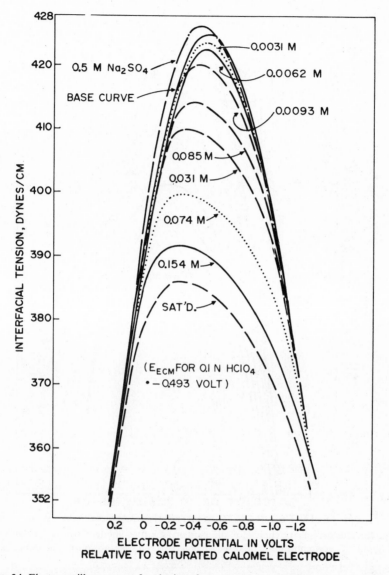

Fig. 24. Electrocapillary curves for the interface mercury-0.1 F HClO$_4$ + amyl alcohol concentration indicated [from (44)]. (By permission of the American Chemical Society.)

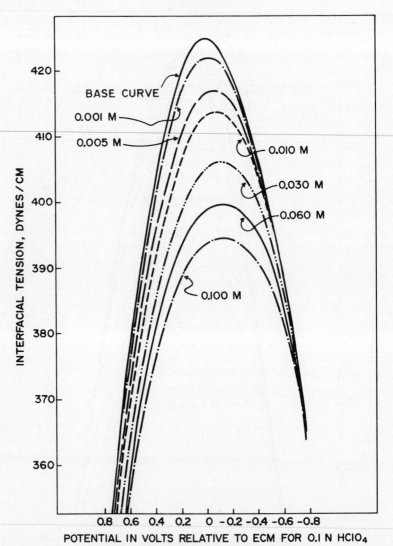

Fig. 25. Electrocapillary curves for the mercury-solution interface with 0.1 F $HClO_4$ plus phenol, concentrations indicated [from (*44*)]. (By permission of the American Chemical Society.)

electrolyte. These figures which are taken from a recent article by Hansen and co-workers (*44*) show the effect of adding various concentrations of *n*-amyl alcohol and phenol, respectively, to a solution which is 0.1 M in $HClO_4$ at an ideal polarized electrode. There is both a lowering of the electrocapillary curve and a shift of the point of zero charge; therefore, one should expect to find a marked Esin and Markov effect (cf. Sec. II.E.4 and below).

The relative surface excess of the neutral component is obtained by making plots of the interfacial tension at constant $E^{\pm}$ versus the logarithm of the concentration (activity) of the neutral substance in solution [cf. Eq. (214)]. Figures 26 and 27 show a series of plots of $\Gamma_{hh'}$ for *n*-amyl alcohol and phenol at various constant values of the electrode potential, which were derived by Hansen et al. from the electrocapillary curves in Figs. 24 and 25. Figure 9 for *n*-butyl alcohol was obtained by Blomgren, Bockris, and Jesch from plots of the same type as shown in Figs. 26 and 27. Such a graph can be derived from the family of $\Gamma - c$ (constant E) curves by plotting values of Γ read along a constant c line (vertical line in Fig. 26 or 27) versus the corresponding potential.

The behavior which is to be noted from Figs. 24 to 27 and Fig. 9 is quite general for the two classes of neutral organic substances, aliphatic and aromatic. These figures show that the adsorption of aliphatic substances is greatest in the vicinity of the point of zero charge and approaches zero at sufficiently negative or positive values of q^M. The adsorption of aromatic neutral substances such as phenol also disappears at sufficiently negative potentials but not at positive potentials. The last behavior, coupled with the fact that the shift of the point of zero charge for substances like phenol is toward more negative potentials with increasing concentration, has led to the conclusion that in the case of adsorption of aromatic compounds there is an interaction between the delocalized π electrons of the aromatic ring and the positively charged metal [cf. Barradas and Conway (*105*)]. At more negative charges this interaction is diminished and eventually disappears, but at positive q^M it is intensified. Aromatic substances are pictured as lying flat on the surface of the metal, whereas aliphatic substances are normally pictured as attached at one end, the other end pointing away from the electrode.

2. Differential Capacity Curves

The presence of a neutral organic substance in solution also affects the shape of the differential capacity curve. In some respects the effect of an organic compound on the morphology of the differential capacity curve

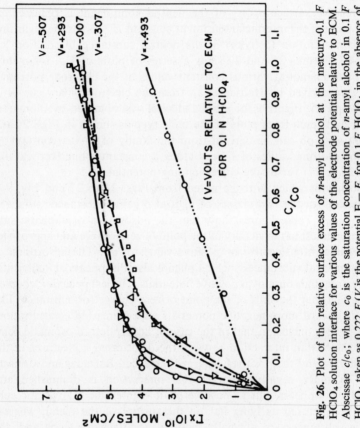

Fig. 26. Plot of the relative surface excess of *n*-amyl alcohol at the mercury-0.1 *F* $HClO_4$ solution interface for various values of the electrode potential relative to ECM. Abscissae c/c_0; where c_0 is the saturation concentration of *n*-amyl alcohol in 0.1 *F* $HClO_4$ taken as 0.222 *F* (*V* is the potential $E - E_z$ for 0.1 *F* $HClO_4$ in the absence of amyl alcohol) [from (*44*)]. (By permission of the American Chemical Society.)

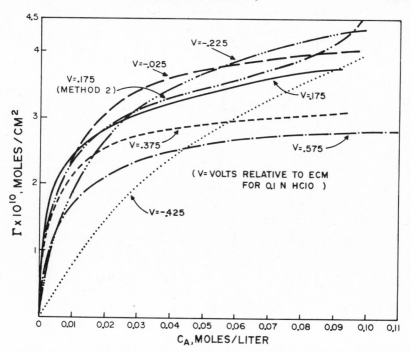

Fig. 27. Relative surface excess of phenol versus concentration for various values of the potential $E - E_z$ for $HClO_4$ in the absence of organic sorbate [from (*44*)]. (By permission of the American Chemical Society.)

is even more striking than on the electrocapillary curve. Figure 28 shows differential capacity curves obtained by Hansen et al. for one of the previously mentioned solutions of amyl alcohol. The dashed line is the differential capacitance curve for the base electrolyte of 0.1 F $HClO_4$. The chief feature to be noticed from these curves is the lowering of the differential capacity from its value in the base electrolyte solution in the vicinity of maximum adsorption. The second general feature is that in the region of desorption, the differential capacity curve shows a peak. The height of these desorption peaks is quite dependent on the frequency v of the ac signal. This phenomenon which is often referred to as the "frequency dispersion" of the differential capacitance does not occur for solutions of simple inorganic salts, at least at mercury electrodes. Quantitative analysis of differential capacity curves for organic substances requires extrapolation to zero frequency, since it is only at this limit that the measured differential capacity equals the thermodynamically defined

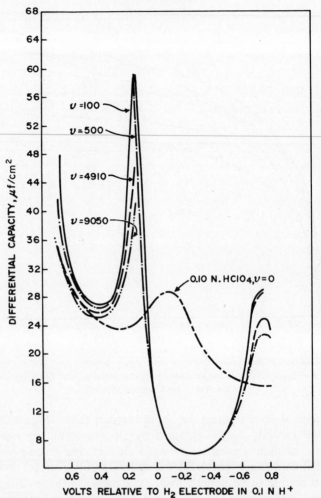

Fig. 28. Differential capacitance for mercury-solution interface with 0.1 F HClO$_4$ + 0.031 M n-amyl alcohol for various frequencies of the ac signal. Dotted line, 0.1 F HClO$_4$ in the absence of organic, extrapolated to zero frequency [from (44)]. (By permission of the American Chemical Society.)

quantity. In general, for aliphatic compounds, there will be two desorption peaks, one on either side of the potential of maximum adsorption, as is shown in Fig. 28. The effect of the adsorption of an aliphatic compound on the differential capacity of a mercury electrode in an aqueous salt solution becomes even more striking as the chain length of the aliphatic compound is increased. Figure 29 shows the results obtained by

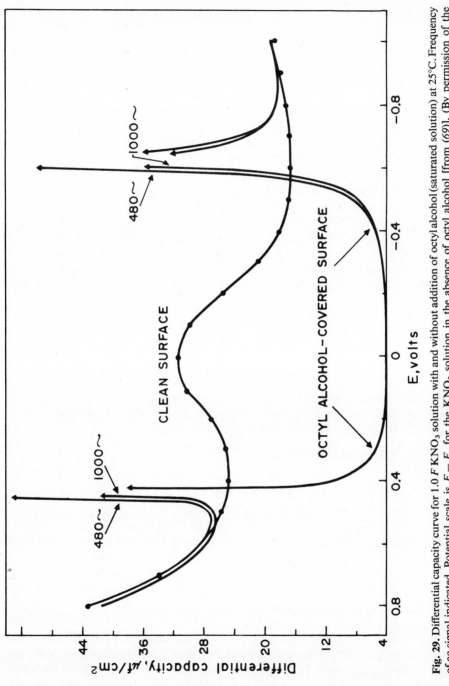

Fig. 29. Differential capacity curve for 1.0 F KNO$_3$ solution with and without addition of octyl alcohol (saturated solution) at 25°C. Frequency of ac signal indicated. Potential scale is $E - E_z$ for the KNO$_3$ solution in the absence of octyl alcohol [from (69)]. (By permission of the American Chemical Society.)

Grahame (69) by saturating a solution containing 1 F KNO$_3$ with octyl alcohol. The frequency of the ac signal is indicated in the figure. It is clear from this figure as well as the previous one that the frequency dependence is greatest in the vicinity of the peaks. In the case of the octyl alcohol the region of lowered differential capacity around the point of zero charge is even broader and flatter than in the case of amyl alcohol. A desorption peak at negative potentials but not at positive potentials is shown in Fig. 30 for the aromatic substance phenol, which shows desorption at negative potentials but no strong desorption at positive potentials within the range of ideal polarizability. The curves shown in Fig. 30 are the extrapolated, that is, zero frequency, differential capacity curves and the concentration of the phenol in the 0.1 F HClO$_4$ solution is indicated. The frequency dependence of the *measured* differential capacity results from adsorption equilibrium not being attained during the measurement. In general, the slow step which inhibits the attainment of adsorption equilibrium is diffusion of the organic material toward the electrode surface. This subject has been discussed by Parsons (9) and will not be considered in detail in this chapter. However, many of the same problems are involved in the treatment of ac polarographic waves in the presence of adsorbed organic substances which are discussed by Smith in this volume. The effect of adsorption on dc polarographic waves has been discussed recently by Maironovskii (106), who gives copious references to related Russian work in this same field. Some elementary ideas on the formulation of the kinetic equations of the adsorption process per se as contrasted with the diffusion of the adsorbing substances is discussed below.

3. Esin and Markov Plots

Since, at constant electrolyte activity, the electrocapillary curves considered above exhibit a shift of the point of zero charge (Esin and Markov effect) with increasing bulk concentration of the adsorbed neutral organic substance, further analysis of the Esin and Markov coefficient is indicated.. The Esin and Markov coefficient for a neutral substance was given by Eq. (81):

$$\left(\frac{\partial E^{\pm}}{\partial \mu_h}\right)_{T,p,q^M,\mu'} = \frac{1}{RT}\left(\frac{\partial E^{\pm}}{\partial \ln a_h}\right)_{T,p,q^M,\mu'} = -\left(\frac{\partial \Gamma_{hh'}}{\partial q^M}\right)_{T,p,\mu}$$

To make an Esin and Markov graph for adsorption of a neutral organic substance one plots the potential $E^{\pm}$ corresponding to any fixed charge density q^M versus the corresponding organic concentration (on a logarithmic scale). Figure 31 shows some Esin and Markov plots prepared from

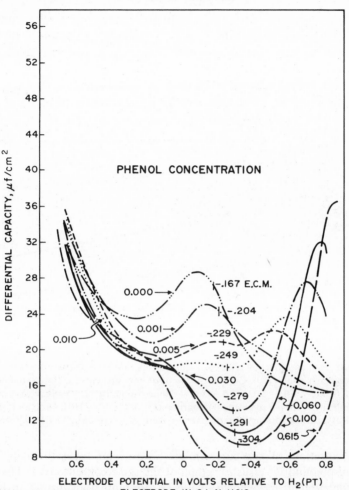

Fig. 30. Differential-capacity curves for mercury-solution interface. Solution is 0.1 *F* HClO₄ + phenol at indicated concentration. Curves are value of differential capacitance extrapolated to zero frequency. Potential measured with respect to hydrogen electrode in 0.1 *F* HClO₄ in the absence of phenol [from (*44*)]. (By permission of the American Chemical Society.)

the data of Blomgren and colleagues (*50*) on the adsorption of *n*-butyl alcohol from aqueous HCl solution (cf. Sec. II.E.3). The most noticeable difference between these Esin and Markov plots and those for specific adsorption of ions (Fig. 20) is the fact that the slope in Fig. 31 varies with charge density. Equation (81) shows that the slope, measured at a particular

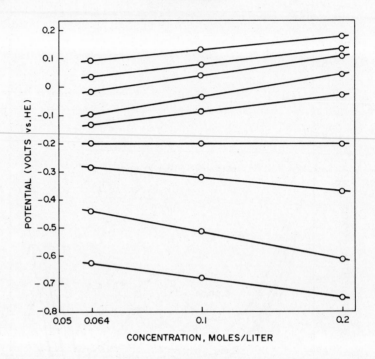

Fig. 31. Esin and Markov plots for n-butyl alcohol in HCl solutions plotted from data of Blomgren et al. (*150*). Constant value of q^M for each curve is (top to bottom): 8.0, 5.0, 3.0, 0.0, -1.0, -2.0, -3.0, -5.0, -8.0 μcoulomb cm^{-2}. Abscissae, concentration in mole liter^{-1} of n-butyl alcohol (logarithmic scale) in 0.1 F HCl. Ordinates: potential of ideal polarized mercury electrode versus hydrogen electrode in 0.1 F HCl [from (*107*)].

concentration, of any curve in Fig. 31 is $-RT$ times the slope of the corresponding Γ versus q^M graph at the same concentration. The slopes of the straight lines shown in Fig. 31 are plotted against q^M in Fig. 32. The corresponding Γ versus q^M plots are shown in Fig. 33. (Actually in Fig. 33, $\theta = \Gamma/\Gamma_s$ is plotted where Γ_s is the saturation, that is, maximum value of Γ.) Comparing Figs. 31 and 32 with Fig. 33, one sees that the Esin and Markov plots can serve as a guideline to the shape of the $\Gamma - q^M$ curves. When the Esin and Markov coefficient is zero, the slope of the Γ versus q^M curve must be zero. When the Esin and Markov coefficient is positive, the slope of the Γ versus q^M curve must be negative. When the Esin and Markov coefficient is negative, the slope of the Γ versus q^M curve must be positive.

It is easy to show (*107*) that, if the Esin and Markov plots for neutral

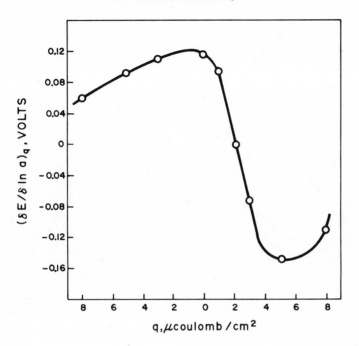

Fig. 32. Esin and Markov coefficient versus q^M for *n*-butyl alcohol solutions in 0.1 *F* HCl. Calculated from Esin and Markov plots in Fig. 31. From data of Blomgren et al. (*50*) [from (*107*)].

substances actually are *truly linear*, then Γ must be the sum of two functions, one of which depends on concentration alone, that is,

$$\Gamma_{hh'} = h(q^M) + g(\mu_h)$$

Of the common isotherms (see Sec. III.E.4.), only the logarithmic Temkin isotherm would fit such a requirement, although the Frumkin isotherm reduces to the logarithmic Temkin isotherm at moderate coverages. More extensive experimental work (*108*) on the adsorption of *n*-amyl alcohol in a series of halide solutions has shown that the Esin and Markov plots appear to be fairly linear in the case of fluoride solutions. Even with these solutions, however, the logarithmic Temkin isotherm was not obeyed. The failure to obey the Temkin isotherm may be readily accounted for by noting that logarithmic plots may give a linear appearance to nonlinear data. Therefore, conclusions as to appropriate isotherm selection based on the linearity of the Esin and Markov plots should be reached only with considerable caution. It can be shown (*107*) that, if the Esin and Markov

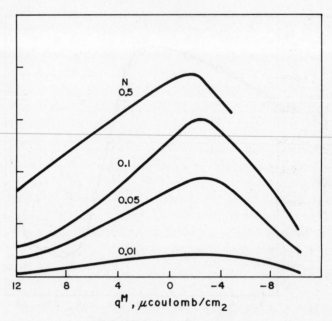

Fig. 33. Variation of coverage $\theta = \Gamma/\Gamma_s$ for n-butyl alcohol versus q^M [from (*21*)]. (By permission of the Royal Society.)

plot is truly linear, a plot of the reciprocal of the differential capacity versus the logarithm of the concentration (activity) of the organic substance at constant electrolyte activity should be linear. Therefore, although the Esin and Markov coefficient provides a valuable qualitative indication of the nature of adsorption, it should be used with caution for quantitative deductions as to that nature.

4. Isotherms

An adsorption isotherm describes the variation, at constant temperature, of Γ_h^i, the amount of adsorbed substance per unit area, with a_h^b, the bulk activity. In nonelectrochemical adsorption, the isotherm is a function only of the concentration. However, at an ideal polarized electrode, one must also take account of the additional electrical variable which determines the state of the system. Thus, an electrochemical adsorption isotherm must show how Γ_h^i varies both with bulk activity and with the electrical state of the system, that is, q^M or E.

a. The Measure of Adsorption. Since only the relative surface excess $\Gamma_{hh'}$ is measurable, the quantity Γ_h^i used in the isotherm, requires further explanation. As in the case of the diffuse layer, the adsorption isotherm is based on a molecular model and Γ_h^i is the measure of adsorption employed in that model. The model which is nearly always used assumes adsorption in a monolayer on the electrode surface. [Melik-Gaikazyan (*109*) has discussed multi-layer adsorption of organic substances.] The symbol Γ_h^i denotes the number of moles of neutral substance *h* adsorbed per square centimeter in the monolayer. The superscript *i* expresses the belief that monolayer adsorption requires the penetration of the adsorbing molecule into the inner layer and that every molecule of substance *h*, which is in the inner layer† is adsorbed. Thus, according to the model, the total amount of substance *h* in the real system, that is, interphase, may be divided into two parts, the number of moles per square centimeter in the inner layer Γ_h^i and the number of moles per square centimeter in the diffuse layer Γ_h^d. The total number of moles in the interphase is given by

$$\Gamma_h = \Gamma_h^i + \Gamma_h^d$$

Because of the assumption that all of substance *h* which is in the inner layer is adsorbed in the monolayer, it is natural and convenient to choose the dividing surface in the Gibbs model, again, at the OHP. Thus, whether we are speaking of ionic adsorption or adsorption of organic neutral molecules the dividing surface is at the OHP. As before, to calculate the surface excess of substance *h* reckoned at the OHP, one must calculate the number of moles of substance *h* in a Gibbs model whose dividing surface is placed at the position of the OHP, that is, x_2. This is given by

$$\Gamma_{hG}^{OHP} = (x_S - x_2)c_h^b$$

where x_s denotes the distance from the metal surface to the bounding surface *WX* of the real system (Fig. 5). The difference $(\Gamma_h - \Gamma_{hG}^{OHP})$ defines the surface excess of *h* reckoned at the OHP, that is,

$$\Gamma_h^{OHP} = \Gamma_h - \Gamma_{hG}^{OHP} = \Gamma_h^i + (\Gamma_h^d - \Gamma_{hG}^{OHP})$$
$$= \Gamma_h^i + {}^d\Gamma_h^{OHP}$$

${}^d\Gamma_h^{OHP}$ is the surface excess reckoned at the OHP of substance *h* as a result of adsorption in the diffuse layer. In the case of neutral substances it is

---

† An adsorbed molecule will always be *counted* as belonging to the inner layer, whether it is entirely there such as an aromatic compound lying flat on the metal surface or whether it is merely attached on one end, the "head," with its "tail" penetrating back into the diffuse layer as might well be the case with a long-chain aliphatic compound.

commonly supposed that the contribution of the diffuse layer to the total adsorption is negligibly small, that is,

$$^{d}\Gamma_{h}^{OHP} = 0$$

Strictly speaking, this assumption is probably incorrect because of the extremely intense electric field in the diffuse layer near the OHP. Nevertheless, we shall neglect the diffuse-layer contribution. Thus

$$\Gamma_{h}^{i} = \Gamma_{h}^{OHP}$$

In comparing experimentally measured quantities and surface concentrations calculated according to a monolayer model, one should, therefore, expect the same type of intrinsic errors as were discussed in the section on the diffuse layer.

b. The Condition of Adsorption Equilibrium. If the monolayer of adsorbed neutral substance h is in equilibrium with the bulk of the solution, one may write

$$^{A}\bar{\mu}_{h} = \mu_{h}^{b} \tag{215}$$

where $^{A}\bar{\mu}_{h}$ is the electrochemical potential of substance h adsorbed in the monolayer and μ_{h}^{b} is the chemical potential of the neutral substance in the bulk of the solution. Formal expansion of Eq. (215) gives

$$^{A}\bar{\mu}_{h}^{0} + RT \ln a_{h}^{A} = \mu_{h}^{0} + RT \ln a_{h}^{b} \tag{216}$$

where $^{A}\bar{\mu}_{h}^{0}$ is the standard electrochemical potential of substance h in the monolayer, a_{h}^{A} is its activity in the monolayer, and μ_{h}^{0} and a_{h}^{b} are the standard chemical potential and bulk activity of h, respectively. Let ΔG^{0}, called the *standard electrochemical free energy of adsorption*, be defined by

$$\Delta G^{0} = {}^{A}\bar{\mu}_{h}^{0} - \mu_{h}^{0} \tag{217}$$

and let [cf. Parsons (*52*)]

$$\beta = e^{-\Delta G^{0}/RT} \tag{218}$$

Combining Eqs. (216) to (218), we have

$$a_{h}^{A} = \beta a_{h}^{b} \tag{219}$$

The last equation is the general form of the adsorption isotherm. It expresses the activity in the adsorbed state as a function of the bulk activity and the electrical state of the system, which is included in β. The problem is now to obtain expressions for a_{h}^{A} in terms either of surface concentration Γ_{h}^{i} or surface pressure. To formulate particular adsorption isotherms, it is useful to think of the interactions of the adsorbed particles in the mono-layer as being divided into two categories: particle-metal interactions and particle-particle interactions. It is reasonable to suppose that the particle-metal interactions are the ones most directly affected by the electrical state of the system. If, at constant temperature, the electrical state is held constant, changes in the amount of adsorption produced by changing the bulk activity of the adsorbing material ought to be determined by the nature of particle-particle interactions. The latter should be determined by essentially the same factors as govern ordinary, that is, nonelectrochemical, adsorption, for example, adsorption of gases on solid or liquid surfaces.

c. Two-Dimensional Equations of State. Just as in nonelectrochemical adsorption, it is convenient to derive electrochemical adsorption isotherms at constant electrical state through the use of two-dimensional equations of state analagous to the various equations of state for three·dimensional gases (real or ideal). A three-dimensional equation of state relates the pressure and the volume of the gas, the temperature and the number of moles. A two-dimensional equation of state relates a two-dimensional pressure, the area, the temperature, and the number of moles. Table III lists corresponding parameters for two- and three-dimensional equations of state.

Parsons (*102*), following the work of Everett (*110*) on adsorption, has examined the form of several possible equations of state for the adsorbed film and has given the form of the corresponding isotherms (Table IV). The isotherm corresponding to each equation of state gives the form of the activity of the adsorbed species a^A, at constant electrical state, as a function of the surface concentration in the monolayer Γ^i or, alternately, the surface pressure. If one is speaking of the adsorption of a neutral substance, the

TABLE III

CORRESPONDING TERMS IN TWO- AND THREE-DIMENSIONAL
EQUATIONS OF STATE

Two-Dimensional		Three-Dimensional	
Name	Symbol	Symbol	Name
Pressure	P	P	Pressure
Area	A	V	Volume
Temperature	T	T	Temperature
Number of moles	n	n	Number of moles
Surface concentration	$\Gamma^i = n/A$	$c = n/V$	Volume concentration
Gas constant	R	R	Gas constant
Molar area	$1/\Gamma^i = A/n$	$v = V/n$	Molar volume

TABLE IV

Two-Dimensional Equations of State and Corresponding Adsorption Isotherms (54,102)

Name	Equation of state	Isotherm — Surface pressure form	Isotherm — Surface concentration form
Henry's law	$p = RT\Gamma^i$	$\beta a^b = p$	$\beta a^b = RT\Gamma^i$
Freundlich	$p = xRT\Gamma^i$	$\beta a^b = p^x$	$\beta a^b = (xRT\Gamma^i)^x$
Volmer	$p = RT\dfrac{\Gamma^i}{(1-b\Gamma^i)}$	$\beta a^b = pe^{bp/RT}$	$\beta a^b = \dfrac{RT\Gamma^i}{(1-b\Gamma^i)}\exp\left[\dfrac{b\Gamma^i}{(1-b\Gamma^i)}\right]$
van der Waals	$p = \dfrac{RT\Gamma^i}{(1-b\Gamma^i)}+g(\Gamma^i)^2$		$\beta a^b = \dfrac{RT\Gamma^i}{(1-b\Gamma^i)}\exp\left[\dfrac{b\Gamma^i}{(1-b\Gamma^i)}\right]\times\exp\left(\dfrac{2g\Gamma^i}{RT}\right)$
Virial	$p = RT\Gamma^i + g(\Gamma^i)^2$	$\beta a^b = \dfrac{RT}{2g}\left[\left(1+\dfrac{4gp}{R^2T^2}\right)^{\frac{1}{2}}-1\right]\times\exp\left[\left(1+\dfrac{4gp}{R^2T^2}\right)^{\frac{1}{2}}-1\right]$	$\beta a^b = \Gamma^i\exp\left(\dfrac{2g\Gamma^i}{RT}\right)$
Langmuir[a]	$p = -RT\Gamma_s\ln\left(1-\dfrac{\Gamma^i}{\Gamma_s}\right)$	$\beta a^b = e^{p/(RT\Gamma_s)}-1$	$\beta a^b = \dfrac{\Gamma^i}{\Gamma_s-\Gamma^i}=\dfrac{\theta}{1-\theta}$
Logarithmic Temkin	$p = g(\Gamma^i)^2$	$\beta a^b = \exp\left[\left(\dfrac{4gp}{R^2T^2}\right)^{\frac{1}{2}}\right]$	$\beta a^b = \exp\left(\dfrac{2g\Gamma^i}{RT}\right)$
Frumkin	$p = -RT\Gamma_s\left(1-\dfrac{\Gamma^i}{\Gamma_s}\right)+g(\Gamma^i)^2$		$\beta a^b = \dfrac{\Gamma^i}{\Gamma_s-\Gamma^i}\exp\left(\dfrac{2g\Gamma^i}{RT}\right)$
"Square root"	$\dfrac{1}{\Gamma^i} = \dfrac{RT}{p}+\left(\dfrac{g}{p}\right)^{\frac{1}{2}}$	$\beta a^b = p\exp\left[\left(\dfrac{4gp}{R^2T^2}\right)^{\frac{1}{3}}\right]$	
Modified H. F. L.[a] [Helfand, Frisch, and Lebowitz, modified by Parsons (54)]	$p = RT\dfrac{\Gamma^i}{(1-b\Gamma^i)^2}+g(\Gamma^i)^2$		$\beta a^b = \dfrac{\theta}{1-\theta}\times\dfrac{b}{0.907}\exp\left[g\theta+\dfrac{1}{1-\theta}+\dfrac{1}{(1-\theta)^2}-2\right]$

$^a\ \Gamma_s$ = saturation value of Γ^i; $\theta = \Gamma^i/\Gamma_s$.

activity referred to is the activity of that neutral substance in the mono-layer. The same formalism may be used for specific ionic adsorption, where a^A corresponds to the activity of the adsorbed ion in the mono-layer and a^b corresponds to the activity of the salt whose concentration is being varied.

Henry's law amounts to assuming that the adsorbed film behaves as a two-dimensional ideal gas, that is, without interaction between the molecules. For all adsorbed substances in the limit of infinite dilution the monolayer will approach Henry law behavior. At higher concentrations in the monolayer, particle-particle interactions must be taken into account.

Since it may not be entirely evident from the table how the isotherm is derived from the equation of state, we shall illustrate how this may be done in a simple way for three equations of state, the Henry law, the van der Waals equation of state, and the Volmer equation of state.

Before we begin the derivation, a few preliminary remarks on the thermodynamics of gases will be in order (*111*). First, since we are com-paring adsorbed particles in a monolayer film to the molecules of a two-dimensional gas and since we are considering the adsorption of only one kind of substance the analogy may be made to a pure gas, that is, a gas composed of a single chemical substance. For any pure chemical substance the Gibbs free energy may be expressed as a function of the temperature and pressure by

$$dG = -S\,dT + V\,dp$$

and

$$\left(\frac{\partial G}{\partial p}\right)_T = V$$

where V is the volume of the sample of the gas containing n moles. For 1 mole we have, by definition

$$\mu = \frac{G}{n}$$

$$v = \frac{V}{n}$$

so that

$$(\partial\mu/\partial p)_T = v \tag{220}$$

where v is the molar volume of the gas at the specified temperature. But the chemical potential of any substance is defined in terms of its fugacity f by an equation of the form

$$\mu = RT \ln f + B(T) \tag{221}$$

where B for a particular substance is a function of the temperature alone. The fugacity is defined in such a manner that it approaches the pressure as the pressure approaches zero, that is,

$$\lim_{p \to 0} f/p = 1 \tag{222}$$

For an ideal gas, the fugacity is equal to its pressure, and for any real gas the fugacity approaches the pressure as the pressure approaches zero, that is, the gas approaches ideality. The activity of any substance is defined as

$$a = f/f^0$$

where f^0 is the fugacity in some chosen standard state. For gases the chosen standard state is the ideal gas at one atmosphere. Thus, for all gases,

$$f^0 = 1 \text{ atm}$$

Hence,

$$a = f/f^0 = f/1 = f \tag{223}$$

Our problem, therefore, reduces simply to finding the appropriate equation for the fugacity, that is, activity, of a gas obeying the particular equation of state under consideration, and then to make the correspondence to the two-dimensional case by means of Table III. Since $\ln f \to -\infty$ as $p \to 0$, it is more convenient for practical derivations to consider the ratio f/p. From Eqs. (220) and (221), we have

$$\left(\frac{\partial \ln f}{\partial p} \right)_T = \frac{v}{RT}$$

Therefore,

$$\left(\frac{\partial \ln\left(\frac{f}{p}\right)}{\partial p} \right)_T = \frac{v}{RT} - \frac{1}{p} = \frac{1}{RT}\left(v - \frac{RT}{p} \right) \tag{224}$$

The derivation of the form of the isotherm for any equation of state may now be accomplished by integrating Eq. (224) from zero pressure to pressure p:

$$\int_0^p \left(\frac{\partial \ln (f/p)}{\partial p} \right)_T dp = \ln f - \ln p = \frac{1}{RT} \int_0^p \left(v - \frac{RT}{p} \right) dp$$

Hence,

$$\ln f = \ln p + \frac{1}{RT} \int_0^p \left(v - \frac{RT}{p} \right) dp \tag{225}$$

The quantity on the left side is the logarithm of the fugacity (activity) of the gas which obeys the given equation of state. Making the correspondence to the two-dimensional case, f becomes the quantity a^A and may be set equal to βa^b to give the form of the isotherm.

EXAMPLES

1. HENRY'S LAW (IDEAL GAS)

Equation of state:

$$pv = RT$$

$$\ln f = \ln p + \frac{1}{RT} \int_0^p \left(\frac{RT}{p} - \frac{RT}{p} \right) dp$$

$$f = p = \frac{RT}{v} \tag{226}$$

Making the correspondence (Table III) to the two-dimensional case, we have, noting that v corresponds to $1/\Gamma^i$,

$$f = a^A = RT\Gamma^i$$

Therefore,

$$\beta a^b = RT\Gamma^i$$

Alternately, if one wishes to express the isotherm in terms of the surface pressure, one has simply from Eq. (226)

$$\beta a^b = p$$

2. VAN DER WAALS EQUATION†

$$p = \frac{RT}{v - b} + \frac{g}{v^2}$$

$$dp = -\frac{RT}{(v - b)^2} dv - \frac{2g}{v^3} dv$$

---

† We have written the van der Waals equation in a somewhat unfamiliar way by replacing $-a/v^2$ by g/v^2. When g is negative, this corresponds to the usual type of van der Waals gas with intermolecular attractions. Positive g would correspond to a gas obeying an equation of the same form but for which the intermolecular forces are repulsive. The notation g was introduced by Delahay (5).

Substituting for dp in Eq. (225), we obtain

$$\ln f = \ln p + \frac{1}{RT} \int_0^p v \left[-\frac{RT}{(v-b)^2} - \frac{2g}{v^3} \right] dv - \int_0^p d \ln p$$

or

$$\ln f = \ln p - \ln[p(v-b)] \Big|_0^p + \frac{b}{(v-b)} \Big|_0^p + \frac{2g}{v} \Big|_0^p \qquad (227)$$

The lower limits may be evaluated by noting

$$\lim_{p \to 0} v = \infty$$

$$\lim_{p \to 0} \left(\frac{b}{v-b} \right) = 0$$

$$\lim_{p \to 0} \left(\frac{2g}{v} \right) = 0$$

$$\lim_{p \to 0} p(v-b) = pv = RT$$

The last limit expresses the fact that the gas approaches ideality as $p \to 0$. Substituting these limits into Eq. (227), we obtain

$$\ln f = \ln \left(\frac{RT}{v-b} \right) + \frac{b}{v-b} + \frac{2g}{RTv}$$

or

$$f = \frac{RT}{v-b} e^{b/(v-b)} e^{2g/RTv} \qquad (228)$$

Now, making the correspondence to the two-dimensional case via Table III, we obtain

$$f = a^A = \frac{RT\Gamma^i}{1 - b\Gamma^i} \exp\left(\frac{b\Gamma^i}{1 - b\Gamma^i} \right) \exp\left(\frac{2g\Gamma^i}{RT} \right)$$

Thus

$$\beta a^b = \frac{RT}{1 - b\Gamma^i} \exp\left(\frac{b\Gamma^i}{1 - b\Gamma^i} \right) \exp\left(\frac{2g\Gamma^i}{RT} \right)$$

By the same technique used in the case of the ideal gas, we could also write the isotherm in terms of surface pressure by substituting for pressure by means of the van der Waals equation into Eq. (228). Unfortunately, in this case the result is very cumbersome, since solving for v involves solving a cubic equation. To illustrate fully how this procedure works for a case where one can obtain either the surface concentration or the surface pressure version of the isotherm in a relatively simple form, we conclude with the example of the Volmer gas.

3. VOLMER GAS

Equation of state:

$$p = \frac{RT}{v - b}$$

Solution for activity:

$$f = \frac{RT}{v - b} e^{b/(v-b)} \tag{229}$$

Expressing in surface concentration:

$$\beta a^b = \frac{RT\Gamma^i}{1 - b\Gamma^i} \exp\left(\frac{b\Gamma^i}{1 - b\Gamma^i}\right)$$

Solving the equation of state for $v - b$ and substituting in Eq. (229), we have

$$f = \beta a^b = p \, e^{bp/RT}$$

d. The Electrical Parameter. To deduce the nature of the particle-particle interactions in an adsorbed monolayer, it is necessary to compare experimental adsorption curves with those predicted by isotherms of the type discussed above. These isotherms are based on the assumption of constant electrical state. The experimental data are derived (in principle) from a family of electrocapillary curves, each of which is obtained at a different constant bulk activity of the adsorbing substance. The comparison of experiment and model is made by selecting points corresponding to the same electrical state from each member of the family of curves. There seems to be little question that the electrical parameter which should be held constant for such correlations is the electric-field operative in the adsorbed monolayer. Parsons (53) has suggested that a reasonable model for this electric field is that resulting from the addition of the field which would result from the charge q^M if the inner layer were vacuum (Gauss' law) and an opposing field produced by orientation of the dipoles of adsorbed solvent and solute molecules. On this basis it would appear that control of q^M would permit control of the effective field. Parsons has uniformly preferred to study adsorption at constant q^M (40,52–54,104,112, 113) and has obtained useful correlations by this method. He has pointed out that the choice of q^M as the electrical variable to hold constant during analysis of adsorption curves corresponds more closely to the case of nonelectrochemical adsorption, since in such systems the value of charge density is always zero. Damaskin (57), on the other hand, has preferred to hold the potential of the electrode constant when dealing with adsorbed neutral organic molecules [cf. Damaskin and Tedoradze (114)]. The question of which of the two practical electrical variables q^M or E should be held constant to keep the electrical state of the monolayer constant is not completely resolved. However, at the present stage it would appear that q^M is the better choice.

The linear Esin and Markov plots which are obtained for specifically adsorbed halide ions (Fig. 20) can be explained readily on the assumption of a Temkin isotherm at constant charge. For neutral organic substances, there is, perhaps, still some room for argument in favor of choosing constant E. However, a recent comparison of the adsorption of n-butyl alcohol in fluoride and chloride solutions by Parsons (53) argues strongly for the rationality of holding q^M constant also for neutral organic substances.

The difference between constant q^M or constant E isotherms is illustrated by using the surface-pressure form of the isotherm. At constant E the surface pressure π resulting from the adsorbed monolayer is the lowering of the surface tension from its value in the base electrolyte solution. Thus

$$\pi = \gamma_0 - \gamma$$

where γ_0 is the interfacial tension at the given value of E of the base electrolyte solution alone and γ is the interfacial tension at the same E with the organic species present. The surface pressure at constant charge Φ is defined in terms of Parsons' function ξ:

$$\Phi = \xi_0 - \xi = (\gamma_0 - \gamma) + q^M(E_0 - E) \qquad (230)$$

where ξ_0 is the value of the function ξ in the absence of organic solute at the given value of q^M. The term $\gamma_0 - \gamma$ in Eq. (230) is not the surface pressure π, because the value of E is not being held constant.

Parsons plotted values of the surface pressure π versus electrode potential at constant concentration of n-butyl alcohol for the KF and KCl base electrolyte solutions and compared them to similar curves obtained by plotting the function Φ again q^M for the same solutions. The results are shown in Figs. 34 and 35. One sees that the $\Phi - q^M$ curves in Fig. 35 for both base electrolyte solutions have essentially the same shape and that their maxima lie at almost the same values of q^M (-2 μcoulomb cm^{-2}). On the other hand the $\pi - E$ curves have a quite different shape in the presence

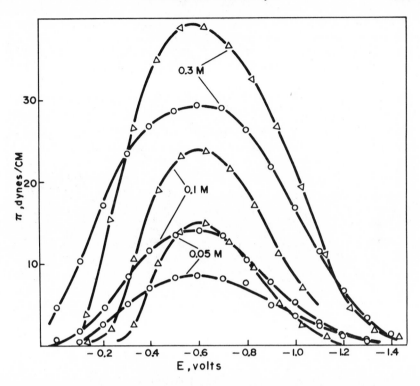

Fig. 34. Surface pressure π versus electrode potential E for n-butyl alcohol. $\bigcirc$, base solution 0.1 F KF; $\triangle$, base solution 3 F KCl [from (53)]. (By permission of Elsevier).

or absence of specifically adsorbed ion. This dependence on base electrolyte may be explained on the following basis. Since the specific adsorption of Cl^- raises the differential capacity of the double layer, a smaller change in E will produce the same change in q^M in the region of strong specific adsorption (more positive potentials). Thus, if q^M is the controlling variable, one would predict that in the region of strong Cl^- adsorption the π versus E curve should rise more rapidly than the corresponding curves for the F^- solution.

5. Effect of Adsorption of Neutral Substances on ϕ_2

Despite its obvious importance in the study of double-layer effects in many electrode reactions which occur in the presence of adsorbed organic inhibitors, very little work has been done on the effect of the adsorption of neutral organic substances on the diffuse layer [however, see Tidwell

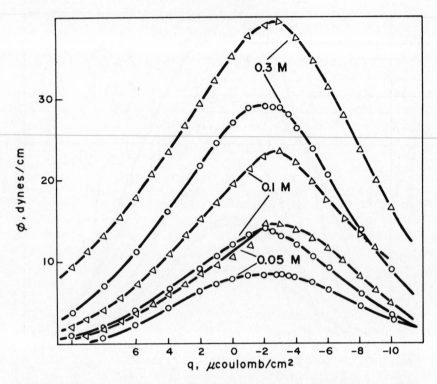

Fig. 35. Surface pressure Φ versus q^M for *n*-butyl alcohol. $\bigcirc$, base solution 0.1 *F* KF; $\triangle$, base solution 3 *F* KCl [from (*53*)]. (By permission of Elsevier.)

(*108*)]. Parsons (*9*) has indicated how the effect of the adsorption of a neutral substance on the potential ϕ_2 of the OHP could be calculated in the absence of specific adsorption. One would simply integrate the differential capacity curve obtained in the presence of the adsorbed neutral species from the potential of interest to the point of zero charge.† Since there is often a considerable potential range over which the differential capacity in the presence of an adsorbed organic substance is low and nearly constant (cf. Figs. 28 to 30), an approximation of the effect can be made by assuming a constant value for the differential capacity and multiplying this value by the *rational potential*,* that is, $E - E_z$, to obtain the charge q^M. Substitution

---

† It should be noted, incidentally, that the difference of potential between the measured value and the point of zero charge in the absence of specific adsorption was called, by Grahame, the *rational potential* and that this quantity is used by many authors in their discussions.

of q^M into the GCS equations will, by the usual methods given above, yield ϕ_2.

Figure 36, which was calculated by Parsons on this basis, assuming a constant differential capacity of the inner layer of 4 μfarads cm^{-2}, shows that the effect would be merely a lowering of the ϕ_2 curves but not a change in their shape. In the presence of specific adsorption the situation would be more complicated. It would be expected that the ϕ_2 curves would still be essentially of the same shape as shown in Fig. 22. However, the variation of ϕ_2 with E would certainly be smaller throughout the range where the differential capacity was lowered by the presence of the organic substance. Aramata and Delahay (*91*) have studied the effect of adsorption of *n*-amyl alcohol on the kinetics of the zinc-zinc amalgam discharge reaction and calculated ϕ_2 via the GCS theory in an exact manner by using the charges obtained from electrocapillary curves measured in the presence of the amyl alcohol.

6. Adsorption Kinetics

a. Introduction. Although most work to date on adsorption of neutral substances at the metal-solution interface has been concerned with the equilibrium situation, that is, thermodynamics, a complete understanding also requires a knowledge of the kinetics of the adsorption process. A knowledge of the rates of adsorption processes is also important in many studies in electrode kinetics, that is, kinetics of charge-transfer processes (cf., for example, Smith, Chapter 1, who discusses the importance of this problem in connection with ac polarography).

Since physical adsorption processes are very rapid, the rate of adsorption of a neutral substance at a metal electrode is usually controlled by mass transfer of the adsorbing substance from the bulk of the solution rather than by the adsorption process itself. Adsorption kinetics with complete or partial control by diffusion has been considered by Delahay and co-workers (*115–117*), Frumkin and Melik-Gaikazyan (*109,118*), and Lorenz and Möckel (*119,120*). For a review of this work, see Parsons (*9*). In none of these works, however, was a particular form of the adsorption rate equation given. Apparently only one attempt has been made to derive a general form of the rate equation for an adsorption process at an ideal polarized electrode (*55*). The last theory has many similarities in its formalism to the classical theory of electrode kinetics [cf. Parsons (*121*)]. It was developed by Delahay and Mohilner only for the particular case of an adsorption process obeying both the logarithmic Temkin isotherm and Temkin kinetics. However, it is capable of much greater generalization.

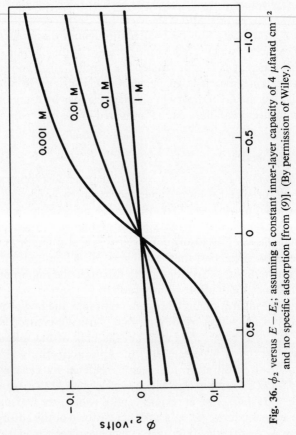

Fig. 36. ϕ_2 versus $E - E_z$; assuming a constant inner-layer capacity of 4 μfarad cm^{-2} and no specific adsorption [from (9)]. (By permission of Wiley.)

The purpose of this section is to indicate the basic ideas behind the theory and to stress those portions of the theory which do not depend on the particular type of isotherm or kinetics under consideration. In principle, nearly any isotherm or form of kinetics should be amenable to this type of treatment (*122*).

b. Formal Expression of Adsorption Reaction. The adsorption of a neutral substance h at the ideal polarized electrode is represented formally by the reaction†

$$h + S \rightleftharpoons A \qquad (231)$$

where h represents a neutral substance in solution, S represents a "site" on the electrode surface, and A represents the adsorbed species in the monolayer. It should be emphasized from the outset that Eq. (231) is merely a formalism and does not necessarily imply the existence of fixed adsorption sites on the surface of the electrode. On a homogenous electrode surface, for example, mercury, S could imply merely the obvious fact that adsorption of neutral substance h probably involves the replacement of an adsorbed molecule or group of molecules of solvent which is then indicated by the symbol S. The condition of adsorption equilibrium under this formalism is given by

$$\mu_h + \bar{\mu}_S = \bar{\mu}_A \qquad (232)$$

where μ_h represents the chemical potential of the neutral substance h in the bulk of the solution. [At equilibrium only the activity (chemical potential) in the bulk of the solution need be considered.] $\bar{\mu}_S$ represents the electrochemical potential of the site on the electrode (or adsorbed solvent in the monolayer), and $\bar{\mu}_A$ is the electrochemical potential of the adsorbed neutral species in the monolayer. Equation (232) appears to be different from the previous equilibrium condition discussed in Sec. III.E.4.b in connection with isotherms, namely,

$$\mu_h^b = {}^A\bar{\mu}_h \qquad (215)$$

However, the two equations are completely equivalent. Insofar as one is discussing only the equilibrium situation, a simpler formalism than Eq. (232) will suffice, and the formal adsorption reaction implied in Sec. III.E.4.b was

$$h \rightleftharpoons A \qquad (233)$$

---

† This formalism was inspired by the order-disorder theory of adsorption (*123*).

which led to the equilibrium condition, Eq. (215). In the discussion of kinetics, however, as we shall see, Eq. (231) has advantages. Comparing Eqs. (215) and (232), one sees that

$$^A\bar{\mu}_h = \bar{\mu}_A - \bar{\mu}_S = \mu_h^b \tag{234}$$

When all species are in their standard states, Eq. (234) becomes

$$^A\bar{\mu}_h^0 = \bar{\mu}_A^0 - \bar{\mu}_S^0 = \mu_h^0$$

Expanding Eq. (234) in terms of standard chemical potentials and activities, one then obtains

$$a_h^A = \frac{a_A}{a_S} \tag{235}$$

where a_A and a_S represent the activities of species A and S on the electrode surface (new formalism), and a_h^A is the activity of adsorbed species h in the monolayer (old formalism). Equation (235) relates in a convenient way the two formalisms for the adsorption process, and, moreover, it enables one to fit the new formalism into the system of equations of state and isotherms considered earlier. Thus

$$\frac{a_A}{a_S} = a_h^A = \beta a_h^b$$

and thus the *ratio* a_A/a_S at constant electrical state is given for any particular type of adsorption equilibrium process by the corresponding entries in Table IV.

While the ratio a_A/a_S must be consistent with the isotherm in question, the individual forms of a_A and a_S do not come from the isotherm but come from the particular mode of adsorption kinetics assumed. Customarily, one associates one particular formulation for a_A and a_S with each type of isotherm; for example, for the Langmuir isotherm, a_S is normally formulated as $1 - \theta$ and a_A as θ, where θ is the coverage of the surface. However, it should be realized that other possibilities exist.

Using Eq. (231), we now postulate that the adsorption reaction takes place in the following stages. First a molecule of neutral substance h moves from the bulk of the solution to a position close to the electrode surface, the *preadsorption state*. The preadsorption state is probably very close to the electrode surface, for example, at the OHP. Thus, one would expect that, strictly speaking, a correction of the Frumkin type, which is always required in electrode kinetics, would have to be made. However, this detail is ignored in the present treatment. (If the interaction of the neutral substance with the electric field in the diffuse layer should have to be taken into account, this consideration would not alter the basic ideas involved in the kinetic treatment.)

The activity of the adsorbing substance in the preadsorption state is, therefore, postulated to be equal to the activity in the bulk solution unless there is some control by mass transfer. In that case the activity (concentration) in the preadsorption state may be calculated by solving Fick's equation. For the case of logarithmic Temkin adsorption, the mass-transfer problem with semiinfinite linear diffusion has been worked out by Delahay (*124*) (cf. Sec. e, below).

Only after the molecule of h is in the preadsorption state can it be adsorbed. We postulate that the adsorption process involves an activation energy and apply absolute-reaction-rate theory. The process of adsorption, then, is represented by the passage of the adsorbing system along a reaction coordinate and over a standard electrochemical free energy barrier, curve G^0 in Fig. 37.

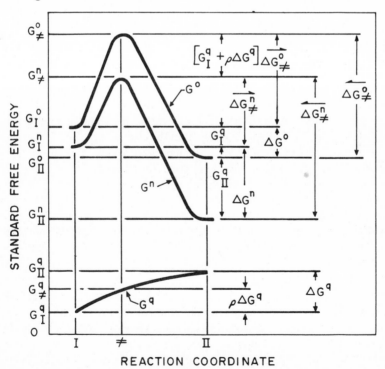

Fig. 37. Standard free energies versus reaction coordinate for adsorption process [from (*55*)]. (By permission of the American Chemical Society.)

In Fig. 37 position I represents the system in the preadsorption state, $\neq$ represents the transition state (activated complex), and II represents the system after the molecule has been adsorbed in the monolayer.

c. Charge Dependence of Rate Constants.

The fundamental idea of this theory of adsorption kinetics is that the rate constants for adsorption and desorption both depend upon the electrical state of the system. The form of this dependence may best be understood by referring to Fig. 37. The standard electrochemical free energy of the adsorbing system, curve G^0, may be written as the sum of two parts, one of which is independent of the electrical state G^n and the other of which depends upon the electrical state, G^q. Following the ideas of Parsons, the independent electrical variable is chosen to be the charge density on the metal q^M. One may then write

$$G^0 = G^n + G^q \tag{236}$$

The electrochemical free energy of the system is represented in Fig. 37 as passing over a free-energy barrier corresponding to the transition state. Curve G^n represents the corresponding change in the charge-independent part of G^0, and the curve G^q represents the corresponding charge dependent part. It will be noted that, although curves G^0 and G^n both have humps at the state corresponding to the activated complex $\neq$, curve G^q does not. The shape of curve G^q expresses for adsorption the same postulate which is made in electrode kinetics, namely, that the electrical part of the standard electrochemical free energy of a system undergoing a reaction at an electrode is a monotonic function along the reaction coordinate. This postulate, which may be called the *intrinsic postulate of electrode kinetics*, implies that the change in the electrical part of the standard electrochemical free energy of adsorption in passing from the preadsorption state I to the transition state $\neq$ is some fraction $(0 < \rho < 1)$, of the total change ΔG^q in passing from the preadsorbed state to the adsorbed state. Thus, the quantity ρ which is called the *charge parameter*, plays the same role in adsorption kinetics as the transfer coefficient α plays in electrode kinetics, and ΔG^q, the charge-dependent part of the standard electrochemical free energy of adsorption ΔG^0, plays the same role in adsorption kinetics as the electrode potential E does in electrode kinetics. ΔG^0 and changes is ΔG^q are accessible from equilibrium experiments.

Applying this adsorption analog of the intrinsic postulate of electrode kinetics, we can express the forward (adsorption) and backward (desorption) rate constants as the product of two factors, one charge in-

dependent, the other charge dependent. Thus, for the adsorption proc.
we write

$$\overrightarrow{k} = k_f \exp\left(-\rho\,\frac{\Delta G^q}{RT}\right) \tag{237}$$

and for the desorption process

$$\overleftarrow{k} = k_b \exp\left[(1-\rho)\,\frac{\Delta G^q}{RT}\right] \tag{238}$$

where k_f and k_b are charge independent.

The last equations were derived (55) by applying the well-known method of Parsons (121) originated for the treatment of electrode kinetics [see also Mohilner (125)].

d. Rate Equations for the Adsorption Process. The rates of the adsorption and desorption processes may now be written as follows:

Adsorption:

$$\overrightarrow{v} = \overrightarrow{k}a_h a_s = k_f \exp\left(-\rho\,\frac{\Delta G^q}{RT}\right)a_h a_s \tag{239}$$

Desorption:

$$\overleftarrow{v} = \overleftarrow{k}a_A = k_b \exp\left[(1-\rho)\,\frac{\Delta G^q}{RT}\right]a_A \tag{240}$$

It is clear from these two equations that, at constant charge density, the rate of the adsorption process varies only with the activities of substance h at the preadsorption state and of the sites S, and that the rate of the desorption process varies only with the activity of the species A.†

---

† Strictly speaking, Eq. (240) which follows from the formal adsorption reaction, Eq. (231), is still an oversimplification. If one assumes that S actually represents a group of n adsorbed solvent molecules, h', which are replaced by one molecule of h, then the formal adsorption process could be written

$$h + S \rightleftharpoons A + nh'$$

Using this equation, the desorption rate equation, Eq. (240), would become

$$\overleftarrow{v} = \overleftarrow{k}a_A a_{h'}^n$$

Thus k_b in Eq. (240) should strictly be considered a psuedo-first-order desorption rate constant. It is also possible that in a detailed treatment of adsorption kinetics one would have to take into account the possibility that n, the number of adsorbed solvent molecules replaced by a molecule of h, would vary with q^M. The last statement might be true, for example, in cases in which the orientation of the adsorbing h molecule varied greatly with q^M.

vation that one applies particular adsorption
ctrochemical surface chemistry. The adsorption
.e form of the ratio a_A/a_S as a function of Γ_h^i at
.rticular form of adsorption kinetics determines the
.ial forms of a_A and a_S as functions of Γ_h^i.
, the equations for gas-adsorption kinetics derived by
, one can express a_S and a_A at any fixed value of q^M by

$$a_S = \exp\left(-\frac{\lambda b\Gamma_h^i}{RT}\right)$$

$$a_A = \exp\left[\frac{(1-\lambda)b\Gamma_h^i}{RT}\right]$$

where $0 < \lambda < 1$ and b is the quantity $2g$ appearing in the isotherm in Table IV.

Substituting for a_S in the adsorption rate equation and for a_A in the desorption rate equation, we then obtain

$$\overrightarrow{v} = k_f a_h \exp\left(-\frac{\lambda b\Gamma_h^i}{RT}\right) \exp\left(-\rho\,\frac{\Delta G^q}{RT}\right) \tag{241}$$

$$\overleftarrow{v} = k_b \exp\left[\frac{(1-\lambda)b\Gamma_h^i}{RT}\right] \exp\left[(1-\rho)\frac{\Delta G^q}{RT}\right] \tag{242}$$

The Temkin isotherm, Table IV, can be written in the form

$$b\Gamma_h^i = RT \ln a_h^b - \Delta G^0$$

At equilibrium the value of Γ_h^i appearing in the rate equation for adsorption and desorption is the same as the value for Γ_h^i given in the adsorption isotherm, and, moreover,

$$\overrightarrow{v} = \overleftarrow{v} = v^0$$

where v^0 is called the *adsorption-exchange rate*, by analogy with the exchange current in electrode kinetics and has dimensions of moles per square centimeter per second. Substituting the equilibrium value of Γ_h^i into Eqs. (241) and (242) and rearranging, one obtains

$$v^0 = k^0 (a_h^b)^{1-\lambda} \exp\left[(\lambda - \rho)\frac{\Delta G^q}{RT}\right] \tag{243}$$

where k^0, the *standard adsorption rate constant*, comparable to the standard rate constant in electrode kinetics, is given by

$$k^0 = k_f \exp\left(\frac{\lambda \Delta G^n}{RT}\right) = k_b \exp\left[-\frac{(1-\lambda)\Delta G^n}{RT}\right]$$

The exchange rate v^0, the coverage parameter λ, and the charge parameter ρ, can be experimentally determined. k^0 cannot be experimentally determined.

If another type of adsorption isotherm and kinetics had been assumed, one would obtain equations analogous to the last equations, but different in their exact forms.

If one now considers an electrode which is initially at adsorption equilibrium, the value Γ_h^i becomes $(\Gamma_h^i)_{eq}$. If one considers further a perturbation, $\delta(\Delta G^q)$, on $(\Delta G^q)_{eq}$ such that

$$\delta(\Delta G^q) = \Delta G^q - (\Delta G^q)_{eq}$$

then a corresponding change in the adsorption will be produced according to the rate equation, the change being $\delta\Gamma_h^i$. For the rate of adsorption, we have

$$v = \frac{d\Gamma_h^i}{dt} = \frac{d(\delta\Gamma_h^i)}{dt} = v^0\left\{\exp\left(-\frac{\lambda b \delta\Gamma_h^i}{RT}\right)\exp\left(-\rho\frac{\delta(\Delta G^q)}{RT}\right)\right.$$
$$\left. -\exp\left[\frac{(1-\lambda)b\delta\Gamma_h^i}{RT}\right]\exp\left[\frac{(1-\rho)\delta\Delta G^q}{RT}\right]\right\}$$
(244)

One notes that $\delta(\Delta G^q)$ plays the same role in adsorption kinetics as the overvoltage η in electrode kinetics. The initial net rate of adsorption $v_{t=0}$ is given by replacing $\delta\Gamma_h^i$ by zero in Eq. (244), since at time, $t = 0$, $\delta\Gamma_h^i$ will be equal to zero, and

$$v_{t=0} = v^0\left\{\exp\left[-\rho\frac{\delta(\Delta G^q)}{RT}\right] - \exp\left[\frac{(1-\rho)\delta(\Delta G^q)}{RT}\right]\right\}$$
(245)

One notes that Eq. (245) is of precisely the same form as the classical equation for electrode kinetics, where v^0 plays the role of the exchange current, ρ plays the role of the transfer coefficient, and $\delta(\Delta G^q)$ plays the role of the overvoltage. A plot of initial rate of adsorption versus $\delta(\Delta G^q)/RT$ will, therefore, have exactly the same shape as a plot of charge-transfer-controlled current versus overvoltage. Such a graph is shown in Fig. 38 for

D. M. Mohilner

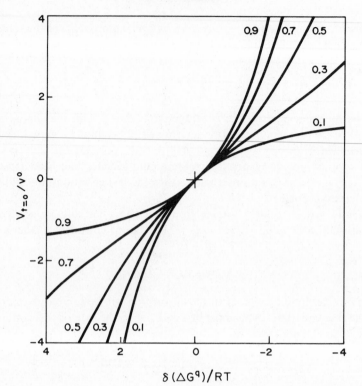

Fig. 38. Variation of $v_{t=0}/v^0$ with $\delta(\Delta G^q)/RT$ according to Eq. (245) [from (55)]. (By permission of the American Chemical Society.)

various values of the charge parameter ρ. The analogy between $\delta(\Delta G^q)$ and overvoltage leads immediately to the idea of a coulostatic method for adsorption (55) as the analog of the potentiostatic method in electrode kinetics.

For small departures from equilibrium, Eq. (244) can be linearized to give

$$v = -\frac{v^0}{RT}\left[b\delta\Gamma_h^i + \delta(\Delta G^q)\right] \qquad (246)$$

For large departures from equilibrium, that is, large values of $\delta(\Delta G^q)$, one concludes that there should be an analogy for adsorption and desorption to ordinary Tafel lines in electrode kinetics, and an example of these is shown in Fig. 39.

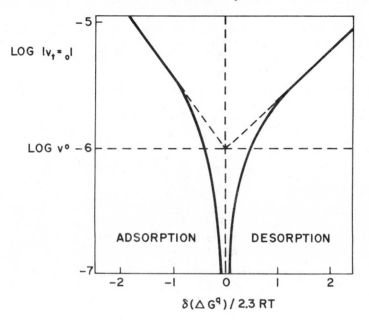

Fig. 39. Adsorption analogy to a Tafel plot. Theoretical plot for $v^0 = 10^{-6}$ mole cm^{-2} sec^{-1} and $\rho = 0.6$ [from (55)]. (By permission of the American Chemical Society.)

The influence of mass transfer is easily handled in a formal way. The equation corresponding to Eq. (244) but including mass transfer is

$$v = v^0 \left\{ \frac{a_h}{a_h^b} \exp\left(-\frac{\lambda b \delta \Gamma_h^i}{RT} \right) \exp\left[-\rho \frac{\delta(\Delta G^q)}{RT} \right] \right.$$

$$\left. -\exp\left[\frac{(1-\lambda)b\delta\Gamma_h^i}{RT} \right] \exp\left[\frac{(1-\rho)\delta(\Delta G^q)}{RT} \right] \right\} \qquad (247)$$

where v^0 is the particular value calculated from Eq. (243) for $a_h = a_h^b$.

For small departures from equilibrium, Eq. (247) may be linearized to give

$$v = v^0 \left\{ \frac{\delta a_h}{a_h^b} - \frac{1}{RT} \left[b\delta\Gamma_h^i + \delta(\Delta G^q) \right] \right\} \qquad (248)$$

In Eqs. (247) and (248), a_h is the concentration at the electrode surface, that is, preadsorption state, and a_h^b is the bulk concentration.

e. Influence of Mass Transfer. Delahay considered the problem of mass transfer by semiinfinite linear diffusion for the particular case of the linearized Temkin adsorption equation, Eq. (248). He further considered a coulostatic experiment in which $\delta(\Delta G^q)$ is a step function going from zero to a given value $\delta(\Delta G^q)$, which then remains constant. The boundary-value problem to be solved was, therefore, Fick's law with the initial and boundary conditions (using concentrations c instead of activities)

$$c = c^b \quad \text{(for } x \geq 0 \text{ and } t = 0\text{)}$$

$$c = c^b \quad \text{(for } x \rightarrow \infty \text{ and } t > 0\text{)}$$

$$D\left(\frac{\partial c}{\partial x}\right)_{x=0} = v^0\left[\frac{c_{x=0}}{c^b} - \frac{bD}{RT}\int_0^t\left(\frac{\partial c}{\partial x}\right)_{x=0}dt - \frac{\delta(\Delta G^q)}{RT} - 1\right]$$

where x is the distance from the electrode surface and D is the diffusion coefficient. The last boundary condition corresponds to the usual conservation of flux at $x = 0$ condition in electrode kinetics [cf. Delahay (126)]. It expresses the fact that the flux of the adsorbing substance h at the electrode surface is equal to the rate of adsorption. The general solution was carried out in the usual way by Laplace transformation, giving the following equation for $\delta\Gamma_h^i$:

$$\delta\Gamma_h^i = v^0\left[\frac{\delta(\Delta G^q)}{RT}\right]\left(\frac{1}{m-p}\right)\left\{\frac{(1/m)[1 - e^{m^2t}\,\text{erfc}(mt^{\frac{1}{2}})]}{-(1/p)[1 - e^{p^2t}\,\text{erfc}(pt^{\frac{1}{2}})]}\right\} \tag{249}$$

where

$$m = \frac{v^0}{2c^bD^{\frac{1}{2}}} + \left[\left(\frac{v^0}{2c^bD^{\frac{1}{2}}}\right)^2 - \frac{bv}{RT}\right]^{\frac{1}{2}}$$

and

$$p = \frac{bv^0}{RT}\left(\frac{1}{m}\right)$$

The variations of the ratio $\delta\Gamma_h^i/\delta(\Delta G^q)$ with time for mixed control by adsorption kinetics and diffusion for different values of adsorption exchange rates are given in Fig. 40. The curve marked ∞ corresponds to pure diffusion control. When there is pure diffusion control,

$$v^0 \gg 4bc^bD/RT$$

and

$$\frac{v^0}{c^bD^{\frac{1}{2}}}t^{\frac{1}{2}} \gg 1 \tag{250}$$

Under this condition, the function

$$e^{m^2t}\,\text{erfc}(mt^{\frac{1}{2}})$$

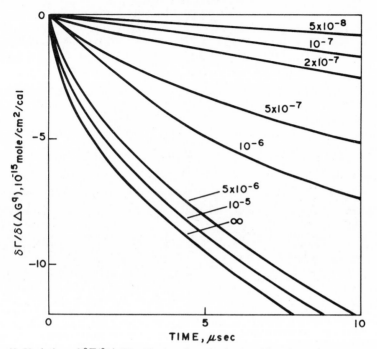

Fig. 40. Variation of $\delta\Gamma/\delta(\Delta G^q)$ with time for mixed control by adsorption kinetics and diffusion for different adsorption exchange rates. Theoretical plot for Temkin kinetics assuming $b = 5 \times 10^{12}\ \text{mole}^{-1}\ \text{cm}^2\ \text{cal}$; $c = 10^{-6}\ \text{mole cm}^{-3}$; $D = 5 \times 10^{-6}\ \text{cm}^2\ \text{sec}^{-1}$; temperature, $25°C$ [from (*124*)]. (By permission of the American Chemical Society.)

in Eq. (249) vanishes and one obtains

$$\delta\Gamma_h^i = -\frac{\delta(\Delta G^q)}{b}\left\{1 - \exp\left[\left(\frac{bc^b D^{\frac{1}{2}}}{RT}\right)^2 t\right]\ \text{erfc}\left[\left(\frac{bc^b D^{\frac{1}{2}}}{RT}\right)t^{\frac{1}{2}}\right]\right\}$$

The magnitude of the exchange rate required for pure diffusion control may be estimated from reasonable values of the diffusion coefficient. Thus if

$$v^0 \geq 10^{-4}\quad \text{mole cm}^{-2}\ \text{sec}^{-1}$$

and if

$$b = 5 \times 10^{12}\quad \text{moles}^{-1}\ \text{cm}^2\ \text{cal}$$

a reasonable value for Temkin adsorption, and if the concentration of substance h in the bulk of the solution c^b is millimolar, that is, 10^{-6} mole cm^{-3}, and

$$D = 5 \times 10^{-6}\quad \text{cm}^2\ \text{sec}^{-1}$$

then condition (250) would be satisfied for any time greater than 10^{-8} sec. Meaningful coulostatic measurements within so short a time do not appear to be feasible with presently available equipment. On the other hand, pure control by adsorption kinetics would imply that

$$v^0 \ll 4bc^b D/RT$$

Under this condition, m and p are imaginary, but it can be shown (*127*) that Eq. (249) reduces to

$$\delta\Gamma_h^i = -\frac{\delta(\Delta G^q)}{b}\left[1 - \exp\left(-\frac{v^0 bt}{RT}\right)\right]$$

This result can also be derived directly from Eq. (246), assuming a coulostatic step $\delta(\Delta G^q)$. Figure 41 illustrates the initial segments of the

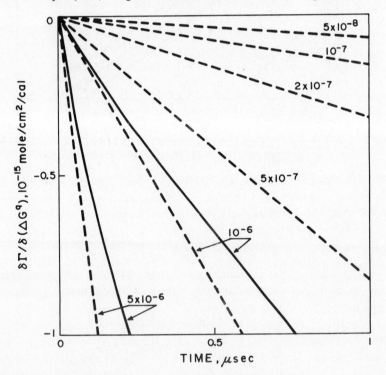

Fig. 41. Initial segments of the curves in Fig. 40 (solid curves), and the corresponding curves for pure control by adsorption kinetics (dotted curves). The curves for mixed control by adsorption and diffusion for the lowest four exchange rates practically coincide with those for pure control by adsorption [from (*124*)]. (By permission of the American Chemical Society.)

curves shown in Fig. 40 for the extremes of mixed control and pure adsorption control. The dotted lines correspond to pure adsorption control. One notes that, for exchange rates less than 5×10^{-7}, pure adsorption ought to be observed if the measurement can be made within 1 μsec. Thus, one concludes that, for very fast adsorption processes with present equipment, it would probably be impossible to detect the presence of any adsorption control, but for slower processes such detection and measurement of v^0 and other adsorption parameters is probably possible.

It should be mentioned that the logarithmic Temkin isotherm is not strictly obeyed for the adsorption of most neutral substances, and a better approximation is probably the Frumkin isotherm. However, the possibility of kinetic analysis along the lines outlined above is not ruled out. The rate equations corresponding to Frumkin adsorption have been derived (*122*).

ACKNOWLEDGMENT

The author is greatly indebted to the Robert A. Welch Foundation of Houston, Texas, for financial support and to Professor Norman Hackerman for providing the necessary freedom and facilities for this work as well as for helpful discussions.

APPENDIX A: EQUIVALENCE OF THE GIBBSIAN AND THE INTERPHASE APPROACHES

Parsons and Devanathan (*27*) derived the general electrocapillary equation using Guggenheim's (*37*) concept of the interphase without employing the Gibbs model as was done in Sec. II. Their final equations are identical in form with our final equations [Eqs. (52)] for constant temperature and pressure. (Parsons and Devanathan did not derive the temperature and pressure dependence of γ.) Since it may not be obvious that the relative surface concentrations appearing in their equations are equivalent to relative surface excesses, this appendix will demonstrate that equivalence.

The interphase of Guggenheim is defined in an identical manner to the real system we have employed. The number of moles per unit area in the real system (interphase) is called the *surface concentration* by Parsons and Devanathan and denoted by the symbol Γ. In this discussion, we shall use the symbol Γ_h^{sc} for their surface concentration to distinguish it from our Γ_h, which is a gibbsian surface excess. Thus, Γ_h^{sc}, the surface concentration of component h, is defined by the equation

$$\Gamma_h^{sc} = n_h/A$$

Here n_h is the total number of moles of substance h in the interphase and A is the area of the interface. In contrast, the gibbsian surface excess is defined as the difference between the number of moles per unit area in the interphase and the number of moles in a Gibbs model with an arbitrarily located dividing surface. Let the number of moles of substance h in the Gibbs model be n_{hG}. Then the Gibbsian surface excess is [cf. Eqs. (8) and (13)]

$$\Gamma_h = \frac{n_h}{A} - \frac{n_{hG}}{A} = \Gamma_h^{sc} - \frac{n_{hG}}{A} \tag{A1}$$

The surface concentration in the interphase of the reference component h' will be denoted by

$$\Gamma_{h'}^{sc} = n_{h'}/A$$

where $n_{h'}$ is the number of moles of h' in the interphase. Let $n_{h'G}$ be the number of moles of the reference component in the Gibbs model. Then the gibbsian surface excess of h' is

$$\Gamma_{h'} = \frac{n_{h'}}{A} - \frac{n_{h'G}}{A} = \Gamma_{h'}^{sc} - \frac{n_{h'G}}{A} \tag{A2}$$

By definition, the relative surface excess we have used [cf. Eq. (41)] is

$$\Gamma_{hh'} = \Gamma_h - \frac{x_h}{x_{h'}}\Gamma_{h'} \tag{A3}$$

The relative surface concentration defined by Parsons and Devanathan is

$$\Gamma_{hh'}^{sc} = \Gamma_h^{sc} - \frac{x_h}{x_{h'}}\Gamma_{h'}^{sc}$$

Substituting by means of Eqs. (A1) and (A2) into (A3), we obtain

$$\begin{aligned}
\Gamma_{hh'} &= \left(\Gamma_h^{sc} - \frac{n_{hG}}{A}\right) - \frac{x_h}{x_{h'}}\left(\Gamma_{h'}^{sc} - \frac{n_{h'G}}{A}\right) \\
&= \left(\Gamma_h^{sc} - \frac{x_h}{x_{h'}}\Gamma_{h'}^{sc}\right) - \frac{n_{hG}}{A} + \left(\frac{x_h}{x_{h'}}\right)\frac{n_{h'G}}{A} \\
&= \Gamma_{hh'}^{sc} - \frac{n_{hG}}{A} + \left(\frac{x_h}{x_{h'}}\right)\frac{n_{h'G}}{A}
\end{aligned} \tag{A4}$$

Remembering that the mole fractions x_h and $x_{h'}$ can be given by

$$x_h = \frac{n_{hG}}{n_{Gs}}$$

$$x_{h'} = \frac{n_{h'G}}{n_{Gs}}$$

where n_{Gs} is the total number of moles in the solution phase of the Gibbs model, we obtain

$$\left(\frac{x_h}{x_{h'}}\right)\frac{n_{h'G}}{A} = \frac{n_{hG}}{A} \tag{A5}$$

Substituting Eq. (A5) into Eq. (A4), we obtain

$$\Gamma_{hh'} = \Gamma_{hh'}^{sc}$$

which proves that the relative surface concentration is the same as the relative surface excess.

The proof of the equivalence of relative surface excess and relative surface concentration for any other component is similar. Thus, the equations of Parsons and Devanathan are identical with ours.

APPENDIX B: EXAMPLES OF THE ELIMINATION OF THE ELECTROCHEMICAL POTENTIAL OF THE INDICATOR ION

Equations (31a) and (31b) can be readily proved for any particular case. We shall consider here two cases, first a cation-reversible indicator electrode and then an anion-reversible indicator electrode.

EXAMPLE 1

Let the solution contain Ba^{2+} and let IN be a hypothetical barium-barium ion electrode. The cell is

$$Cu \,|M|\, Ba^{2+} \text{ solution } |Ba|\, Cu'$$

The equation expressing the electrochemical equilibrium at the indicator electrode-solution interface is

$$\bar{\mu}_{Ba^{2+}} + 2\bar{\mu}_e^{Ba} = \mu_{Ba}^{Ba}$$

Since electrons in the barium metal are in equilibrium with electrons in terminal Cu', we have

$$\bar{\mu}_e^{Cu'} = \bar{\mu}_e^{Ba}$$

Thus

$$d\bar{\mu}_{Ba^{2+}} = -2d\bar{\mu}_e^{Cu'} + d\mu_{Ba}^{Ba}$$

Since the indicator electrode is supposed to be made of pure barium metal, μ_{Ba}^{Ba} is equal to the molar Gibbs free energy of Ba; that is,

$$\mu_{Ba}^{Ba} = \mu_{Ba}^{\bullet}$$

where the superscript $^{\bullet}$ ("spot") indicates a pure substance at any given temperature and pressure. Thus

$$d\bar{\mu}_{Ba^{2+}} = -2d\bar{\mu}_e^{Cu'} + d\mu_{Ba}^{\bullet} \tag{B1}$$

At constant temperature and pressure, $\mu_{Ba}^{\bullet}$ is a constant and

$$d\bar{\mu}_{Ba^{2+}} = -2d\bar{\mu}_e^{Cu'}$$

The last equation would be the form of Eq. (31a) required to derive the electrocapillary equation if one did not wish to be able to consider variations of temperature and pressure [cf. Parsons and Devanathan (27)]. However, if the temperature and/or pressure dependence of the interfacial tension is to be taken into account, it is necessary to realize that the indicator electrode will also be subject to the same changes of temperature or pressure. (It might be thought that the problem could be simplified by keeping the indicator electrode at constant temperature and pressure. Far from simplifying the problem, this procedure would necessitate taking account of irreversible changes due to temperature and pressure gradients within the system.) The molar Gibbs free energy $\mu^{\bullet}$ of any pure substance is a function of the temperature and pressure. For example, in the case of barium metal,

$$d\mu_{Ba}^{\bullet} = -s_{Ba}^{\bullet} dT + v_{Ba}^{\bullet} dp$$

where $s_{Ba}^{\bullet}$ is the molar entropy of pure Ba at the temperature and pressure under consideration and $v_{Ba}^{\bullet}$ is the corresponding molar volume. Thus Eq. (B1) may also be written

$$d\bar{\mu}_{Ba^{2+}} = -2d\bar{\mu}_e^{Cu'} + d\mu_{Ba}^{\bullet}$$
$$= -2d\bar{\mu}_e^{Cu'} - s_{Ba}^{\bullet} dT + v_{Ba}^{\bullet} dp \tag{B2}$$

Equation (B2) is the particular form of Eq. (31a) for a hypothetical Ba^{2+}/Ba indicator electrode.

For any other indicator electrode reversible to a cation j' of the solution, an equation of the same form as Eq. (B2) would be obtained for $d\bar{\mu}_{j'}$. Therefore,

$$d\bar{\mu}_{j'} = -z_{j'} d\bar{\mu}_e^{Cu'} + d\mu_+^{\bullet}$$
$$= -z_{j'} d\bar{\mu}_e^{Cu'} - s_+^{\bullet} dT + v_+^{\bullet} dp \tag{31a}$$

The symbols $\mu_+^{\cdot}$, $s_+^{\cdot}$, and $v_+^{\cdot}$ in Eq. (31a) refer to the pure component of the cation-reversible indicator electrode. Thus if the indicator electrode is a reversible H^+/H_2 electrode, $\mu_+^{\cdot}$, $s_+^{\cdot}$, and $v_+^{\cdot}$ refer to the chemical potential, molar entropy, and molar volume of pure hydrogen gas in equilibrium with the solution when the pressure of the gas is p and the absolute temperature is T.

EXAMPLE 2

Let the solution contain SO_4^{2-} ions and let IN be a mercury-mercurous sulfate electrode. The cell is

$$Cu \mid M \mid SO_4^{2-} \text{ solution} \mid Hg_2SO_4 \mid Hg \mid Pt \mid Cu'$$

The equation expressing the electrochemical equilibrium at the indicator electrode is

$$\bar{\mu}_{SO_4^{2-}} = 2\bar{\mu}_e^{Hg} + \mu_{Hg_2SO_4}^{\cdot} - 2\mu_{Hg}^{\cdot}$$

where the $\mu^{\cdot}$'s are the chemical potentials of pure Hg_2SO_4 and pure Hg, respectively. Since the electrons in Hg are in equilibrium with the electrons in terminal Cu',

$$\bar{\mu}_e^{Cu'} = \bar{\mu}_e^{Pt} = \bar{\mu}_e^{Hg}$$

Thus, solving for the electrochemical potential of the indicator ion and taking the differential, we obtain

$$d\bar{\mu}_{SO_4^{2-}} = 2d\bar{\mu}_e^{Cu'} + d\mu_{Hg_2SO_4}^{\cdot} - 2d\mu_{Hg}^{\cdot} \tag{B3}$$

At constant temperature and pressure, it would take the simpler form

$$d\bar{\mu}_{SO_4^{2-}} = 2d\bar{\mu}_e^{Cu'}$$

For any other indicator electrode reversible to an anion, a similar equation would be obtained. Let the symbol $\mu_-^{\cdot}$ indicate the summation of the $\mu^{\cdot}$'s for the pure substances of any anion indicator electrode, the order of summation to be determined by the form of Eq. (B3). For example, in the case of the $SO_4^{2-}/Hg_2SO_4/Hg$ electrode just considered,

$$d\mu_-^{\cdot} = d\mu_{Hg_2SO_4}^{\cdot} - 2d\mu_{Hg}^{\cdot}$$

Similarly, we define $s_-^{\cdot}$ and $v_-^{\cdot}$. Thus for the same electrode

$$s_-^{\cdot} = s_{Hg_2SO_4}^{\cdot} - 2s_{Hg}^{\cdot}$$

and

$$v_-^{\cdot} = v_{Hg_2SO_4}^{\cdot} - 2v_{Hg}^{\cdot}$$

Thus, in general, for any anion-reversible indicator electrode,

$$d\bar{\mu}_{k'} = |z_{k'}|\, d\bar{\mu}_e^{Cu'} + d\mu_-^-$$

$$= |z_{k'}|\, d\bar{\mu}_e^{Cu'} - s_-^- \, dT + v_-^- \, dp \qquad (31b)$$

APPENDIX C: PERMITTIVITY AND ELECTRICAL UNITS

The permittivity of a dielectric ε is related to the (dimensionless) dielectric coefficient K_D (often called the dielectric constant) by the equation

$$\varepsilon = K_D \varepsilon_0$$

where ε_0 is the *permittivity of free space*. The value of ε_0 depends on the particular system of electrical units chosen; its dimensions are always charge^2 × force^{-1} × length^{-2}, which may be reduced to capacitance × length^{-1}. In the *rationalized mks* (meter, kilogram, second) system of units,

$$\varepsilon_0 = 8.85 \times 10^{-12} \quad \text{coulomb}^2 \; \text{newton}^{-1} \; \text{m}^{-2}$$

$$= 8.85 \times 10^{-12} \quad \text{coulomb volt}^{-1} \; \text{m}^{-1}$$

$$= 8.85 \times 10^{-12} \quad \text{farad m}^{-1}$$

where the electrical units, coulomb, volt, and farad, are the familiar *practical units*.

In the rationalized mks system Coulomb's law of force between two point charges q and q' separated by a distance r is

$$F = \frac{qq'}{4\pi\varepsilon_0 r^2}$$

where F is the force in newtons (kg. × m × sec^{-2}) and q and q' are in ordinary coulombs. If r is given in meters, $\varepsilon_0 = 8.85 \times 10^{-12}$ farad m^{-1}. If r is given in centimeters, $\varepsilon_0 = 8.85 \times 10^{-14}$ farad cm^{-1}. The factor 4π is included in the denominator in the expression of Coulomb's law in the rationalized system of electrical units in order to eliminate its appearance in other, more commonly used equations such as Poisson's equation and Gauss' law.

Grahame (6) chose to employ an *unrationalized* system of electrical units in which the quantity

$$D_0 = 4\pi\varepsilon_0 = 1.112 \times 10^{-10} \quad \text{farad m}^{-1}$$

Using D_0, called the *diabbativity of free space*, Coulomb's law takes on the simpler form

$$F = \frac{qq'}{D_0 r^2}$$

However, with this system, a factor 4π now appears in the statement of Poisson's equation and Gauss' law. Regardless of the system of units employed, the value of the dielectric coefficient K_D is the same. The force exerted between two point charges q and q' immersed in a dielectric of coefficient K_D is

$$F = \frac{qq'}{4\pi K_D \varepsilon_0 r^2} = \frac{qq'}{K_D D_0 r^2}$$

For a dieletric with constant dielectric coefficient K_D, Poisson's equation in one dimension, expressed in the rationalized mks system of units, is given by

$$\frac{d^2\phi}{dx^2} = -\frac{\rho(x)}{\varepsilon} = -\frac{\rho(x)}{K_D \varepsilon_0}$$

whereas in the unrationalized system of units it becomes

$$\frac{d^2\phi}{dx^2} = -\frac{4\pi\rho(x)}{K_D D_0} \tag{C1}$$

It is the appearance of the factor 4π on the right side of Eq. (C1) which accounts for the difference in form between the diffuse-layer equations in this chapter and those in the works of Grahame and others. It may be noted that the old *cgs electrostatic system* of units was an unrationalized system, and therefore the same factor 4π appeared in the statement of Poisson's equation and Gauss' law in that system. The early workers in double-layer theory (for example, Gouy) employed the cgs electrostatic system. Grahame (6) stated in 1947 that his reason for employing D_0 instead of ε_0 was simply to make his equations resemble more closely those which had appeared in previous treatments of the electrical double layer. Although this practice undoubtedly did prevent a certain amount of confusion in 1947, it would appear to offer less pedagogical advantage now, for today, at least in American universities, electrostatics is almost always taught in terms of the rationalized mks system.

APPENDIX D: EFFECT OF VARYING IONIC SIZE ON THE PREDICTIONS OF DIFFUSE-LAYER THEORY

A pictorial representation of the idea of two OHP's for the mixed electrolyte solution considered by Joshi and Parsons (80) discussed in Sec. III.C.7.c will now be given. Figure 42(a) represents the concentration-distance profile of Ba^{2+}; Fig. 42(b) of H^+, and Fig. 42(c) of water. It is assumed that q^M is sufficiently negative that Cl^- may be neglected in this

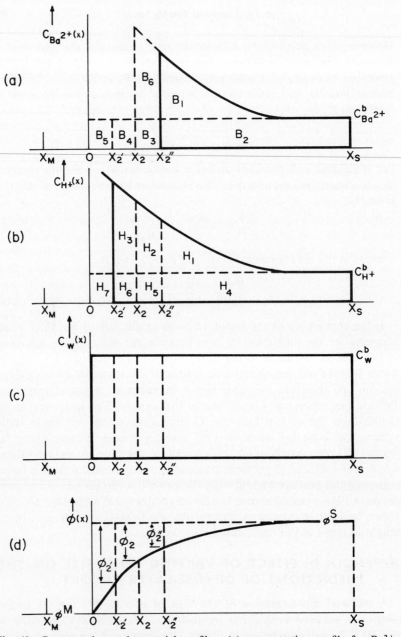

Fig. 42. Concentration and potential profiles. (a) concentration profile for Ba^{2+}; (b) concentration profile for H^+; (c) concentration profile for water; (d) potential profile.

discussion. The position of the metal surface is indicated on each diagram at $x = 0$. The position of the actual H^+ OHP and Ba^{2+} OHP, as well as the position of the "average" OHP calculated from the GCS theory are indicated. The distances between the metal surface and these planes is, of course, considerably exaggerated to make the figure clearer. No attempt has been made to draw these purely qualitative figures to scale nor are the concentration scales employed in the figure necessarily the same. The distance scales on the three profiles, however, do coincide precisely. Part (c) of the figure represents the water profile assumed for the purpose of locating the Gibbs surface ($\Gamma_w = 0$) in parts (a) and (b) (cf. Sec. III.B.1).

For simplicity in this picture, it is assumed that the water concentration remains uniform up to the metal surface, in which case the Gibbs surface is the metal surface itself. It is possible that the Gibbs surface could lie behind the metal surface if there were an increased concentration of water in the inner layer. Such an increase in concentration might result from breaking the normal water structure in the inner layer [cf. (129)]. It seems unlikely that the Gibbs surface could lie very far from the metal surface into the solution, especially not as far as x'_2 (in the absence of specific adsorption). The discussion below, based on the assumed water profile, will not be affected by any of the likely errors in the assumption of the location of the Gibbs surface.

In Fig. 42(a) the number of moles of Ba^{2+} in the real system (interphase), $n_{Ba^{2+}}$, is indicated by the area $B_1 + B_2$. The dotted curve proceeding from the true Ba^{2+} OHP to the "average" OHP indicates the additional amount of Ba^{2+} adsorption which would occur if Ba^{2+} could actually adsorb at the "average" OHP when the potential at that plane is the average value ϕ_2 given by the GCS theory. Thus, the number of moles of Ba^{2+} in the real interphase is less than the number of moles of Ba^{2+} according to the GCS model.

Figure 42(d) represents the electric-potential profile in the double layer. ϕ^M is the inner potential of the metal, ϕ^S represents the inner potential of the solution and, since we are assuming that the charge density on the metal is negative, ϕ^M is negative with respect to ϕ^S. The electrical potential controlling the actual adsorption of Ba^{2+} is ϕ''_2. ϕ_2 represents the difference of potential between the "average" OHP and the bulk of the solution. ϕ'_2 represents the potential difference between the true H^+ OHP and the bulk of the solution. Since the electrical potential ϕ''_2 is smaller in absolute value than ϕ_2, one sees immediately from the Boltzmann equation [Eq. (105)] that the concentration of Ba^{2+} at the true Ba^{2+} OHP will be less than the concentration of Ba^{2+} at the "average" OHP. Conversely, the concentration of H^+ at the true H^+ OHP will be greater than the concentration of H^+ at the "average" OHP.

In Fig. 42(b) the area $H_1 + H_2 + H_3 + H_4 + H_5 + H_6$ represents the actual number of moles of H^+ in the real interphase per unit area. Areas $H_3 + H_6$ represent the additional amount of H^+ which is adsorbed in the real system over that which would be calculated on the basis of the GCS theory, assuming that the controlling potential is ϕ_2 instead of ϕ_2'.

The number of moles of Ba^{2+} in a Gibbs model, with dividing surface taken at the location of the Gibbs surface ($\Gamma_w = 0$), is given by areas $B_2 + B_3 + B_4 + B_5$. Similarly, the number of moles of H^+ in the same Gibbs model is given by areas $H_4 + H_5 + H_6 + H_7$. The relative surface excess of Ba^{2+} which is actually measured is equal to the number of moles of Ba^{2+} in the real system minus the number of moles of Ba^{2+} in the Gibbs model with dividing surface at the Gibbs surface. The relative surface excess of H^+ is calculated in a similar way. Thus the relative surface excesses of Ba^{2+} and H^+ are represented by the areas:

$$\Gamma_{Ba^{2+},w} = B_1 - (B_3 + B_4 + B_5)$$

and

$$\Gamma_{H^+,w} = H_1 + H_2 + H_3 - H_7$$

In contrast, since the number of moles of Ba^{2+} which are supposedly present according to the GCS theory is given by $B_1 + B_2 + B_3 + B_6$, the nonrelativized surface excess reckoned at the "average" OHP is given by the area $B_1 + B_6$. In the case of H^+, since the number of moles of H^+ which are supposedly in the real system using the value of ϕ_2 calculated from the GCS theory is given by $H_1 + H_2 + H_4 + H_5$, one concludes that the surface excess of H^+ reckoned at the "average" OHP is given by areas $H_1 + H_2$. Comparing these two expressions with the previously derived equations for the relative surface excesses, we conclude that the discrepancy between the measured relative surface excess and the surface excess reckoned at the "average" OHP for Ba^{2+} is

$$\Gamma_{Ba^{2+},w} - \Gamma_{Ba^{2+}}^{OHP_{av}} = -B_3 - B_4 - B_5 - B_6 \tag{D1}$$

Thus, the measured surface excess of Ba^{2+} must be less than that calculated by the GCS theory. Figure 16 (taken from Joshi and Parsons) supports this conclusion. Similarly, the measured versus calculated discrepancy for H^+ is

$$\Gamma_{H^+,w} - \Gamma_{H^+}^{OHP_{av}} = H_3 - H_7 \tag{D2}$$

Equation (D2) predicts that, if the measured surface excess of H^+ is greater than that calculated on the basis of the simple GCS theory, the area H_3 must be greater than area H_7. Since the data of Joshi and Parsons (cf. Fig. 16) show that the measured surface excess of H^+ is larger than that predicted by the GCS theory, we conclude that the excess quantity of H^+ contained between the "average" OHP and the H^+ OHP, that is, area H_3, is greater than the number of moles of H^+ which would be contained in a volume of unit area of thickness x_2' if the concentration of H^+ were bulk concentration. Such behavior is reasonable in view of the remarkable increase of concentration which can be produced by even moderately high values of ϕ_2 (cf. Sec. III.C.6).

The difference between the nonrelativized surface excesses calculated on the basis of the "average" OHP and those calculated on the basis of the true OHP, assuming that Ba^{2+} is the larger and H^+ is the smaller ion, are clearly given by

$$\Gamma_{Ba^{2+}}^{OHP_{av}} - \Gamma_{Ba^{2+}}^{OHP\,Ba^{2+}} = +B_6$$

and

$$\Gamma_{H^+}^{OHP_{av}} - \Gamma_{H^+}^{OHP\,H^+} = -H_3$$

Thus, the improved-model calculations will necessarily shift the calculated surface excesses in the direction of the measured surface excesses, as found by Joshi and Parsons (cf. Fig. 17).

GLOSSARY OF SYMBOLS

In the text, subscripts or superscripts are sometimes omitted when detailed specification is not required at that point in the discussion. For example, E_z is defined below simply as the potential of the point of zero charge. In the context in which it is used it will always be clear whether E_z is on an $E^{\pm}$ or an E_{ref} scale.

A	area of interface
A	constant $= (2RT\varepsilon c^b)^{\frac{1}{2}}$
A	an adsorbed species
C	differential capacity; superscript i or d denotes differential capacity due to inner or diffuse layer, respectively; subscript $+$ or $-$ denotes component of differential capacity attributable to cations or anions, respectively
D	magnitude of electric-displacement vector
D	diffusion coefficient
$\mathscr{E}$	magnitude of electric-field-strength vector

E	electrode potential measured with respect to
	superscripts: $\pm$ an unspecified indicator electrode
	$+$ indicator electrode reversible to a cation of the solution
	$-$ indicator electrode reversible to an anion of the solution
	subscripts: ref a constant-potential reference electrode
	H a hydrogen electrode
	NCE a normal calomel electrode
E_z	potential of the point of zero charge
F	the faraday
G	Gibbs free energy
G^0	standard electrochemical free energy of adsorption
G^n	charge independent part of G^0
G^q	charge dependent part of G^0
IN	indicator electrode
K	integral capacity
K_D	dielectric coefficient ("constant")
M	metallic phase of electrode system
O	generalized oxidized species
R	generalized reduced species
R	gas constant
S	entropy; subscripts x, MG, SG indicate excess entropy, entropy of the metallic, and solution phases of the Gibbs model, respectively
S	solution phase of electrode system
S	a site on the electrode surface
T	absolute temperature
U	internal energy; subscripts G, MG, SG indicate internal energy of the Gibbs model, and its metallic and solution phases, respectively; subscript x indicates excess internal energy
V	volume, subscripts MG and SG indicate volume of metallic and solution phases of the Gibbs model, respectively
a	number of anionic species in solution phase of electrode system
a	activity; subscript indicates species, superscript indicates location of species (A indicates species in adsorbed monolayer)
$a_\pm$	mean ionic activity
b	number of neutral species in solution phase of electrode system
b	parameter appearing in van der Waals, Volmer, and other equations of state
b	alternate expression for $2g$ (used only in discussions of adsorption kinetics)
b	superscript denoting bulk of solution
c	number of cationic species in solution phase of electrode system

c	concentration; subscript indicates species, superscript indicates location of species
d	superscript denoting diffuse layer
f	fugacity; f^o fugacity in standard state
g	interaction parameter in van der Waals, Temkin, and other isotherms
h	a neutral species in solution; h' reference substance in solution
i	substance in metallic phase; i' reference substance in metallic phase (as used in Sec. I and II)
i	ionic species in solution phase (as used in Sec. III)
i	superscript denoting inner layer
j	cationic species in solution phase; j' cation of indicator salt
k	anionic species in solution phase; k' anion of indicator salt
$\vec{k}, \overleftarrow{k}$	adsorption-rate constants including electrical dependence
k^o	standard adsorption-rate constant
k_f, k_b	adsorption-rate constants independent of electrical state
m	number of metallic species in metallic phase of electrode system
n	number of electrons in charge-transfer reaction
n	(with single subscript) number of moles of indicated component in real system
n	(with second subscript G) number of moles of indicated component in Gibbs model; superscript, if any, specifies location of dividing surface
n	(with second subscript x) excess number of moles of indicated component
p	pressure; subscripts M or S indicate metallic or solution phase, respectively
q	charge density; for clarification of particular q's see Eqs. (172–174)
q^M	total excess charge density on metallic side of double layer
q^s	total excess charge density on solution side of double layer
q_+	component of q^s equivalent to relative surface excess of cation
q_-	component of q^s equivalent to relative surface excess of anion
q_+^{OHP}	component of q^s equivalent to surface excess of cation reckoned at OHP
q_-^{OHP}	component of q^s equivalent to surface excess of anion reckoned at OHP
q_+^d	component of q_+^{OHP} due to diffuse layer
q_-^d	component of q_-^{OHP} due to diffuse layer
q_+^i	component of q_+^{OHP} due to specifically adsorbed cations
q_-^i	component of q_-^{OHP} due to specifically adsorbed anions
s	mean molar entropy; subscript S denotes solution phase, M, metallic phase; superscript $\cdot$ denotes molar entropy of pure component at specified temperature and pressure

t	time
v	mean molar volume; subscript S denotes solution phase, M, metallic phase; superscript $\cdot$ denotes molar volume of pure component at specified temperature and pressure
v	rate of adsorption; $\overrightarrow{v}, \overleftarrow{v}$ adsorption and desorption rates, respectively
v^0	adsorption exchange rate
w	subscript denoting water
x	distance from metal surface: x_1 to IHP, x_2 to OHP, x_S to bounding surface of real system located in homogeneous solution
x	mole fraction, subscript indicates species, superscript indicates in bulk of solution
z	charge of indicated ion; no subscript, $z = z_+ = |z_-|$, subscripts indicate species (+ for cation, − for anion)
Γ	surface excess
	single subscript: nonrelativized surface excess of indicated component
	two subscripts: surface excess of component indicated by first subscript relative to component indicated by second subscript
	superscript: location of dividing surface in Gibbs model at which nonrelativized surface excess is reckoned
	presuperscript: portion of surface excess in inner (i) or diffuse (d) layer
Γ_s	saturation value of Γ_h^i in complete monolayer, cf. Table IV
Γ_h^i, Γ_h^d	surface excess of indicated component in inner (i) or diffuse (d) layer
Φ	surface pressure at constant q^M
β	$= \exp(-\Delta G^0/RT)$
γ	interfacial tension
ε	dielectric permittivity
ε_0	permittivity of free space
η	overvoltage
θ	coverage; $\theta = \Gamma_h^i/\Gamma_s$ (cf. Table IV)
κ	Debye reciprocal length
λ	coverage parameter in Temkin adsorption kinetics
μ	chemical potential, subscript indicates species, superscript b indicates in bulk of solution, ^{0} indicates standard chemical potential; superscript $\cdot$ denotes chemical potential (molar Gibbs free energy) of pure component at specified temperature and pressure
$\bar{\mu}$	electrochemical potential, superscripts and subscripts as for chemical potential
ν	stoichiometric coefficient, superscript: +, cation; −, anion; subscript designates cation and anion of salt in question

$\xi^{\pm}$	$= \gamma + q^M E^{\pm}$
π	surface pressure at constant electrode potential
ρ	charge parameter in adsorption kinetics
$\rho(x)$	volume charge density
σ	nonrelativized surface excess of entropy
$\sigma_{i'h'}$	relative surface excess of entropy
τ	nonrelativized thickness of the interphase
$\tau_{i'h'}$	relative thickness of the interphase
$\tau_{99.99}$	effective thickness of diffuse layer
ϕ^M	inner potential of metallic phase
ϕ^S	inner potential of solution phase
$\phi(x)$	electric potential at distance x from metal surface
ϕ_2	potential drop across diffuse layer
ϕ	(without super- or subscript) $= \phi(x) - \phi^S$

REFERENCES

1. J. F. Dewald, "Semiconductor Electrodes," in *Semiconductors* (N. B. Hannay, ed.), Reinhold, New York, 1964, Chap. 17.
2. H. Gerischer, "Semiconductor Electrode Reactions," in *Advances in Electrochemistry and Electrochemical Engineering*, Vol. 1 (P. Delahay, ed.), Wiley (Interscience), New York, 1961, Chap. 4.
3. M. Green, "Electrochemistry of the Semiconductor-Electrolyte Interface," in *Modern Aspects of Electrochemistry*, No. 2 (J. O'M. Bockris, ed.), Academic Press, New York, 1959, Chap. 5.
4. P. J. Holmes, ed., *The Electrochemistry of Semiconductors*, Academic Press, New York, 1961.
5. P. Delahay, *Double Layer and Electrode Kinetics*, Wiley (Interscience), New York, 1965.
6. D. C. Grahame, *Chem. Rev.*, **41**, 441 (1947).
7. D. C. Grahame, *Ann. Rev. Phys. Chem.*, **6**, 337 (1955).
8. R. Parsons, "Equilibrium Properties of Electrified Interphases," in *Modern Aspects of Electrochemistry*, No. 1 (J. O'M. Bockris, ed.), Butterworths, London, 1954, Chap. 3.
9. R. Parsons, "The Structure of the Electrical Double Layer and its Influence on the Rates of Electrode Reactions," in *Advances in Electrochemistry and Electrochemical Engineering*, Vol. 1 (P. Delahay, ed.), Wiley (Interscience), New York, 1961, Chap. 1.
10. P. Delahay, "The Study of Fast Electrode Processes by Relaxation Methods," in *Advances in Electrochemistry and Electrochemical Engineering*, Vol. 1 (P. Delahay, ed.), Wiley (Interscience), New York, 1961, Chap. 5.
11. M. Breiter, "Some Problems in the Study of Oxygen Overvoltage," in *Advances in Electrochemistry and Electrochemical Engineering*, Vol. 1 (P. Delahay, ed.), Wiley (Interscience), New York, 1961, Chap. 3.
12. A. N. Frumkin, "Hydrogen Overvoltage and Adsorption Phenomena: Part I. Mercury," in *Advances in Electrochemistry and Electrochemical Engineering*, Vol. 1 (P. Delahay, ed.), Wiley (Interscience), New York, 1961, Chap. 2.

13. A. N. Frumkin, "Hydrogen Overvoltage and Adsorption Phenomena: Part II," in *Advances in Electrochemistry and Electrochemical Engineering*, Vol. 3 (P. Delahay, ed.), Wiley (Interscience), New York, 1963, Chap. 5.

14. A. N. Frumkin and B. B. Damaskin, "Adsorption of Organic Components at Electrodes," in *Modern Aspects of Electrochemistry*, No. 3 (J. O'M. Bockris, ed.), Butterworths, Washington, D.C., 1965, Chap. 3.

15. J. R. Macdonald and C. A. Barlow, Jr., "Equilibrium Double-Layer Theory," in *Electrochemistry, Proceedings of the First Australian Conference on Electrochemistry*, Pergamon, New York, 1964.

16. R. B. Whitney and D. C. Grahame, *J. Chem. Phys.*, **9**, 827 (1941).

17. D. C. Grahame, *Proc. 3rd Meeting CITCE, Berne, 1951*, Carlo-Manfredi-Editore, Milan, 1952, pp. 330–345.

18. O. Stern, *Z. Elektrochem.*, **30**, 508 (1924).

19. P. Van Rysselberghe, *Electrochemical Affinity* (No. 1237 in the series, *Actualities Scientifiques et Industrielles*), Hermann, Paris, 1955.

20. M. A. V. Devanathan, *Trans. Faraday Soc.*, **50**, 373 (1954).

21. J. O'M. Bockris, M. A. V. Devanathan, and K. Müller, *Proc. Roy. Soc. (London)*, **A274**, 55 (1963).

22. D. C. Grahame, M. A. Poth, and J. I. Cummings, *J. Am. Chem. Soc.*, **74**, 4422 (1952).

23. S. Levine, G. M. Bell, and D. Calvert, *Can. J. Chem.*, **40**, 518 (1962).

24. T. N. Andersen and J. O'M. Bockris, *Electrochim. Acta*, **9**, 347 (1964).

25. A. N. Frumkin, B. B. Demaskin, and N. V. Nikolayeva-Fedorovich, *Dokl. Akad. Nauk SSSR*, **115**, 751 (1957).

26. D. C. Grahame and R. B. Whitney, *J. Am. Chem. Soc.*, **64**, 1548 (1942).

27. R. Parsons and M. A. V. Devanathan, *Trans. Faraday Soc.*, **49**, 404 (1953).

28. F. O. Koenig, *J. Phys. Chem.*, **38**, 339 (1934).

29. L. Meites, "Voltammetry at the Dropping Mercury Electrode (Polarography)," in *Treatise on Analytical Chemistry*, Part I, Vol. 4 (I. M. Kolthoff and P. J. Elving, eds.), Wiley (Interscience), New York, 1963, Chap. 46.

30. L. Meites, *J. Am. Chem. Soc.*, **73**, 2035 (1951).

31. G. Lippmann, *Compt. Rend.*, **76**, 1407 (1873); *Pogg. Ann.*, **149**, 561 (1873); *J. Phys.*, [1] **3**, 41 (1874).

32. G. Lippmann, *Ann. Chim. Phys.*, [5] **5**, 494 (1875); [5] **12**, 265 (1877).

33. M. A. V. Devanathan and P. Peries, *Trans. Faraday Soc.*, **50**, 1236 (1954).

34. J. W. Gibbs, *Collected Works*, Vol. 1, Yale Univ. Press, New Haven, Conn., 1928, p. 219 ff.

35. N. K. Adam, *The Physics and Chemistry of Surfaces*, 3rd ed., Oxford, New York, 1941, p. 107.

36. R. Defay and I. Prigogine, *Tension Superficielle et Adsorption*, Editions Desoer, Liege, 1951.

37. E. A. Guggenheim, *Trans. Faraday Soc.*, **36**, 397 (1940); cf. Ref. (*35*) pp. 404–7; See also E. A. Guggenheim, *Thermodynamics*, 3rd ed., North-Holland, Amsterdam, 1957, pp. 46–59.

38. D. M. Mohilner and N. Hackerman, *Electrochim. Acta* (in press).

39. I. M. Klotz, *Chemical Thermodynamics*, Prentice-Hall, New York, 1950, p. 12.

40. R. Parsons, *Trans. Faraday Soc.*, **51**, 1518 (1955).

41. S. J. Kline and F. O. Koenig, *J. Appl. Mech.*, **24**, 29 (1957).

42. (a) R. Parsons, *Can. J. Chem.*, **37**, 308 (1959); (b) W. Anderson and R. Parsons, *Proc. Intern. Congr. Surface Activity 2nd London*, **1957**, Vol. 3; 45–52.

43. J. E. B. Randles and K. S. Whiteley, *Trans. Faraday Soc.*, **52**, 1509 (1956).

44. R. S. Hansen, D. J. Kelsh, and D. H. Grantham, *J. Phys. Chem.*, **67**, 2316 (1963).

45. R. G. Barradas, Ph.D. Dissertation, University of Ottawa, Ottawa, Can., 1960.

46. M. A. V. Devanathan and S. G. Canagaratna, *Electrochim. Acta*, **8**, 77 (1963).

47. R. Parsons and M. A. V. Devanthan, *Trans. Faraday Soc.*, **49**, 673 (1953).

48. D. C. Grahame, R. P. Larsen, and M. A. Poth, *J. Am. Chem. Soc.*, **71**, 2978 (1949).

49. D. C. Grahame, E. M. Coffin, J. I. Cummings, and M. A. Poth, *J. Am. Chem. Soc.*, **74**, 1207 (1952).

50. E. Blomgren, J. O'M. Bockris, and C. Jesch, *J. Phys. Chem.*, **65**, 2000 (1961).

51. R. Parsons, *Proc. Intern. Congr. Surface Activity 2nd London*, **1957**, Vol. 3, 38–44.

52. R. Parsons, *Trans. Faraday Soc.*, **55**, 999 (1959).

53. R. Parsons, *J. Electroanal. Chem.*, **8**, 93 (1964).

54. R. Parsons, *J. Electroanal. Chem.*, **7**, 136 (1964).

55. P. Delahay and D. M. Mohilner, *J. Phys. Chem.*, **66**, 959 (1962); *J. Am. Chem. Soc.*, **84**, 4247 (1962).

56. D. M. Mohilner and P. Delahay, *J. Phys. Chem.*, **67**, 588 (1963).

57. B. B. Damaskin, *J. Electroanal. Chem.*, **7**, 155 (1964).

58. O. A. Esin and B. F. Markov, *Acta Physicochim, URSS*, **10**, 353 (1939).

59. D. M. Mohilner, *J. Phys. Chem.*, **66**, 724 (1962).

60. F. O. Koenig, W. H. Wohlers, and D. Bandini, *Abstracts, CITCE Meeting*, Moscow, 1963.

61. H. L. F. von Helmholtz, *Ann. Physik.*, [2] **89**, 211 (1853); [3] **7**, 337 (1879); *Wiss. Abhandl. physik. tech. Reichsanstalt I*, p. 925 (1879); *Monatsh. Preuss, Akad. Sci.* Nov. 1881.

62. G. Quincke, *Ann. Physik*, [2] **113**, 513 (1861).

63. J. A. V. Butler, *Electrocapillarity*, Chemical Publishing Co., New York, 1940, Chap. 4.

64. G. Gouy, *Ann. Chim. Phys.*, [7] **29**, 145 (1903); [8] **8**, 291 (1906); **9**, 75 (1906).

65. G. Gouy, *J. Phys.*, [4] **9**, 457 (1910); *Compt. Rend.*, **149**, 654 (1910).

66. D. L. Chapman, *Phil. Mag.*, [6] **25**, 475 (1913).

67. G. Gouy, *Ann. Phys. (Paris)*, **7**, 163 (1917).

68. D. C. Grahame, *J. Am. Chem. Soc.*, **63**, 1207 (1941).

69. D. C. Grahame, *J. Am. Chem. Soc.*, **68**, 301 (1946).

70. D. C. Grahame and R. Parsons, *J. Am. Chem. Soc.*, **83**, 1291 (1961).

71. F. W. Sears, *Principles of Physics II, Electricity and Magnetism*, Addison-Wesley, Reading, Mass., 1946.

72. A. F. Kip, *Fundamentals of Electricity and Magnetism*, McGraw-Hill, New York, 1962.

73. R. Payne, *J. Electroanal. Chem.*, **7**, 343 (1964).

74. D. C. Grahame, *J. Am. Chem. Soc.*, **76**, 4819 (1954).

75. D. C. Grahame, *J. Am. Chem. Soc.*, **79**, 2093 (1957).

76. C. D. Russell, *J. Electroanal. Chem.*, **6**, 440 (1963).

77. A. N. Frumkin, *Trans. Faraday Soc.*, **36**, 117 (1940).

78. H. S. Harned and B. B. Owen, *The Physical Chemistry of Electrolyte Solutions*, Reinhold, New York, 1950, Chap. 2.

79. D. C. Grahame, *J. Chem. Phys.*, **21**, 1054 (1953).
80. K. M. Joshi and R. Parsons, *Electrochim. Acta*, **4**, 129 (1961).
81. J. J. Bikerman, *Phil. Mag.*, [7] **33**, 884 (1942).
82. T. Grimley and N. F. Mott, *Discussions Faraday Soc.*, **43**, 3 (1947).
83. V. Freise, *Z. Elektrochem.*, **56**, 822 (1952).
84. M. Eigen and E. Wicke, *Z. Elektrochem.*, **55**, 354 (1951); **56**, 836 (1952).
85. M. Spaarnay, *Rec. Trav. Chim.*, **77**, 382 (1958).
86. H. Brodowsky and H. Strehlow, *Z. Elektrochem.*, **63**, 262 (1959).
87. I. Prigogine, P. Mazur, and R. Defay, *J. Chim. Phys.*, **50**, 146 (1953).
88. H. D. Hurwitz, A. Sanfeld, and A. Steinchen-Sanfeld, *Electrochim. Acta*, **9**, 929 (1964).
89. D. C. Grahame, *J. Chem. Phys.*, **18**, 903 (1950).
90. A. Aramata and P. Delahay, *J. Phys. Chem.*, **66**, 2710 (1962).
91. A. Aramata and P. Delahay, *J. Phys. Chem.*, **68**, 880 (1964).
92. W. J. Argersinger, Jr. and D. M. Mohilner, *J. Phys. Chem.*, **61**, 99 (1957).
93. G. McDougall and C. W. Davies, *J. Chem. Soc.*, **1935**, 1416.
94. P. Delahay and A. Aramata, *J. Phys. Chem.*, **66**, 1194 (1962).
95. F. P. Buff and F. H. Stillinger, *J. Chem. Phys.*, **39**, 1911 (1963).
96. V. G. Levich and V. S. Krylov, *Dokl. Akad. Nauk SSSR*, **141**, 1403 (1961).
97. V. S. Krylov and V. G. Levich, *Zh. Fiz. Khim.*, **37**, 106, 2273 (1963).
98. A. N. Frumkin, in *Transactions of the Symposium on Electrode Processes* (E. Yeager, ed.), Wiley, New York, 1961.
99. J. W. Mellor, *Higher Mathematics for Students of Chemistry and Physics*, 4th ed., Dover, New York, 1955, p. 349.
100. J. W. Mellor, *Higher Mathematics for Students of Chemistry and Physics*, 4th ed., Dover, New York, 1955, p. 613.
101. D. C. Grahame and B. A. Soderberg, *J. Chem. Phys.*, **22**, 449 (1954).
102. R. Parsons, *Soviet Electrochemistry* (Proc. 4th Conf. of Electrochem., 1956, USSR), Vol. 1, Consultants Bureau, New York, 1961, p. 18.
103. N. F. Mott and R. J. Watts-Tobin, *Electrochim. Acta*, **4**, 79 (1961).
104. J. M. Parry and R. Parsons, *Trans. Faraday Soc.*, **59**, 241 (1963).
105. R. G. Barradas and B. E. Conway, *Electrochim. Acta*, **5**, 349 (1961).
106. S. G. Mairanovskii, *Electrochim. Acta*, **9**, 803 (1964).
107. D. M. Mohilner and P. Delahay, unpublished investigation.
108. T. H. Tidwell, Jr., Ph.D. Dissertation, Louisiana State University, 1964; T. H. Tidwell, Jr., D. M. Mohilner, and P. Delahay, unpublished investigation.
109. V. I. Melik-Gaikazyan, *Zh. Fiz. Khim.*, **26**, 560 (1952).
110. D. H. Everett, *Trans. Faraday Soc.*, **46**, 453, 942 (1950).
111. I. M. Klotz, Ref. (*39*), Chap. 15, 16.
112. R. Parsons, *J. Electroanal. Chem.*, **5**, 397 (1963).
113. R. Parsons, *Proc. Roy. Soc.* (*London*) **A261**, 79 (1961).
114. B. B. Damaskin and G. A. Tedoradze, *Electrochim. Acta*, **10**, 529 (1965).
115. T. Berzins and P. Delahay, *J. Phys. Chem.*, **59**, 906 (1955).
116. M. Senda and P. Delahay, *J. Am. Chem. Soc.*, **83**, 3763 (1961).
117. R. Delahay and I. Trachtenberg, *J. Am. Chem. Soc.*, **79**, 2355 (1957).
118. A. N. Frumkin and V. I. Melik-Gaikazyan, *Dokl. Akad. Nauk SSSR*, **77**, 855 (1951).
119. W. Lorenz, *Z. Physik. Chem.* (*Frankfurt*), **18**, 1 (1958); *Z. Elektrochem.*, **62**, 192 (1958).

120. W. Lorenz and F. Möckel, *Z. Elektrochem.*, **60**, 507 (1956).

121. R. Parsons, *Trans. Faraday Soc.*, **47**, 1332 (1951).

122. D. M. Mohilner, unpublished investigation.

123. J. M. Honig, *J. Chem. Educ.*, **38**, 538 (1961).

124. ʀ. Delahay, *J. Phys. Chem.*, **67**, 135 (1963).

125. D. M. Mohilner, *J. Phys. Chem.*, **68**, 623 (1964).

126. P. Delahay, *New Instrumental Methods in Electrochemistry*, Wiley (Interscience), New York, 1954.

127. H. S. Carslaw and J. C. Jaeger, *Conduction of Heat in Solids*, Oxford, New York, 1947, p. 372.

128. M. I. Temkin, *Zh. Fiz. Khim.*, **15**, 296 (1941).

129. H. S. Frank and M. W. Evans, *J. Chem. Phys.*, **13**, 507 (1945).

130. P. Delahay, "Double Layer Structure and Polarographic Theory" in *Progress in Polarography* (P. Zuman, ed.), Wiley (Interscience), New York, Vol. 1, 1962, pp. 65–80.

131. C. N. Reilley and W. Stumm, "Adsorption in Polarography," in *Progress in Polarography* (P. Zuman, ed.), Wiley (Interscience), New York, Vol. 1, 1962, pp. 81–121.

132. M. A. V. Devanathan, *Chem. Rev.*, **65**, 635 (1965).

133. G. J. Hills and R. Payne, *Trans. Faraday Soc.*, **63**, 326 (1965).

AUTHOR INDEX

Numbers in parentheses are reference numbers and indicate that an author's work is referred to although his name is not cited in the text. Numbers in italics give the page on which the complete reference is listed.

A

Abrahamson, E. W., 206(5), *237*
Acheson, R. M., 225(145), *239*
Adachi, K., 138(284), *155*
Adam, N. K., 254(35), *406*
Adams, R. N., 5(26), 7(26), *149*, 177 (51, 52), 179(59), 183(74), *194*
Adler, E., 216(84), *238*
Agarwal, H. P., 8(58, 59), *149*
Aikin, C. L., 186(92), *195*
Airey, L., 180(64), *194*
Akamatsu, H., 213(65), *237*
Alberts, G. S., 71(169), *152*
Alden, J. R., 5(26), 7(26), *149*
Andersen, T. N., 245(24), *406*
Anderson, S., 227, *240*
Anderson, W., 279, *407*
Anson, F. C., 87(232), *154*, 164(19), 169(26), 171(26), 172(26), 174 (47), 182(19), 184(75, 76), 185 (68, 77, 81, 82), 186(19, 94, 95, 100, 103, 109), 187(105, 109), 188, 189(75, 117, 120), 190, 191 (81, 82, 100), 192(19), *193, 194, 195*
Aramata, A., 12(86), *150*, 328, 331, 377, *408*
Arbit, H. A., 219(101), 220(101), 221, *238*
Argersinger, W. J., Jr., 328(92), *408*
Armstrong, G., 160, 164(2), *193*
Aten, A. C., 8(80), 10(80), *150*
Audubert, R., 215(75), 224(131), *237, 238, 239*
Auerbach, C., 187(108), 189(108), *195*

Ayabe, Y., 93(249), 94(249), 100, 138 (281–284), *154, 155*
Aylward, G. H., 30(122, 123, 124, 125), 33(122, 123, 124, 125), 44, 50, 51(150), 71(122), 92(122–125), 93, *151, 152*

B

Bäckström, H. L. J., 204(18), *236*
Bader, J., 228, *240*
Baker, W. P., 204(22), *236*
Balchin, L. A., 7(44), *149*
Bancroft, W. D., 198, *235*
Bandini, D., 293(60), *407*
Bannister, L. C., 192(139), *196*
Bard, A. J., 72(174), *152*, 171(33, 34, 35), 174(44, 45), 176(33), 177 (44), 178, 185(34, 35, 83, 90, 97), 186(35, 44, 45, 97), 187(34), 189 (34), 191(132), *193–196*, 228, *240*
Bardeen, J., 220(111), 222(111), *238*
Barker, G. C., 7(52, 53), 8(52, 53, 61, 62), 9(53, 61, 62), 10(61), 85 (217), 86, 102, 132, *149, 150, 153, 155*, 215(81), 216(81), *238*
Barlow, C. A., Jr., 243, 244, 301, 318, 327, 351, *406*
Barradas, R. G., 279, 355, *407, 408*
Bartok, W., 204(33), *236*
Bauer, H. H., 2(9), 3(9, 19), 4(9, 19), 6(9, 31), 7(9, 45), 8(9, 45, 76–78), 10(76–78), 11, 25(78, 78a, 109, 110), 30(9), 32, 35(78), 39, 71(154), 80(9), 86, 100, 103(9),

411

111(78), 114(45), 116(45), *148–152, 154*

Baumeiser, L., 219(89), 220(89), *238*

Beckett, A., 204(23, 31), *236*

Becquerel, E., 198, *235*

Beevers, J. R., 71(154), *152*

Beilby, A. L., 178(57), *194*

Bell, G. M., 245(23), *406*

Berg, H., 201, 212(14), 215(55, 57, 58, 79, 80), 216(79, 80), 218, 232, *236, 237*

Bernanose, A., 230(164), *240*

Berzins, T., 19, 71(159), 79(178), *151, 152,* 162(13), 163(16), 164(16), 165, 166(13), 167(23), 168(16, 23), 169(16), 185(23), 186(13, 16), 187(16), 189(13), *193,* 377 (115), *408*

Bianchi, G., 186(98), *195*

Biegler, T., 96(248a), *154*

Bikerman, J. J., 326(81), *408*

Black, O. D., 198(2), 206(2), *235*

Blocker, J. M., Jr., 215(76), 219(76), *237*

Blomgren, E., 288, 289(50), 290(50), 361, 363, *407*

Bockris, J. O'M., 244(21), 245(21, 24), 246(21), 288(50), 289(50), 290 (50), 361(50), 363(50), 364(21), *406, 407*

Bolland, J. L., 206(50), *237*

Booman, G. L., 7(36), 103(257), 104, 108, *149, 155*

Booth, D., 223(117), *239*

Bowden, F. P., 215, 219(72), *237*

Bowen, E. J., 225(138), *239*

Bowers, R. C., 192(138), *196*

Brattain, W. H., 220(110, 111, 112), 222(111, 112), *238*

Braun, F., 226, *239*

Brdicka, R., 43(132, 135, 136), 44(132, 135, 136, 138), 45(153), 46(132), 69(132), *151, 152,* 190(125), *195*

Breiter, M., 243, *405*

Bremer, T., 230(164), *240*

Breyer, B., 2(9), 3(9, 14–16, 18), 4 (1–3, 9, 14–16, 18), 6(1–3, 9, 14–

16, 18, 29–31), 7(9), 8(9), 11, 21 (30), 25(30), 30(9), 32, 71, 80 (9), 96(30), 103(9), *148, 149, 152*

Bridge, N. K., 206(53), *237*

Brodowsky, H., 326(86,) *408*

Brooks, W., 178(57), *194*

Brown, E. R., 104(259), 108(259), 114 (259), 116(266), *155*

Brown, W. L., 220(110), 222(110), *238*

Browne, R. J., 225, *239*

Bruckenstein, S., 179(60a), 187(106), *194, 195*

Bublitz, D. E., 185(86), *194*

Buchanan, G. S., 21(105), *151*

Buck, R. P., 45(152), 46(152), *152,* 181(114), 185(88), 189(88, 114), *195*

Buff, F. P., 331, *408*

Burington, R. S., 115(265), *155*

Bush, M. B., 206(47), *237*

Butler, J. A. V., 160, 164(2), 173(42), *193,* 300, 318, *407*

C

Calvert, D., 245(23), *406*

Campbell, W. E., 192(140), *196*

Canagaratna, S. G., 282(46), 306(46), 348, *407*

Carslaw, H. S., 16(98), *150,* 390(127), *409*

Chaki, S., 84(201), *153*

Chambers, L. M., 191, *195*

Chance, D. A., 219(100), 221(100), *238*

Chandross, E. A., 225(147), 227, 232, *239, 240*

Chapman, D. L., 298, 300, *407*

Chapman, O. L., 204(25), 206(25), *236*

Chaykowsky, O. C., 118(268), *155*

Christensen, C. R., 184(75), 189(75, 117), *194*

Ciamician, G., 204(16), *236*

Cizek, J., 97, 98, *154*

Clark, P. E., 215(77), 219(77), *237*
Coffin, E. M., 285(49), *407*
Connolly, T. W., 119(270), *155*
Conway, B. E., 355, *408*
Cooke, W. D., 7(51), *149,* 179(61), *194*
Cooper, H. R., 206(50), *237*
Copeland, A. W., 198, 206, *235*
Cottrell, F. G., 160(4), *193*
Cummings, J. I., 244(22), 284(22), 285(22, 49), 286(22), *406*

D

Dainton, F. S., 223(122), *239*
Damaskin, B. B., 3(17), 6(17), *148,* 243, 245(25), 291, 374, *406–408*
Davies, C. W., 331, *408*
Davis, D. G., 164(17), 167(17), 179 (60), 182(67, 69, 70), 185(17, 60, 69), 186(17, 93), 187(17, 69, 70), 188(60, 69), *193, 194, 195*
Defay, R., 254(36), 327(87), *406, 408*
DeFord, D. D., 7(37), 8(56), 108(37), 111(37), 116(266), 122(56), *149, 155,* 192(138), *196*
deGroot, M. S., 205(35), *236*
Delahay, P., 2(10, 12), 3(12), 4(10), 6(12), 8(10, 57, 65–69, 71, 72), 9(10, 57, 65, 67–69, 71, 72), 12 (10, 84–88), 15(12), 18(12), 19 (10), 21(12), 23(10), 26(10, 115), 28(12, 117), 30(10, 12), 44(146), 45(12), 69, 71(159), 79, 83, 84, 85, 86, 88(10, 236), 90, 91(236), 102, 139, 145(10), *148– 154,* 160, 162(12, 13), 163(8, 12, 16), 164(16), 165(8), 166(8, 13), 167(8, 23), 168(16, 23), 169(16), 174(9, 46), 181(9), 183(72), 185 (12, 23), 186(12, 13, 16), 187(12, 16), 189(8, 13), *193–195,* 231, *240,* 242(5), 243, 287, 291(55, 56), 318, 321(10), 326, 327, 328, 331, 340, 351, 362(107), 363 (107, 108), 371, 376(108), 377

(55), 381(55), 383(55), 386(55), 387(55), 388, 389(124), 390 (124), *405, 407–409*
Delmastro, J. R., 39(131), 40(131), 41 (131), 44(137), 49(131, 137), 65 (137), 93(250), 95(248b, 250), 96(248b, 250), 99(250), *151, 152, 154*
de Maeyer, L., 90(241), *154*
Demars, R. D., 179(63), *194*
De Mayo, P., 206(42), *236*
Devanathan, M. A. V., 243, 244(20, 21), 245(20, 21), 246(21), 249 (27), 251(27), 252, 253(33), 255 (27), 259, 263(27), 264(27), 266 (27), 272(27), 280(33), 282(33, 46), 283(47), 306(46), 345, 348, 364(21), 391, 394, *406, 407, 409*
Dewald, J. F., 220(114), 222, *238,* 242 (1), *405*
Dickinson, R. G., 223, *239*
Dopo, R. F., 187(107), *195*
Dorfman, L. M., 219(87), *238*
Doss, K. S. G., 8(58, 59), 30(120), 44(147), 89, 90(237), *149, 151, 152, 154*
Dračka, O., 170(27), 189(118, 119), *193, 195*
Duclaux, J. P. E., 215(78), *237*

E

Egan, J. J., 192(136), *196*
Eigen, M., 90(241), *154,* 326(84), *408*
Eisenberg, M., 199, 213(13, 62), 214 (62), *236, 237*
Elving, P. J., 3(19), 4(19), 8(77), 10 (77), 86(225), *148, 150, 154,* 178 (55), 185(55), 186(55), *194*
Enke, C. G., 2(13), *148,* 182(66), *194*
Epstein, B., 230(165, 166), *240*
Erbelding, W., 7(51), *149*
Ermolaev, V., 205, *236*
Ershler, B., 2, *148*
Esin, O. A., 292, *407*
Evans, D. H., 185(80), 186(102), *194, 195*

Evans, M. W., 399(129), *409*
Evans, U. R., 192(139), *196*
Everett, D. H., 367, *408*
Everett, G. W., 160(10), 162(10), 163 (10), 164(10), 175(10), 176(10), 185(10), 186(10), 187(10), *193*
Everhart, M. E., 179(60), 185(60), 188 (60), *194*
Eyring, H., 26(116), *151,* 216(86), *238*

F

Faircloth, R. L., 8(62), 9(62), 132, *150, 155*
Feldberg, S. W., 187(108), 189(108), *195,* 233, *240*
Ferguson, W. S., 192(133), *196*
Fernandez-Biarge, J., 86(222, 223), *154*
Ferrett, D. J., 7(43), 8(43), *149*
Fike, C. T., 79(177), 84(177), *152*
Findl, E., 213(64), *237*
Fischer, O., 189(118, 119), *195*
Fischerová, E., 189(118, 119), *195*
Fisher, D. J., 7(38), 111(38), 132(38), *149*
Fleisher, H., 111(263), *155*
Florence, T. M., 30(124), 33(124), 92 (124), 93, *151*
Foo, D. C. S., 25, *151*
Forejt, J., 84(210), *153*
Foss, R. P., 204(21), *236*
Fournier, M., 21(104), *151*
Frank, H. S., 399(129), *409*
Freise, V., 326(83), *408*
Frumkin, A. N., 3(17), 6(17), 79, 84 (196–198), 86, 87, *148, 153, 154,* 243, 245(25), 318, 334, 377, *405–408*
Furlani, C., 170(28), 171(28), 189 (28), *193*
Furman, N. H., 185(78, 89), 186(78), 188(89), *194, 195*

G

Galus, Z., 179(59), *194*
Ganchoff, J., 164(17), 167(17), 185 (17), 186(17), 187(17), *193*

Gardner, A. W., 8(62), 9(62), *149,* 215 (81), 216(81), 217, *238*
Garrett, A. B., 198(2), 206(2), 215(76, 77), 219(76, 77), *235, 237*
Garrett, C. G. B., 220(110, 112), 222 (110, 112), *238*
Gaur, H. C., 192(134), *196*
Gerischer, H., 4(25), 12(83), 14, 44, 69, 84(215), 86, 93, 103(256), 133(279), 138, *148, 150,153–155,* 179, *194,* 220(113), 222, *238,* 242 (2), *405*
Geske, D. H., 170, *193*
Gibbs, J. W., 254(34), 256(34), *406*
Gierst, L., 90, 91(247), *154,* 160(7), 165, 172(36), 176(50), 185(50), 186(50), 189(7), *193, 194*
Ginzburg, V. I., 215(73), 219(92, 108), 220(92), *237, 238*
Glasstone, S., 26(116), *151,* 216(86), *238*
Gobrecht, H. R., 220(115), 222, *239*
Goldfinger, P., 230(164), *240*
Grahame, D. C., 2, 3(23), 14(92), 21 (106), 79, 81(106), *148, 150–152,* 243, 244, 249(6, 26), 251, 254, 259, 263(6), 284(22), 285(48, 49), 286(22), 292, 293, 298, 300, 305(26), 310(6), 311(6), 315, 316, 318, 319(6), 322–324, 327, 341, 343, 347, 348, 349(101), 352 (101), 359(69), 396, 397, *405–408*
Grantham, D. H., 279(44), 353(44), 354(44), 355(44), 356(44), 357 (44), 358(44), 361(44), *407*
Gray, T. S., 140(287), *155*
Green, M., 242(3), *405*
Greenberg, S. A., 213(69), *237*
Griffith, L. R., 185(88), 189(88), *195*
Grimley, T., 326(82), *408*
Grube, G., 219(89), 220, *238*
Guggenheim, E. A., 255, 272, 391, *406*
Guoy, G., 298, 300, 332, *407*
Guterman, M. S., 84(206), *153*
Gutman, F., 2, 4(1–3), 6(1–3, 29), *148, 149*

H

Hackerman, N., 255(38), 256(38), *406*

Hacobian, S., 3(14, 15, 18), 4(14, 15, 18), 6(14, 15, 18, 29–31), 21 (30), 25(30), 96(30), *148, 149*

Hamm, R. E., 8(55), *149*

Hammond, G. S., 199(10), 204(20, 21, 22, 26, 29), 206(26), *236*

Haneda, Y., 224(129), *239*

Hansen, R. S., 279, 353(44), 354(44), 355, 356(44), 357(44), 358(44), 361(44), *407*

Hanus, V., 43(132), 44(132, 138, 145), 46(132), 69(132), *151, 152*

Harned, H. S., 319(78), 325(78), 328, *407*

Harnwell, G. P., 85(219), 140(286), *154, 155*

Hartmann, G., 225(134), *239*

Harvey, E. N., 198, 226, 230(152), *235, 240*

Hatchard, C. G., 205(39), *236*

Hayami, N., 219(95), 220(95), *238*

Hayes, J. W., 7(45), 8(45), 30(122, 123, 125), 33(122, 123, 125), 44 (122, 123, 150), 50, 51(150), 71 (122), 92(122, 123, 125), 93, 114 (45), 116(45), *149, 151*

Hercules, D. M., 206, 209(49), 212 (49), 227, 234(171), *237, 240* -

Herman, H. B., 72(174), *152*, 171(33, 34, 35), 176(33), 185(34, 35), 186(35), 187(34), 189(34), 191 (132), *193, 196*

Heyrovsky, J., 84(208, 209, 210), *153*

Heyvrosky, M., 215, *237*

Hickling, A., 182(65), *194*, 226, 230, *240*

Hillson, P. J., 215(74), 216, 222, *237*

Hochstrasser, R. M., 204(24), 206(24), *236*

Hoh, G. L. K., 185(86, 87), *194, 195*

Holbrook, W. B., 103(257), 104, 108, *155*

Holleck, L., 71(170), 84(214), *152, 153*

Honig, J. M., 379(123), *409*

Horiuti, J., 88, *154*

Howe, R., 119(269), *155*

Hubbard, A. T., 184(76), *194*

Hume, D. N., 2(7), *148*

Hung, H. L., 30(121, 122), 33(121, 122), 34(121), 35(129), 44(121, 122, 123, 137), 49(137), 50(122), 65(137), 71(122, 129), 73(129), 74(129), 77(129), 81(129), 82 (129), 92(121, 122, 129), 93 (122), 137(129, 137), *151, 152*

Hurd, R. M., 71(164), *152*

Hurwitz, H., 90, 91(247, 248), *154, 172(36–38)*, 173(38), *193*

Hurwitz, H. D., 187(111), *195*, 327, *408*

Hutchinson, C. A., 205(34), *236*

Hutton, E., 205(40), *236*

I

Ibl, N., 160(6), *193*

Imai, H., 8(69, 70, 71, 72), 9(70, 71, 72), 84(201), 102(69–72), *150, 153*

Ingram, M. D., 226, 230(151, 162), *240*

Ives, D. J. G., 219(99), 221(99), *238*

Ivey, H. F., 198, 226, *235*

Iwaki, R., 225(146), *239*

Iwamoto, R. T., 183(71, 74), *194*

J

Jaeger, J. C., 16(98), *150*, 390(127), *409*

Jenkins, I. L., 7(52), 8(52), *149*

Jesch, C., 288(50), 289(50), 290(50), 361(50), 363(50), *407*

Jessop, G., 7(41, 42), 114(41, 42), *149*

Johns, R. H., 160(10), 162(10), 163 (10), 164(10), 175(10), 176(10), 185(10), 186(10), 187(10), *193*

Johnson, F. H., 224(129), *239*

Johnson, H. W., Jr., 204(27), 206(27), *236*

Jones, H. C., 7(38), 111(38), 132(38), *149*
Jortner, J., 223(123, 125), *239*
Joshi, K. M., 85(218), *153,* 324, 327, 328, 329(80), 330(80), 331, 397, *408*
Juliard, A., 160(7), 165, 189(7), *193*
Juston-Coumat, F., 213(61), *237*

K

Kalab, D., 212(56), 213(56), *237*
Kallman, H. P., 213(66), *237*
Kambara, T., 16(100), 26(119), 100 (254), *150, 151, 154,* 183(73), *194*
Kamiya, I., 225(146), *239*
Karaoglanoff, Z., 160(11), 161, *193*
Karchmer, J. H., 186(96), *195*
Karpukhin, O. N., 227(157), *240*
Kasgkarev, V. E., 219(97), *238*
Kasha, M., 225, *239*
Kastening, B., 84(214), *153*
Kaufman, F., 224(126), *239*
Kelley, M. T., 7(38), 111(38), 132 (38), *149*
Kelsh, D. J., 279(44), 353(44), 354 (44), 355(44), 356(44), 357(44), 358(44), 361(44), *407*
Kern, D. M. H., 71(157, 158), *152*
Khan, A. U., 225, *239*
King, D. M., 170(29), 186(109), 187 (109), *193, 195*
Kip, A. F., 307(72), *407*
Kisiakowsky, G. B., 223, *239*
Kleinberg, J., 185(87), *195*
Kleinerman, M., 44(146), *152*
Kline, S. J., 270(41), *406*
Klotz, I. M., 260(39), 369(111), *406, 408*
Koenig, F. O., 251, 270(41), 272, 293, *406, 407*
Kolthoff, I. M., 84(202), 138(285), *153, 155*
Koryta, J., 43(133), 44(133), 71(155, 156), 79(186), 84(186), 138, *152, 153, 155*
Kosonogoya, K. M., 219(97), *238*

Koutecky, J., 14, 43(132), 44(132, 140, 144, 145), 45(153), 46(132), 69(132), 71(155, 156), 79, 84 (185–188), 93, 97, 98, *150–154*
Krause, M., 12(83), *150*
Kruger, G., 79(183, 184), *153*
Kruger, J., 219(98), 221(98), *238*
Krylov, V. S., 331, *408*
Kryukova, A. A., 84(199, 204, 205), *153*
Kuhnkies, R., 220(115), 222(115), *239*
Kuta, J., 14(93, 94), 44(139), 79 (187), 84(187), *150, 152,153*
Kuwana, T., 177(51, 52), 185(86), *194,* 204(27), 205(36), 206(27, 52), 208(52), 209(52), 212(52), 227(154), 228, 230(154, 165), *236, 237, 240*

L

Laidler, K. J., 26(116), *151,* 216(86), *238*
Laitinen, H. A., 3(22), 30(22), 71 (160), 86, 96(248a), *148, 152, 154,* 178(56), 182(66), 190, 191, 192(133, 134), *194–196*
Lansbury, R. C., 234, *240*
Larsen, R. P., 285(48), *407*
Lawrence, G. L., 178(57), *194*
Lawrence, J. B., 87(233), *154*
Lee, J., 231(167), *240*
Lee, W. B., 213(64), *237*
Leermakers, P. A., 204(29), *236*
Letsinger, R. L., 204(19), *236*
Levich, V. G., 90, *154,* 331, *408*
Levin, I., 206, *237*
Levine, S., 245(23), *406*
Levkovtseva, L. V., 198(6), *235*
Lingane, J. J., 164(18, 19), 178(54), 182(19), 185(85), 186(19, 54, 91, 99, 101, 102, 104), 192(19), *193, 194, 195*
Lingane, P. J., 186(91), *195*
Lippmann, G., 252, *406*
Llopis, J., 86, *154,* 160(5), *193*
Lloyd, R. A., 225(138), *239*
Longworth, J. W., 232(169), *240*

Lorenz, W., 79, 83, *153,* 189, 190(121), 192(121), *195,* 377, *408, 409*
Los, J. M., 44(142, 143), *152*
Losev, V. V., 84(203), *153*
Loshkarev, M. A., 84(204, 205), *153*
Lott, H., 183(74), *194*
Lucchesi, P. J., 204(33), *236*

M

McCord, T., 35(130), 48(130), 60 (130), 70(130), *151*
Macdonald, J. R., 243, 244, 301, 318, 327, 351, *406*
McDougall, G., 331, *408*
McElroy, W. D., 224(130), *239*
McEwen, W. E., 185(87), *195*
McKee, W. E., 213(60, 63, 64), *237*
McKeon, M. G., 187(110), *195*
McKeown, E., 225, *239*
McKinney, P. S., 7(40), 103, *149*
Mairanovskii, S. G., 44(145), *152,* 360, *408*
Malmstadt, H. V., 2(13), *148*
Mamantov, G., 160(9), 174(9), 181 (9), *193*
Mangrum, B. W., 205(34), *236*
Mann, C. K., 187(107), *195*
Margerum, J. D., 213(63, 64), *237*
Maricle, D. L., 234, *240*
Mark, H. B., 178, 186(92, 103), 187 (105), 188, *194, 195*
Markov, B. F., 292, *407*
Martin, R. B., 204(19), *236*
Martirosyan, A. P., 84(199), *153*
Mather, W. B., 185(77), *194*
Matsuda, H., 8(65), 9(65), 10(65), 14 (91), 15, 16(91, 99), 21(91), 25, 44, 88(236), 90, 91(236), 93 (249), 94(249), 100, 138, *149, 150, 152, 154, 155*
Mattax, C. C., 162(12), 163(12), 167 (23), 168(23, 25), 185(12, 23), 186(12), 187(12, 113), *193, 195*
Matyas, M., 84(209, 211), *153*
Mauser, H., 71(165), *152*
Mazur, P., 327(87), *408*
Meadows, L. F., 223(118), *239*

Mechelynck, P., 176(50), 185(50), 186 (50), *194*
Mehl, W., 85(218), *153*
Meites, L., 251, 252(30), *406*
Melik-Gaikazyan, V. I., 79(190), *153,* 365, 377, *408*
Mellor, J. W., 338(99, 100), *408*
Mikho, V. V., 198(6), *235*
Miller, L. J., 213(60), *237*
Milner, G. W. C., 7(43), 8(43), *149*
Möckel, F., 79(180), 83, *153,* 377, *409*
Mohilner, D. M., 2(11), 12(87), 71 (162), 79(175), 83, 86, *148, 150, 152,* 255(38), 256(38), 291(55, 56), 293, 328(92), 362(107), 363 (107, 108), 376(108), 377(55), 379(122), 381(55), 383(55), 386 (55), 387(55), 391(122), *406–409*
Momyer, W. R., 213(62), 214(62), *237*
Montgomery, H. C., 220(110), 222 (110), *238*
Moore, R. D., 118(268), *155*
Moore, W. N., 204(20, 21, 22), *236*
Moorhead, E. D., 185(89), 188(89), *195*
Morpurgo, G., 170(28), 171(28), 189 (28), *193*
Mott, N. F., 326(82), 348, *408*
Mülhberg, H., 189(122, 123), *195*
Müller, K., 244(21), 245(21), 246(21), 364(21), *406*
Munson, R. A., 191(129), *195*
Murayama, T., 189(115), *195*
Murray, R. W., 6(28), *149,* 172(39, 40), 173(39, 40, 41), 176(40), 191, *193, 195,* 199, *236*
Mussini, T., 186(98), *195*

N

Nagel, E. H., 8(56), 122(56), *149*
Narayanan, K., 90(245), *154*
Neeb, R., 8(74, 75), 10(74, 75, 82), 111(82), *150*
Newns, G. R., 230(161), *240*
Nicholson, M. N., 186(96), *195*

Nicholson, R. S., 71(172), *152*
Nikelley, J. G., 179(61), *194*
Niki, E., 7(46, 47, 48, 49), *149*
Nikolayeva-Fedorovich, N. V., 245 (25), *406*
Nisbet, A. R., 185(97), 186(97), *195*
Nobe, K., 219(100, 101), 220(101), 221(100, 101), *238*
Norrish, R. G. W., 215(72a), *237*
Noyes, R. M., 223(117, 118), *239*

O

Ogryzol, E. A., 225, *239*
Okamoto, K., 100, *154*
Okinaka, V., 84(202), *153*
Oldham, K. B., 8(60, 64), *150*
Onsager, L., 90, *154*
Oomen, J. J. C., 86(227), 132(227), 140(227), *154*
Orleman, E. F., 71(157, 158), *152*
Osteryoung, R. A., 87(232), 120(271), *154, 155,* 186(94), 191(131), 192 (94, 131), *195*
Ottolenghi, M., 223(123), *239*
Owen, B. B., 319(78), 325(78), 328 (78), *407*

P

Paldus, J., 44(144), *152*
Palke, W. E., 169(26), 171(26), 172 (26), *193*
Panik, I., 206(51), *237*
Parker, C. A., 205(39), *236*
Parry, E. P., 120(271), *155*
Parry, J. M., 350, 373(104), *408*
Parsons, J. S., 185(84), 186(84), 187 (84), *194*
Parsons, R., 26(114), 83(114), 85 (218), 88(114), *151, 153,* 243, 244, 249(27), 251(27), 255(27), 258(8), 259, 263(27), 264(8, 27, 40), 266(27), 272(27), 279, 283 (47), 291, 292, 293(51), 305, 315 (9), 316(8), 318, 324, 326, 327, 328, 329(80), 330(80), 331, 335 (8), 339(51), 344, 345(51), 348,
350, 351, 360, 366, 367, 368(54, 102), 373, 374, 375(53), 376(53), 377, 378(9), 383, 391, 394, 397, *405-409*
Paszyc, S., 213(68), *237*
Paterson, W. L., 118(267), *155*
Patterson, J. M., 204(19), *236*
Pawlak, Z., 213(67), *237*
Payne, R., 314, 342, *407, 408*
Paynter, J., 8(73), 10(73), 25(73, 112), 35, 39, 69(73), 102(73), 135(73), 139(73), *150, 151*
Perez Fernandez, M., 86(222, 223), *154*
Peries, P., 252, 253(33), 280(33), 282 (33), 345, 348, *406*
Perone, S. P., 132(276), *155*
Peters, D. G., 178(54), 186(54), *194*
Peterson, J. M., 188(112), *195*
Phillips, L. F., 224(128), *239*
Piette, L. H., 205(36), *236*
Pittman, R. W., 219(105–107), *238*
Pitts, J. N., Jr., 199, 204(19, 27), 205 (36), 206(27), 213(63), 230, *237*
Polanyi, M., 88, *154*
Polo, S., 160(5), *193*
Pope, M., 213(66), *237*
Porter, G., 204(17, 23, 24, 31), 206 (24, 53), *236, 237*
Postnikov, I. M., 227(157), *240*
Poth, M. A., 244(22), 284(22), 285 (22, 48, 49), 286(22), *406, 407*
Potter, A. F., Jr., 213(59), *237*
Prager, S., 187(106), *195*
Price, G., 216, *238*
Prigogine, I., 254(36), 327, *406, 408*
Putseiko, E. K., 213(70), *237*

Q

Quincke, G., 298, *407*

R

Rabinowitch, E., 199, 213(12), *236*
Rampazoo, L., 84(212), *153*
Randles, J. E. B., 2, 3(6, 21, 22), 25 (21), 30(6, 21, 22), 33(127), 84

(6, 21), 86, 96(21), 103(21), 132, 148, 151, 154, 279, 407

Rangarajan, S. K., 30(120), 89, 90, 91, 151, 154

Ravitz, S. F., 223, 239

Rawson, A. E., 219(99), 221(99), 238

Recktenwald, G., 204(19), 236

Reddy, A. K. N., 44(147), 152

Reece, J. M., 111(262), 155

Rehbach, M., 86(228, 229), 132(228, 229), 140(228, 229), 144, 154

Reid, C., 206(43), 236

Reid, R. W., 224(127), 239

Reid, S. T., 206(42), 236

Reilley, C. N., 6(28), 80(193), 84, 104, 105(260), 111(260), 149, 153, 155, 160, 162(10), 163(10), 164(10), 170(29), 172(39), 173 (39, 43), 175(10), 176(10, 49), 185(10), 186(10), 187(10), 192 (137), 193, 196, 199, 236, 243, 409

Reinmuth, W. H., 2(8), 6(32), 7(32, 50), 8(8, 81), 9, 10(8, 81), 12 (88–90), 13, 19, 25(81, 108, 109, 112), 29(32), 33(32), 35(128), 39, 45(151), 46(151), 71(167, 168), 79, 84, 85(108), 89, 96(81), 102(8), 111(81), 125(32), 139 (8), 148–154, 162, 163(15), 164 (15), 166(21), 167(20, 22), 168 (24), 170(30), 171(32), 172(32), 174(15, 48), 181(15), 185(30), 187(14), 189(20, 22), 190, 193, 194

Remick, A. E., 189(116), 195

Rhodes, P. R., 178(56), 194

Riddiford, A. C., 71(166), 152

Rideal, E. K., 215(74), 216, 222, 223, 237

Rius, A., 160(5), 193

Rivas, C., 204(28), 236

Roe, D. K., 234(171), 240

Rogers, A. E., 119(270), 155

Ross, J. W., 179, 194

Rouse, T. O., 179(60a), 187(106), 194, 195

Russell, C. D., 169(26), 171(26), 172 (26), 188(112), 193, 195, 314, 407

S

Salie, G., 79(181), 153

Sancier, K. W., 198, 235

Sand, H. J. S., 159, 160(1), 161, 193

Sandros, K., 204(18), 236

Sanfeld, A., 327(88), 408

Sano, M., 213(65), 237

Santhanan, K. S. V., 228, 240

Sarkaneis, K. V., 206(51), 237

Satyanarayana, S., 44, 152

Schaap, W. B., 7(40), 103, 149

Schindler, R., 71(170), 152

Schmidkunz, H., 224(134, 135), 225 (134), 239

Schmidt, H., 3(20), 148

Schultz, F. A., 186(100), 191(100), 195

Schwarz, W. M., 104, 108(261), 155

Schweiss, H., 201(15), 212(15, 58), 215(80), 216(80), 218, 232, 236, 237

Scribner, W. G., 192(137), 196

Sears, F. W., 307(71), 407

Seiger, H. N., 222, 239

Seliger, H. H., 224(130), 225, 231 (167), 239, 240

Senda, M., 8(57, 66–69), 9(57, 66–69), 10(57, 67–69), 16(66), 21(66), 25, 30(126), 79(67, 68), 86, 92 (126), 100(254), 149–151, 154, 183(73), 194, 377(116), 408

Seo, E. T., 230(165, 166), 240

Shain, I., 5(27), 7(27, 39), 25(27), 71 (169, 172), 96(27), 98, 100, 104, 108(261), 149, 152, 155, 179(63), 194

Shalgosky, H. I., 7(43), 8(43), 149

Sharp, J. H., 205(36), 236

Shlyapintokh, V. Ya., 225(137), 227, 239, 240

Sie, E. H. C., 224(129), 239

Sihvonen, V., 215, 238

Silber, P., 204(16), 236

Sills, S. A., 223(122), *239*

Silverman, H. P., 199, 213(13, 62, 69), 214(62), *236, 237*

Silvestroni, P., 84(212, 213), *153*

Simons, J. P., 206(41), *236*

Simpson, L. B., 44(143), *152*

Slee, L. J., 7(43), 8(43), *149*

Sluyters, J. H., 86(226–231), 132(231), 140, 143, 144(226), *154, 155*

Sluyters-Rehbach, M., 86(230, 231), 132(230, 231), 140(230, 231), *154*

Smales, A. A., 180(64), *194*

Smith, D. E., 6(32, 33, 34), 7(32, 34, 50), 8(34, 81), 10(34, 81), 25 (34, 81, 108, 109, 111), 29(32, 33), 30(121, 122), 33(32, 33, 111, 121, 122), 34(33, 111, 121), 35 (128–130), 39(131), 40(131), 41 (131), 44(122, 137), 45(148), 46 (148), 47(148, 149), 48(130, 148, 149), 49(131, 137), 50(122), 51 (33, 111), 60(130), 62(148), 63 (148), 64(148), 65(137), 66(33), 67(33), 69(111), 70(130), 71 (122, 129), 73(129), 74(129), 77 (111, 129), 81(129), 82(129), 85 (108), 92(121, 122, 129), 93 (122, 250), 95(248b, 250), 96(81, 111, 248b, 250), 99(250), 101(149), 104(259), 108(34, 259), 111(33, 34, 81), 113(34), 114(259), 116 (266), 118(34), 119(34), 120 (34), 121(34), 125(32–34), 128 (34), 129(34), 137(129, 137, 148), 138(33), *149–152, 155*

Smith, D. L., 86(225), *154,* 178(55), 185(55), 186(55), *194*

Smoler, I., 14(93, 94), *150*

Smutek, M., 26(118), *151*

Snead, W. K., 189(116), *195*

Snider, N. S., 204(33), *236*

Soderberg, B. A., 341, 343, 347, 349 (101), 352(101), *408*

Solon, E., 185(83), *194*

Somerton, K. W., 3(21), 25(21), 30 (21), 84(21), 86, 96(21), 103 (21), *148*

Sonntag, F. I., 227(155), *240*

Šorm, F., 84(210), *153*

Spaarnay, M., 326(85), *408*

Srinivasan, V. S., 231, *240*

Stauff, J., 224(134, 135), 225, *239*

Stein, G., 223(123, 124, 125), *239*

Steinchen-Sanfeld, A., 327(88), *408*

Stern, O., 243, 298, 332(18), *406*

Stevens, B., 205(40), *236*

Stillinger, F. H., 331, *408*

Strehlow, H., 43(134), 44(134), *151,* 326(86), *408*

Stromberg, A. G., 84(206, 207), *153*

Stuart, R. D., 115(264), *155*

Stumm, W., 80(193), 84, *153,* 243, *409*

Sugden, T. M., 224(127, 128), *239*

Suginov, N. P., 198(4), 226, *235*

Suppan, P., 204, *236*

Surash, J. J., 206, 209(49), 212(48, 49), *237*

Susuki, M., 71(171), *152*

Szychlinski, J., 214(67), *237*

T

Tachi, I., 8(66), 9(66), 16(66, 100), 21(66), 25, 26(119), 100(254), *150, 151, 154,* 183(73), *194*

Takahashi, T., 7(49), *149*

Takemori, Y., 100, *154,* 183(73), *194*

Tamamushi, R., 4(24), 29, 30(123), 33(24, 123), 44(123), 84(200), 86, 92(123), 93(123), *148, 151, 153*

Tanaka, N., 4(24), 29, 33(24), 86, *148,* 189(115), *195*

Tatwawadi, S. V., 191(132), *195, 196*

Taub, I. A., 219(87), *238*

Tausend, A., 220(115), 222(115), *239*

Taylor, R. P., 204(19), *236*

Tedoradze, G. A., 374, *408*

Temkin, M. I., 83, *153,* 384, *409*

Terenin, A., 205, *236*

Testa, A. C., 71(167, 168), 89, *152, 154,* 167(20, 22), 168(24), 170 (30), 171(32), 172(32), 185(30), 189(20, 22), *193*

Thaller, L. H., 213(59), *237*
Theml, P., 79(184), *153*
Thiemke, G., 71(161), *152*
Thomas, U. B., 192(140), *196*
Tidwell, T. H., Jr., 363(108), 375, 376, *408*
Timmer, B., 86(231), 132(231), 140 (231), *154*
Tordai, L., 84, *153*
Toren, E. C., Jr., 2(13), *148*
Trachtenberg, I., 79(176), 84(176), 85, *152*, 377(117), *408*
Traini, C., 186(98), *195*
Trümpler, G., 160(6), *193*
Turro, N. J., 204(26), 206(26), *236*

U

Underkofler, W. L., 5(27), 7(27, 39), 25(27), 96(27), 98, 100, *149*

V

van Cakenberghe, J., 8(79), 10(79), 39, *150*
van der Waals, J. H., 205(35), *236*
Van Dijck, W. J. D., 219(96), *238*
VanNorman, J. D., 192(135, 136), *196*
Van Rysselberghe, P., 244, 264(19), 265(19), 294, *406*
Vasil'ev, R. F., 224(132, 133), 225 (133), *239*
Vdovin, I. A., 8(63), *150*
Veretil'nyi, A. Ya., 227(157), *240*
Veselovskii, V. I., 215(73), 219(88, 90–94, 102–104, 108), 220(88, 90–92), *237, 238*
Vetter, K. J., 71(161, 163), *152*
Vichutinskii, A. A., 224, 225, *239*
Vielstich, W., 90, *154*
Visco, R., 227(156), 232(169), 234, *240*
Vlcek, A. A., 71(173), *152*
Vojir, V., 230, *240*
von Helmholtz, H. L. F., 298, *407*
von Stackelberg, M., 3(20), *148*
Voorhies, J. D., 185(78, 84), 186(78, 84), 187(84), *194*

W

Walker, D. E., 5(26), 7(26), *149*
Walker, M. S., 205(40), *236*
Ward, A. F. H., 84, *153*
Ward, G. A., 185(79), 192(138), *194, 196*
Waters, W. A., 225, *239*
Watts-Tobin, R. J., 348, *408*
Wawzonek, S., 71(160), *152*
Weber, J., 79, 84(185–188), *153*
Weis, C. H., 8(57), 9(57), 10(57), *149*
Weiser, H. B., 198, *235*
Wells, C. F., 210, *237*
Werner, R. L., 21(105), *151*
White, C. E., 206(47), *237*
White, E. H., 225(145), *239*
Whiteley, K. S., 279, *407*
Whitney, R. B., 243(16), 249(26), 254, 259, 293, 300(16), 305(26), *406*
Wicke, E., 326(84), *408*
Wiebush, J. R., 206(47), *237*
Wiesner, K., 44(141, 142, 143), *152*, 220(109), 222, *238*
Wilkinson F., 199(10), 204(17, 30), *236*
Williams, D. I., 7(44), *149*
Williams, E. G., 223, *239*
Williams, R., 214, 220, 221(71), *237*
Wilson, C. E., 12(90), *150*
Wilson, C. M., 192(138), *196*
Wilson, W., 182(65), *194*
Wohlers, W. H., 293(60), *407*
Woodbury, J. R., 128(273), *155*

Y

Yamanaka, T., 84(200), *153*
Yang, N. C., 204(28), *236*
Yu, Y. P., 128(274), *155*

Z

Zagainova, L. S., 84(207), *153*
Zakharov, I. V., 225(137), *239*

SUBJECT INDEX

A

AC chronopotentiometry, 100
AC polarography, 1–155
  adsorption processes and, 77–87
  amplifiers, 104–108, 111–114
  controlled-potential configuration,
    108–110
  conventional rectifiers, 119
  data analysis, 132–146
  differentiation of wave, 132
  double-layer effects, 87–92
  electrode growth and geometry, 93–
    98
  experimental aspects, 3–13
  faradaic rectification and intermodu-
    lation polarography, 101–102
  fundamental-harmonic type, 23, 101
    instrumentation, 121
  holding circuits, 120
  impedance-bridge method and, 3–4
  instrumentation, 102–132
    future trends, 128–132
    multireadout, 130
  nonfaradaic impedance, 83–85
  notation, 146–148
  operational amplifier, 104–107
  phase-angle measurement, 125–128
  phase-selective type, 7–8
    instrumentation for, 121–122, 125
  phase-sensitive detectors, 114–118
  phase shifters for, 119
  related techniques, theory, 100–101
  reversible polarographic wave and,
    15–26
    fundamental-harmonic type, crite-
      ria, 23
    quasi- 26–42
  second-harmonic type, 10, 24
    instrumentation, 122–125

  signal sources for, 110–111
  with stationary electrodes, 98–100
  in systems with coupled chemical re-
    actions, 42–71
    first-order, 44–69
    second-order, 69–71
  in systems with multistep charge
    transfer, 71–77
  tensammetric process, 83
  theory, 13–102
  time dependence of wave, 92–93
  trigger circuits, 119–120
  undesirable current components in,
    130–132
Acetic anhydride-acetic acid, chronopotentiometry, 185
Acetophenone photopotentials, 207, 209
  derivatives, 209
Alizarin Red S, chronopotentiometry,
  191
Amino acids, photoeffects, 213
p-Aminophenol, chronopotentiometry,
  185
Amplifiers for AC polarography, 104–
  108, 111–114
Anthracene, chemiluminescence, 228,
  229, 233
  chronopotentiometry, 185
Anthraquinone, photoeffects, 213
  photopotentials, 206–210
Antioxidants, chronopotentiometry, 185
Aromatic hydrocarbons, chemiluminescence, 229, 232–235
Ascorbic acid, chronopotentiometry,
  185

B

Barium chloride electrical double-layer
  studies, 328–331, 397–401

Becquerel effect, 198
  definition of, 199
Benzaldehyde, chronopotentiometry,
    185
  photopotentials, 207
Benzoin photopotentials, 207, 209
Benzophenone, photopotentials, 207,
    209, 210
  photoreduction, 204
Bromide ion, chronopotentiometry, 185
Bromine, chronopotentiometry, 185
n-Butyl alcohol electrode-adsorption
    studies, 288–291

C

Cadmium ion reaction with EDTA, ac
    polarography, 51
Cadmium sulfide irradiation, 220–221
Calcium ion reaction with EDTA, ac
    polarography, 51
Carbon monoxide, chronopotentiometry,
    191
Cerium(IV), chronopotentiometry, 185
Chemiluminescence, mechanism, 223–
    226
Chloride ion, chronopotentiometry, 185
Chlorophyll photoeffects, 214
Chromium(VI), chronopotentiometry,
    185
Chronopotentiometry, 157–192
  adsorption effects, 189–192
  apparatus, 159, 175–176
  applications, 184–192
  cells and electrodes, 177–180
  in chemical kinetics, 188–189
  in concentration measurements, 184–
    186
  current-reversal and cyclic type, 168–
    172
  current supplies for, 175–176
  definition, 157–158, 201
  differential and ac type, 183
  in electrode kinetics, 187–188
  electrode pretreatment, 182–183
  experimental methods, 175–184
  in fused-salt studies, 192

  in metal-film studies, 192
  programmed current type, 172–173
  recorders, 176
  Sand equation, 159–160
  techniques, 180–184
  theory, 161–174
    in consecutive and stepwise pro-
      cesses, 163–164
    in coupled chemical reactions, 164–
      168
    in current-reversal and cyclic type,
      168–172
    in double-layer charging, 173–174
    in irreversible processes, 162–163
    in programmed current type, 172–
      173
    in reversible processes, 162–163
  thin-layer type, 183–184
  transition time, 159–160
    measurement, 180–181
Chrysene, chemiluminescence, 229
Cobalt(II) and (III), chronopotentiom-
    etry, 185, 191
Copper(II), chronopotentiometry, 185
Copper crystals, photopotentials, 220–
    221
Copper oxide films, chronopotentiom-
    etry, 192
Coronene, chemiluminescence, 229
Cyanide ion, chronopotentiometry, 185

D

Decacyclene, chemiluminescence, 229
1,2,5,6-Dibenzanthracene, chemilumines-
    cence, 229
Diphenylpicrylhydrazyl, chronopoten-
    tiometry, 185

E

Electrical double layer, 241–409
  adsorption, 298–391
    isotherms, 364–375
    measures, 301–306
    of neutral compounds, 352–391
    specific ionic type, 331–352

surface excesses of nonspecifically absorbed ions, 321–325
types, 306
definition, 242
differential and integral capacitance, 284–287
diffuse layer, 243, 306–331
differential capacitance, 314–318
electrical charge, 245, 247–248
equations, 306–314, 325–331
potential profile and effective thickness, 318–321
statistical mechanical theories, 331
varying ionic size effects, 397–401
electrocapillary curves, 251–254
electrocapillary equations, 275
derivation, 254–263
for ideal polarized electrodes, 263–275
physical implications, 275–293
for reversible charge-transfer electrodes, 293–298, 299
electrode classification, 249–251
charge-transfer type, 250–251
faradaic and nonfaradaic processes and, 251
ideal polarized type, 249–250
Esin and Markov coefficients, 291–293, 335–340
for neutral compound adsorption, 360–364
for specific ionic adsorption, 343–345
Gibbs adsorption equation, 261
derivatives of electrocapillary equations, 254–263
interphase approach, 391–393
Gibbs surface, 305–306
Helmholtz plane, 244, 301
outer, ionic concentrations, 325
Henry's law in isotherm derivation, 371
inner layer, 243, 244
metallic phase, 243–244
model, 246, 299–301
neutral ion adsorption, 352–391
differential curves, 355–360

electrocapillary curves, 352–355
kinetics, 377–391
parts, 243
permittivity and electrical units, 396–397
qualitative description, 243–248
schematic diagram, 243
specific ionic adsorption, 331–352
criteria for detection, 333–343
differential curves, 341–343
effect on ϕ_2, 349–350
electrocapillary curves, 334–340
Esin and Markov coefficient in presence, 343–345
methods to determine amount, 346–349
studies of, usefulness, 248–249
thermodynamics, 249–298
Gibbs adsorption equation, 254–263
van der Waals equation in isotherm derivation, 371–373
Volmer gas in isotherm derivation, 373
Electrocapillary effects, in electrical double layer, 251–254
Electrodes, metal, irradiation, 215–219
Electroluminescence, 223–231
of aromatic hydrocarbons, 229, 232–235
classification of types, 226–227
historical aspects, 198
photoelectrochemistry and, 197–240
Esin and Markov coefficients, in electrical double-layer thermodynamics, 291–293
Euflavin, photoeffects, 214

F

Faradaic impedance measurement, 4
Faradaic rectification measurement, 9
Ferric ion, chronopotentiometry, 185
Ferricyanide ion, chronopotentiometry, 158–159, 185
Ferrocene, chronopotentiometry, 185

Ferrocyanide ion, chronopotentiometry, 185
Ferrous ion, chronopotentiometry, 185
Formaldehyde, chronopotentiometry, 185
Formic acid, chronopotentiometry, 185
Fused salts, chronopotentiometry, 192

G

Gallium(III), chronopotentiometry, 185
Galvanoluminescence, definition, 200
Gibbs-Duhem equations, 270–272
Gouy plane in electrical double layer, 245

H

Helmholtz plane of electrical double layer, 244, 301
Hydrazine, chronopotentiometry, 185
Hydrogen, ion, chronopotentiometry, 185
  peroxide, chronopotentiometry, 186
Hydroquinone, chronopotentiometry, 186
Hydroxylamine, chronopotentiometry, 186

I

Intermodulation polarography, 9–10
Iodate, chronopotentiometry, 186
Iodide ion, chronopotentiometry, 186
Iodine, chronopotentiometry, 186, 191, 192
  photodissociation, 223

L

Lead(II), chronopotentiometry, 186
Lippmann equation, 275–276
Lithium chemiluminescence studies, 230
Luminol electroluminescence, 225, 230–231

M

Malachite green leuconitrile, photoeffects, 214
Manganese(VII), chronopotentiometry, 186
Mercaptobenzothiazole, chronopotentiometry, 186
Mercury electrodes irradiation studies, 216–217, 231–232
Metal, electrodes, irradiation, 215–219
  thin films, chronopotentiometry, 192

N

Naphthalene, chemiluminescence, 229
Nickel, chronopotentiometry, 186
Nitrogen oxides, chronopotentiometry, 186
Nitrosyl chloride, photoeffects, 213

O

Organic compounds, photoeffects, 213–215
Oxalate ion, chronopotentiometry, 186
Oxalic acid, chronopotentiometry, 191
Oxygen, chronopotentiometry, 186

P

Perylene, chemiluminescence, 229, 233
Phase-selective ac polarography, 7
Phenosafranine, photoeffects, 214
Phenylenediamines, chronopotentiometry, 186
Phloxine B, photoeffects, 214
Phosphorescence, definition, 202
Photochronoamperometry, definition, 200–201
Photoconduction, definition, 200
Photocurrent, definition, 200
Photodepolarizer, definition, 211
Photoelectrochemistry, coated electrodes, 213–215
  electroluminescence, 197–240
  excited states and energy transfer, 204–206
  fundamental aspects, 201–206

metal-electrode surface irradiation, 215–219, 231–232
  hydrated electrons, 216–219
  mechanism, 215–216
metal–metal oxide surfaces and semiconductors, 219–223
nomenclature, 198–201
photopotentials in organic systems, 206–215
  mechanism, 206–208
  in presence of acceptor molecules, 210
  in presence of oxygen, 209–210
thermal effects, 222–223
Photokinetic current, definition, 212–213
Photopolarography, definition of, 201
  terms used in, 210–213
Photopotential effect, definition, 199–200
Photoreaction-altered diffusion current, 211
Photoresidual current, 211
Phototransfer reaction, definition, 211–212
Photovoltaic effect, definition, 199
Phthalocyanine, photoeffects, 214
Platinum(II) and (IV), chronopotentiometry, 186
Platinum–platinum oxide, irradiation, 219–220
Potassium bromide, electrocapillary curves, 253
  thermodynamic analysis, 279–283
Proflavin-ascorbic acid, photoeffects, 213
Proflavin sulfate, photoeffects, 214
Purines, photoeffects, 213
Pyranthrene, photoeffects, 213
Pyrene, chemiluminescence, 229
  photoeffects, 213
Pyrimidines, photoeffects, 213

R

Rf polarography, 9
Riboflavin, chronopotentiometry, 191

photoeffects, 214
Rubrene, chemiluminescence, 225, 233, 234–235

S

Sand equation, in chronopotentiometry, 159–160
Second-harmonic ac polarography, 10
Silver(I), chronopotentiometry, 186
Sodium fluoride electrical double-layer studies, 314–318
Square-wave polarography, 7–8
Sulfanilamide, chronopotentiometry, 186
Sulfate, peroxide, chronopotentiometry, 187

T

Tetracene, photoeffects, 213
Thallium(I), chronopotentiometry, 187
Thionine-iron system, photoeffects, 213
Tin, photopotential, 221
Titanium(IV), chronopotentiometry, 187
Titanium ion, reduction, ac polarography of, 51 ff.
Toluene-2,4-diamine, chronopotentiometry, 187
Triethylamine, chronopotentiometry, 187

U

Uranium(IV) and (V), chronopotentiometry, 187

V

Vanadium(V), chronopotentiometry, 187
Victoria Blue B, photoeffects, 214

Z

Zinc, chronopotentiometry, 187